Topics in Current Chemistry

Recently Published and Forthcoming Volumes

Anion Sensing
Volume Editor: Stibor, I.
Vol. 255, 2005

Organic Solid State Reactions
Volume Editor: Toda, F.
Vol. 254, 2005

DNA Binders and Related Subjects
Volume Editors: Waring, M.J., Chaires, J.B.
Vol. 253, 2005

Contrast Agents III
Volume Editor: Krause, W.
Vol. 252, 2005

Chalcogenocarboxylic Acid Derivatives
Volume Editor: Kato, S.
Vol. 251, 2005

New Aspects in Phosphorus Chemistry V
Volume Editor: Majoral, J.-P.
Vol. 250, 2005

Templates in Chemistry II
Volume Editors: Schalley, C.A., Vögtle, F., Dötz, K.H.
Vol. 249, 2005

Templates in Chemistry I
Volume Editors: Schalley, C.A., Vögtle, F., Dötz, K.H.
Vol. 248, 2005

Collagen
Volume Editors: Brinckmann, J., Notbohm, H., Müller, P.K.
Vol. 247, 2005

New Techniques in Solid-State NMR
Volume Editor: Klinowski, J.
Vol. 246, 2005

Functional Molecular Nanostructures
Volume Editor: Schlüter, A.D.
Vol. 245, 2005

Natural Product Synthesis II
Volume Editor: Mulzer, J.
Vol. 244, 2005

Natural Product Synthesis I
Volume Editor: Mulzer, J.
Vol. 243, 2005

Immobilized Catalysts
Volume Editor: Kirschning, A.
Vol. 242, 2004

Transition Metal and Rare Earth Compounds III
Volume Editor: Yersin, H.
Vol. 241, 2004

The Chemistry of Pheromones and Other Semiochemicals II
Volume Editor: Schulz, S.
Vol. 240, 2005

The Chemistry of Pheromones and Other Semiochemicals I
Volume Editor: Schulz, S.
Vol. 239, 2004

Orotidine Monophosphate Decarboxylase
Volume Editors: Lee, J.K., Tantillo, D.J.
Vol. 238, 2004

Long-Range Charge Transfer in DNA II
Volume Editor: Schuster, G.B.
Vol. 237, 2004

Long-Range Charge Transfer in DNA I
Volume Editor: Schuster, G.B.
Vol. 236, 2004

Spin Crossover in Transition Metal Compounds III
Volume Editors: Gütlich, P., Goodwin, H.A.
Vol. 235, 2004

Spin Crossover in Transition Metal Compounds II
Volume Editors: Gütlich, P., Goodwin, H.A.
Vol. 234, 2004

Spin Crossover in Transition Metal Compounds I
Volume Editors: Gütlich, P., Goodwin, H.A.
Vol. 233, 2004

245
Topics in Current Chemistry

Editorial Board:
A. de Meijere · K. N. Houk · H. Kessler · J.-M. Lehn
S. L. Schreiber · J. Thiem · B. M. Trost · F. Vögtle

Functional Molecular Nanostructures

Volume Editor: A. Dieter Schlüter

With contributions by
M. Ballauf · R. E. Bauer · M. A. Balbo Block · C. M. Drain · A. Falber
I. Goldberg · A. C. Grimsdale · C. J. Hawker · S. Hecht · A. Hirsch
V. Kalsani · C. Kaiser · A. Khan · C. N. Likos · K. Müllen
A. D. Schlüter · M. Schmittel · I. Sylvain · O. Vostrowsky · K. L. Wooley

 Springer

The series *Topics in Current Chemistry* presents critical reviews of the present and future trends in modern chemical research. The scope of coverage includes all areas of chemical science including the interfaces with related disciplines such as biology, medicine and materials science. The goal of each thematic volume is to give the nonspecialist reader, whether at the university or in industry, a comprehensive overview of an area where new insights are emerging that are of interest to a larger scientific audience.

As a rule, contributions are specially commissioned. The editors and publishers will, however, always be pleased to receive suggestions and supplementary information. Papers are accepted for *Topics in Current Chemistry* in English.

In references *Topics in Current Chemistry* is abbreviated Top Curr Chem and is cited as a journal.

Visit the TCC content at springerlink.com

Library of Congress Control Number: 2004113558

ISSN 0340-1022
ISBN 3-540-21926-9 **Springer Berlin Heidelberg New York**
DOI 10.1007/b79606

This work is subject to copyright. All rights are reserved, whether the whole or part of the material is concerned, specifically the rights of translation, reprinting, reuse of illustrations, recitation, broadcasting, reproduction on microfilms or in any other ways, and storage in data banks. Duplication of this publication or parts thereof is only permitted under the provisions of the German Copyright Law of September 9, 1965, in its current version, and permission for use must always be obtained from Springer-Verlag. Violations are liable to prosecution under the German Copyright Law.

Springer is a part of Springer Science+Business Media
springeronline.com
© Springer-Verlag Berlin Heidelberg 2005
Printed in The Netherlands

The use of general descriptive names, registered names, trademarks, etc. in this publication does not imply, even in the absence of a specific statement, that such names are exempt from the relevant protective laws and regulations and therefore free for general use.

Cover design: KünkelLopka, Heidelberg/design & production GmbH, Heidelberg
Typesetting: Fotosatz-Service Köhler GmbH, Würzburg

Printed on acid-free paper 02/3141 xv – 5 4 3 2 1 0

Volume Editors

Prof. Dr. A. Dieter Schlüter
Chair of Polymer Chemistry
Swiss Federal Institute of Technology
Department of Materials, Institute of Polymers
ETH-Hönggerberg, HCI J 541
8093 Zürich, Switzerland
dieter.schluter@mat.ethz.ch

Editorial Board

Prof. Dr. Armin de Meijere
Institut für Organische Chemie
der Georg-August-Universität
Tammannstraße 2
37077 Göttingen, Germany
ameijer1@uni-goettingen.de

Prof. Dr. Horst Kessler
Institut für Organische Chemie
TU München
Lichtenbergstraße 4
85747 Garching, Germany
kessler@ch.tum.de

Prof. Steven V. Ley
University Chemical Laboratory
Lensfield Road
Cambridge CB2 1EW, Great Britain
svl1000@cus.cam.ac.uk

Prof. Dr. Joachim Thiem
Institut für Organische Chemie
Universität Hamburg
Martin-Luther-King-Platz 6
20146 Hamburg, Germany
thiem@chemie.uni-hamburg.de

Prof. Dr. Fritz Vögtle
Kekulé-Institut für Organische Chemie
und Biochemie der Universität Bonn
Gerhard-Domagk-Straße 1
53121 Bonn, Germany
voegtle@uni-bonn.de

Prof. Kendall N. Houk
Department of Chemistry and Biochemistry
University of California
405 Hilgard Avenue
Los Angeles, CA 90024-1589, USA
houk@chem.ucla.edu

Prof. Jean-Marie Lehn
Institut de Chimie
Université de Strasbourg
1 rue Blaise Pascal, B.P.Z 296/R8
67008 Strasbourg Cedex, France
lehn@chimie.u-strasbg.fr

Prof. Stuart L. Schreiber
Chemical Laboratories
Harvard University
12 Oxford Street
Cambridge, MA 02138-2902, USA
sls@slsiris.harvard.edu

Prof. Barry M. Trost
Department of Chemistry
Stanford University
Stanford, CA 94305-5080, USA
bmtrost@leland.stanford.edu

Prof. Hisashi Yamamoto
Arthur Holly Compton Distinguished
Professor
Department of Chemistry
The University of Chicago
5735 South Ellis Avenue
Chicago, IL 60637
773-702-5059, USA
yamamoto@uchicago.edu

Topics in Current Chemistry
also Available Electronically

For all customers who have a standing order to Topics in Current Chemistry, we offer the electronic version via SpringerLink free of charge. Please contact your librarian who can receive a password for free access to the full articles by registration at:

springerlink.com

If you do not have a subscription, you can still view the tables of contents of the volumes and the abstract of each article by going to the SpringerLink Homepage, clicking on "Browse by Online Libraries", then "Chemical Sciences", and finally choose Topics in Current Chemistry.

You will find information about the

– Editorial Board
– Aims and Scope
– Instructions for Authors
– Sample Contribution

at springeronline.com using the search function.

Contents

Functional, Discrete, Nanoscale Supramolecular Assemblies
M. Schmittel · V. Kalsani 1

Synthesis and Applications of Supramolecular Porphyrinic Materials
C. M. Drain · I. Goldberg · I. Sylvain · A. Falber 55

Discrete Organic Nanotubes Based on a Combination of Covalent
and Non-Covalent Approaches
M. A. Balbo Block · C. Kaiser · A. Khan · S. Hecht 89

A Covalent Chemistry Approach to Giant Macromolecules
with Cylindrical Shape and an Engineerable Interior and Surface
A. D. Schlüter 151

Functionalization of Carbon Nanotubes
A. Hirsch · O. Vostrowsky 193

Equilibrium Structure of Dendrimers – Results and Open Questions
C. N. Likos · M. Ballauff 239

Functionalised Polyphenylene Dendrimers and Their Applications
R. E. Bauer · A. C. Grimsdale · K. Müllen 253

Nanoscale Objects: Perspectives Regarding Methodologies
for their Assembly, Covalent Stabilization and Utilization
K. L. Wooley · C. J. Hawker 287

Author Index Volumes 200–245 307

Subject Index 323

Contents of Volume 226
Colloid Chemistry I

Volume Editor: Markus Antonietti
ISBN 3-540-00415-7

Amphiphilic Block Copolymers for Templating Applications
S. Förster

Nanocasting of Lyotropic Liquid Crystal Phases for Metals and Ceramics
L. Göltner-Spickermann

Nanoparticles Made in Mesoporous Solids
L. M. Bronstein

Nanocasting and Nanocoating
R. A. Caruso

Mineral Liquid Crystals from Self-Assembly of Anisotropic Nanosystems
J.-C. P. Gabriel · P. Davidson

Monodisperse Aligned Emulsions from Demixing in Bulk Liquid Crystals
J. C. Loudet · P. Poulin

Templating Vesicles, Microemulsions and Lyotropic Mesophases by Organic Polymerization Processes
H.-P. Henze · C. C. Co · C. A. McKelvey · E. Kaler

Rational Material Design Using Au Core-Shell Nanocrystals
P. Mulvaney · L. Liz-Marzan

Contents of Volume 227
Colloid Chemistry II

Volume Editor: Markus Antonietti
ISBN 3-540-00418-1

Molecular Recognition and Hydrogen-Bonded Amphiphiles
C. M. Paleos · D. Tsiourvas

Dendrimers for Nanoparticle Synthesis and Dispersion Stabilization
K. Esumi

Organic Reactions in Microemulsions
M. Häger · F. Currie · K. Holmberg

Miniemulsions for Nanoparticle Synthesis
K. Landfester

Molecularly Imprinted Polymer Nanospheres as Fully Affinity Receptors
I. Kräuter · C. Gruber · G. E. M. Tovar

Hollow Inorganic Capsules via Colloid-Templated Layer-by-Layer Electrostatic Assembly
F. Caruso

Biorelevant Latexes and Microgels for the Interaction with Nucleic Acids
A. Elaissari · F. Ganachaut · C. Pichot

Preparation of Monodisperse Particles and Emulsions by Controlled Shear
V. Schmitt · F. Leal-Calderon · J. Bibette

Functional, Discrete, Nanoscale Supramolecular Assemblies

Michael Schmittel (✉) · Venkateshwarlu Kalsani (✉)

Center of Micro and Nanochemistry and Engineering, Organische Chemie I,
Universität Siegen, Adolf-Reichwein-Strasse, 57068 Siegen, Germany
schmittel@chemie.uni-siegen.de

1	Introduction	2
2	Supramolecular Nanoscale Structures	3
3	Hydrogen-Bond-Driven Supramolecular Nanoscale Assemblies	4
3.1	2D and 3D Motifs	4
4	Coordination-Driven Supramolecular Nanoscale Assemblies	8
4.1	Design	8
4.2	Nanoscale Self-Assemblies Built Using Monodentate Ligands	8
4.2.1	2D Nanoscaffolds	9
4.2.2	3D Nanoscale Architectures (Polyhedra)	11
4.3	Nanoscale Self-Assemblies Built from Bidentate and Tridentate Ligands	13
4.3.1	2D Assemblies (Homo- and Heteroleptic Aggregation)	14
4.3.2	3D (Heteroleptic) Assemblies	14
5	Functional Devices	20
5.1	Supramolecular Catalysis	20
5.2	Photoactive Assemblies	27
5.2.1	Homoleptic Devices	28
5.2.2	Heteroleptic Devices	30
5.3	Molecular Recognition	30
5.3.1	Metallosupramolecular Systems	31
5.3.2	Hydrogen-bonded Aggregates	36
5.4	Switching the Molecular Shape of Assemblies	37
5.4.1	Metallosupramolecular Structures	38
5.4.2	Hydrogen-Bonded Structures	41
5.5	Electroactive and Magnetic Assemblies	42
6	Conclusions	45
References		46

In the schemes, several nanostructures have been depicted without unnecessary alkyl or other groups for clarity purposes.

Abstract The last decade has witnessed an unprecedented pursuit of discrete, nanoscale supramolecular aggregates, built by modern methods of self-assembly strategies. Several efficient new synthetic methods have been developed for engineering spectacular multicomponent supramolecular aggregates. Amongst all the techniques explored, metal coordination and hydrogen-bonding motifs are the most celebrated means of producing structurally rich supramolecular architectures. While a truly biomimetic approach would typically employ a balanced mixture of weak interactions (hydrogen-bonding, π-π interactions, etc.), stronger non-covalent interactions (such as the coordinative metal ligand bond) have equally proven their high utility in the preparation of nanoscale assemblies. The time has now come to install functional elements to nanoscale aggregates in order to build nanoscale devices that exhibit non-linearity, interdependence and emergence, i.e. typical characteristics of more complex systems. Currently, functional model designing is still in its early stages, and lags far behind the progress made in structural engineering. Hence, in the present article some recent advances in the structural design of nanoscale assemblies are shown, along with examples from the following areas: supramolecular catalysis, photoactive assemblies, molecular recognition and switches, and electroactive assemblies.

Keywords Nanoscale · Supramolecular · Hydrogen bond · Metal coordination · Molecular recognition · Catalysis · Photoactive aggregates (functional aggregates)

1
Introduction

For millions of years, nature has capitalised on self-assembly strategies based on non-covalent interactions, such as hydrogen bonds, salt bridges, solvation forces and even metal coordination, to organise biological systems. Hence, such forces had been exploited long before the terms "supramolecular chemistry" and "self-assembly" were introduced [1]. It is well-known that protein function largely depends on the global conformation of the protein, and the folding process is governed by a multitude of reversible non-covalent interactions [2]. The folding is guided by several elements of control, such as recognition and self-sorting, which lead the way to the desired shape (correct folding). By learning these lessons from biology, chemists are now starting to compose highly complex chemical systems from components that interact via noncovalent intermolecular forces.

The *de novo* preparation of complex and large structures relying on covalent synthesis is often a very difficult and time-consuming chore. In contrast, supramolecular chemistry offers a convergent entrée to the creation of nanoscale systems for a wide range of applications.

The construction of nanostructures is of great interest not only because of their biological archetypes, but also because of their potential to revolutionise novel technologies such as the development of molecular-level devices. Over the last few decades, numerous supramolecular aggregates have been studied, thus revealing the principles that guide the necessary critical balance of weak interactions. As described by Lehn [1], the supramolecular architecture is a sort of molecular sociology in which non-covalent interactions and the individual properties of the molecules define the intermolecular bond. In this article, we highlight the most often used synthetic strategies to multicomponent [3] nanoscale assemblies (>2 nm), and their chemical and physical properties in view of their potential applications as molecular devices. The vast area of inorganic and organic/inorganic cluster chemistry with strong M-M or M-ligand bonds will not be covered here, although also nanoscopic aggregates have equally been realised by such strategies [4–6].

2
Supramolecular Nanoscale Structures

The self-assembly process, driven by non-covalent interactions that are prominent in biological systems (electrostatic, hydrogen-bonding, π-π stacking, etc.), offers a great tool for engineering nanoscale structures. Using developed non-covalent protocols several groups have created sparkling architectures, such as rosette aggregates [7, 8], self-assembled capsules (for reviews see [9]), and ordered hydrogen-bonded arrays [7, 10] (Scheme 1).

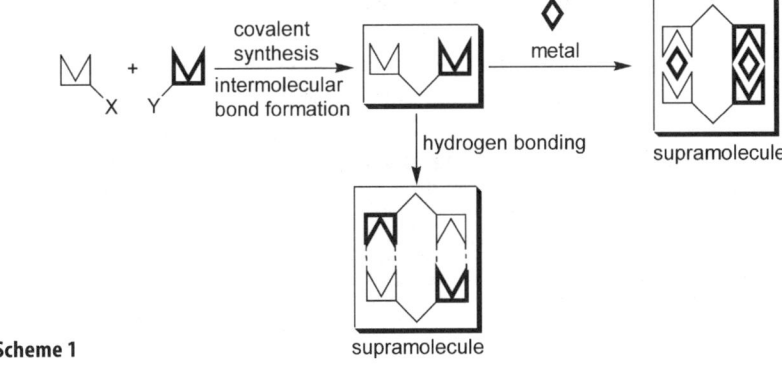

Scheme 1

Formation of nanoscale supramolecular arrays was also achieved using coordination chemistry. This protocol has been adopted by several groups to yield spectacular discrete architectures, such as grids, squares, nanoboxes, ring-in-ring structures, catenanes and rotaxanes, etc., which will be discussed below.

A detailed analysis of synthetic protocols leading to discrete nanoscopic self-assembled systems (>2 nm) reveals clearly that two motifs dominate the scene: hydrogen-bonding and metal-coordination-driven approaches. This review will therefore highlight some recent developments in this field, including a discussion about functional devices.

3
Hydrogen-Bond-Driven Supramolecular Nanoscale Assemblies

As indicated above, hydrogen-bonding motifs play an important role in biological systems, which adds a biomimetic flavour to all artificial hydrogen-bonded 2D and 3D assemblies. A thorough analysis of the size of known aggregates reveals that relatively few discrete hydrogen-bonded assemblies with dimensions above 2 nm, to which this article is restricted, are known.

3.1
2D and 3D Motifs

All larger (>2 nm) hydrogen-bonded assemblies are based on a multitude of hydrogen-bonding interactions in order to compensate entropic losses by enthalpic gains. Sessler et al. described the assembly of artificial dinucleotide modules to yield homoleptic 2D-scaffolds [11]. The stability of these homodimers is improved with respect to their monomer units when a rigid linker is used. A later report by the same group further illustrated the importance of structural rigidity and cooperativity [12]. Along these lines Fenniri et al. utilised heteroaromatic bases possessing both the Watson-Crick DDA pattern of guanine (where A is acceptor and D is donor) and the AAD design of cytosine to mastermind homoleptic rosette aggregates in water (Scheme 2, left). These homoleptic rosette-assemblies further aggregate into nanotubes through hierarchical self-assembly [8a]. Analogously, the ADA hydrogen-bonding arrays of isocyanuric acid are mutually complementary with DAD arrays of melamine. A detailed study of these interactions initiated the engineering of nanoscale scaffolds based on these two building blocks [13]. Hence, the combination of

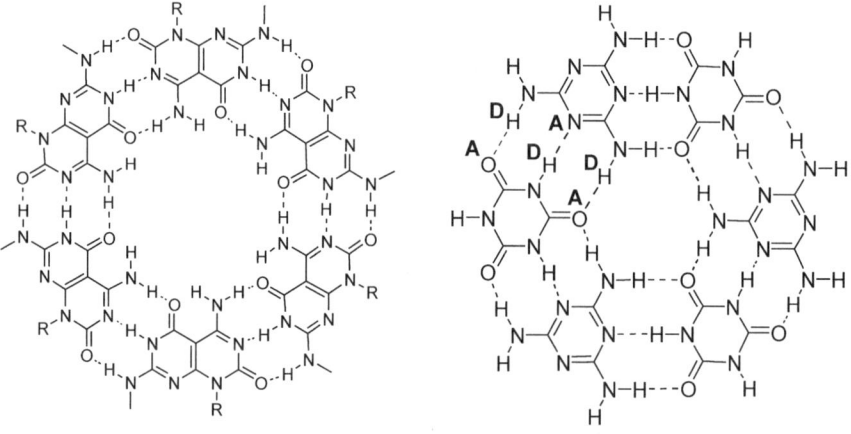

Scheme 2 Strategies used to construct homoleptic and heteroleptic 2D-rosette aggregates [7, 8a]. *A* Acceptor, *D* donor

isocyanuric acid and melamine led to several sparkling architectures such as 2D and 3D rosette motifs (Scheme 2, right) [7]. Timmerman and Reinhoudt et al. reported on the self-assembly of a [2×2] heteroleptic grid architecture through the cooperative formation of 24 hydrogen bonds [14].

Whitesides et al. developed two different approaches to nanoscale 2D and 3D rosette motifs: peripheral crowding [15] and covalent preorganisation [16]. In the first approach, due to the increasing size of substituents at the melamine unit the formation of 2D rosettes over linear tapes is favoured, as repulsions between substituents are reduced in rosette structures, whereas the covalent preorganisation approach uses covalently linked melamine or cyanurate units, thus generating a much more favourable entropic situation for rosette scaffolds. The latter approach was further explored to create nanoscaffolds with internal cavities [17]. A combination of the two approaches was used to obtain two extremely stable rosette assemblies [18]. Calix[4]arene-linked melamine and cyanurate units were used by Reinhoudt et al. to build a family of 3D double-rosette assemblies [19]. Their equilibria were investigated [7] and their growth on gold surface was studied [20].

In a remarkable study by the same group, the quantitative formation of a 15-component 3D tetra-rosette nanoscale assembly held together by 72 hydrogen bonds was demonstrated [21]. Self-sorting, well-known from metallo-supramolecular aggregates [22], is an important phenomenon in these assemblies. Spontaneous generation of discrete nanoscale motifs was further explored to generate 3D heteroleptic hexa- [23] and octa-rosette assemblies (Scheme 3) [24]. One equivalent of the octamelamine ligand 1 forms well-defined assemblies with eight equivalents of 5,5-diethylbarbituric acid (DEB): $(1)_3(DEB)_{24}$. The spontaneous formation of this 27-component octa-rosette structure (~20 kDa) demonstrates the huge potential of this approach in engineering nanoscale 3D heteroleptic aggregates. The resultant assemblies are expected to have a height of 5.5 nm, which makes them like in size to DNA oligomers used for conductivity measurements [25].

Capsule assemblies have received special attention due to their high potential in the field of host-guest chemistry and catalysis (*vide infra*). As a result of the difficulties in constructing capsules by covalent synthesis [26] several groups have explored the fabrication of capsules relying on non-covalent synthesis. Most of the self-assembled capsules reported to date are homoleptic in nature [9a,c, 27] with only a couple of heteroleptic capsules being known. As shown in Fig. 1, six strategies to achieve capsule-like assemblies are explored [7]. While calix[4]arenes, resorcinarenes and cavitands are amongst the most extensively used building blocks, homoleptic capsule formation from calix[4]arenes is hampered due to its high conformational flexibility. Calix[4]arenes, with four urea moieties at the upper rim that allow for sideway-directed urea hydrogen bonds, form well-defined homoleptic capsules in solution (Fig. 1d) [28]. A similar strategy was used to construct homodimers from cyclocholates [29], cyclotriveratrylenes [30] and a heterodimer from complementary cyclodextrin and porphyrin moieties [31]. Rebek et al. prepared homoleptic capsules of

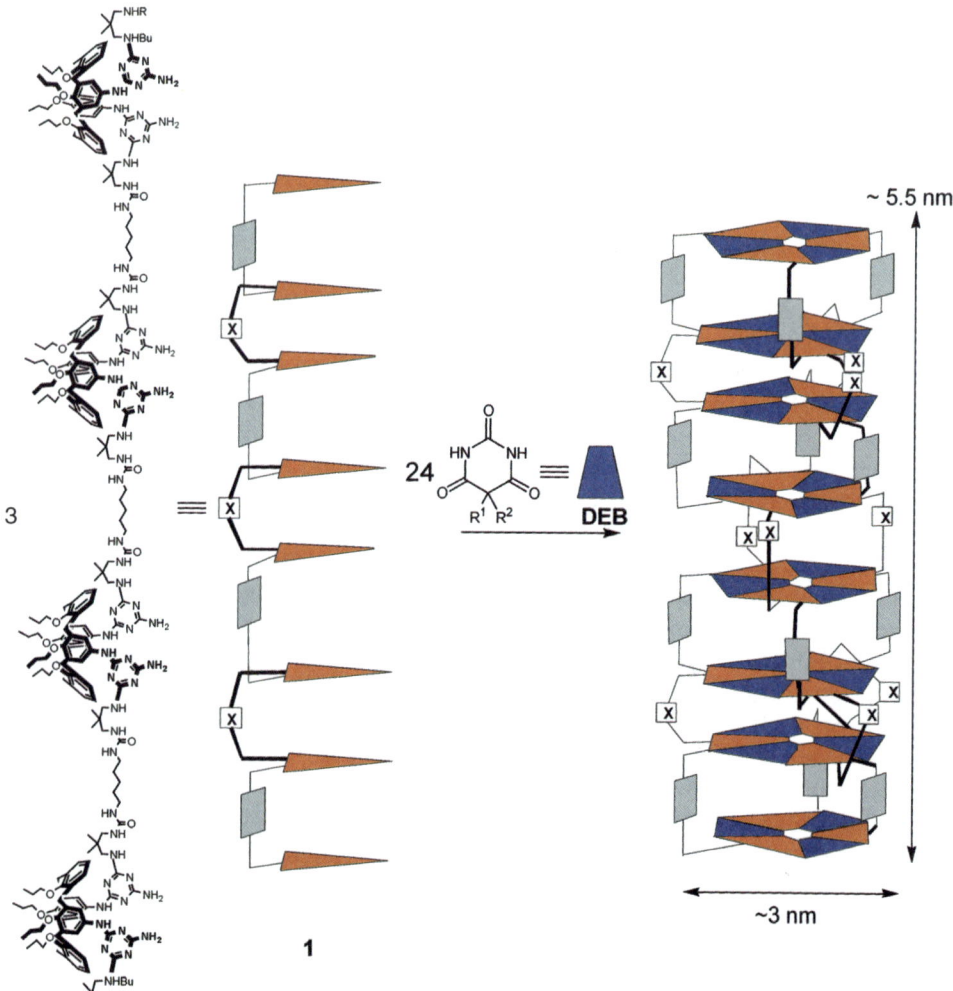

Scheme 3 Hydrogen bond self-assembly of 27 components leading to a heteroleptic rosette aggregate [24]. *DEB* 5,5-Diethylbarbituric acid

tennis ball shape using the hydrogen-bonding properties of glycoluril moieties (Fig. 1b and f) [32]. A similar strategy was explored to construct capsules using glycoluril and sulphamides as complementary hydrogen-bonding blocks (Fig. 1a) and self-assembly of the tetramide block (Fig. 1c). For heteroleptic hydrogen-bonded capsules, the versatile approach (Fig. 1d) was used by Reinhoudt et al. [33], Rebek et al. [34] and others [35]. To construct heteroleptic capsules, two cavitands with carboxylic acid groups serve as end-caps and four 2-aminopyridine molecules act as connectors [36]. These heteroleptic capsules are of particular interest because one can readily introduce multiple functionalities. All these capsules range from ~2 to 3 nm in size.

Functional, Discrete, Nanoscale Supramolecular Assemblies

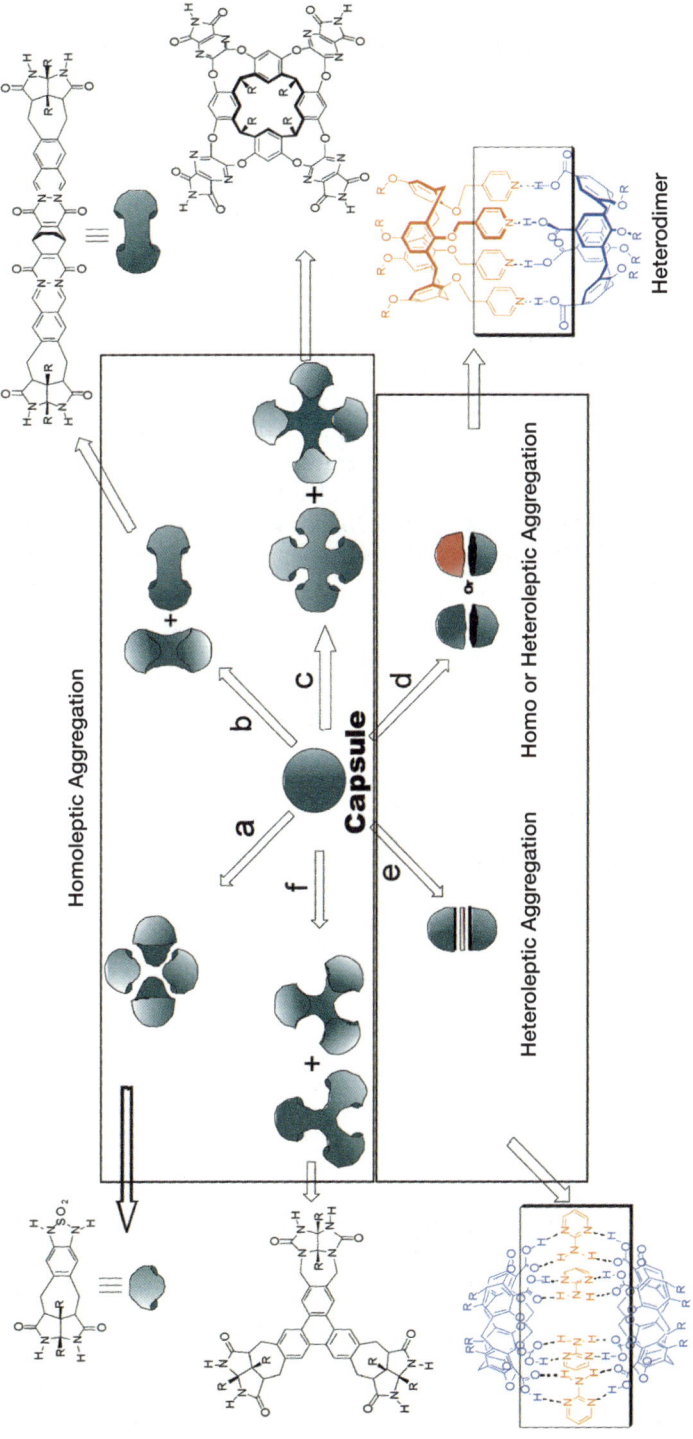

Fig. 1 Strategies used to construct homo and heteroleptic capsule like nanoscale assemblies [28–36]

4
Coordination-Driven Supramolecular Nanoscale Assemblies

4.1
Design

The metal coordination geometry and the information stored in the ligands should provide the construction manual for any self-assembly. Therefore, the selection of appropriate metal ion(s) and ligand(s) is crucial, as witnessed in a multitude of publications, reviews, and books. In this chapter, we will concentrate on nanoscale architectures built from monodentate (pyridine), bidentate (bipyridine, phenanthroline, catechol) and tridentate (terpyridines) ligands. The dentate term is here used purely to describe the interaction of the ligand with one single coordination centre. As depicted in a simplified way in Scheme 4, most self-assemblies described in the literature are constructed about Pd(II) or Pt(II) ions in a square planar arrangement using monodentate ligands, at Cu(I) or Ag(I) ions in a tetrahedral fashion making use of bidentate ligands, and at Co(II)/Cu(II)/Fe(II)/Zn(II)/Hg(II) in an octahedral grouping by employing terpyridine chelating motifs.

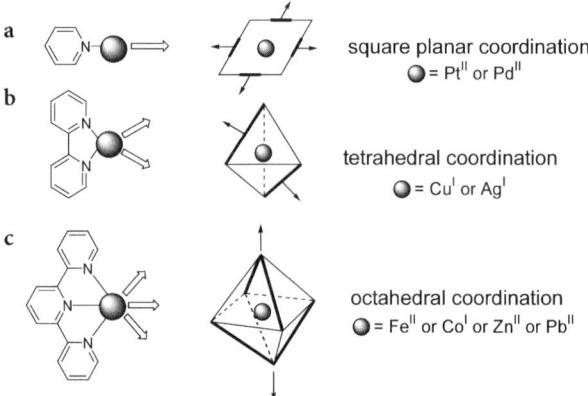

Scheme 4a–c Cartoon representation of preferred coordination geometry of **a** monodentate, **b** bidentate, and **c** tridentate ligands

4.2
Nanoscale Self-Assemblies Built Using Monodentate Ligands

Stang et al. [37], Fujita et al. [38] and Hupp et al. [39] have shown the utility of monodentate ligands in developing nanoscale supramolecular architectures. As emphasised by Stang et al. [37a], the important factors to be considered are the coordination angles at the metal ion equipped with a kinetically inert ligand and at the incoming labile multitopic ligand. This design is usually termed the directional bonding approach. Metal ions can readily provide free *cis* (90°) or *trans* coordination sites depending on the inert ligand. The careful selection and design of labile ligand(s) is therefore a prerequisite for a rational approach to any desired supramolecular assembly.

4.2.1
2D Nanoscaffolds

A molecular square requires 90° metal corners (Scheme 5, left) to define its specific shape. Pioneering work in this area by Fujita et al. described the construction of tetranuclear squares, the structures of which were confirmed by NMR, ESI-MS and the solid-state structure [40]. When the square motif was expanded to the nanoscale by incorporating phenylene or acetylene spacers into the bis-pyridine back bone [40d], an equilibrium resulted between squares and triangles depending on the concentration. Later, Hong et al. could control this equilibrium by induced-fit molecular recognition [41]. Over the years, a large variety of molecular squares have been reported [37–39] and even the gas phase chemistry of these assemblies has been investigated [42].

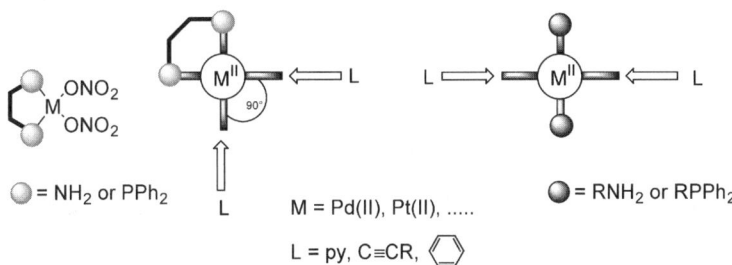

Scheme 5 Cartoon representation of the *cis*- and *trans*-coordinating metal ions (Pt(II) or Pd(II))

After the initial stage had been set, Stang et al. modified the structure of the metal ion corners by using Pt(II) or Pd(II)-phosphane units in the directional bonding approach [43, 44]. This allowed dissolving of the molecular nanoscale squares in organic solvents and analysis of the structures by ^{31}P NMR [45] (Scheme 6).

Larger nanoscale polygons, such as a heteroleptic hexagon of about 5 nm diameter, were accessible by using *trans*-coordinating metal fragments. Twelve individual building blocks were assembled in a cooperative manner to yield the desired macrocycle as a single entity [44a, 45a]. Such authoritative examples illustrate the utility of the metal-driven self-assembly approach to 2D nanoscale structures (Scheme 7).

Analogous self-assembly strategies were employed to produce a large variety of 2D nanoscale polygons {using Pt(II), Pd(II), Hg(II), Ru(II), Re(I) [39a, 46]

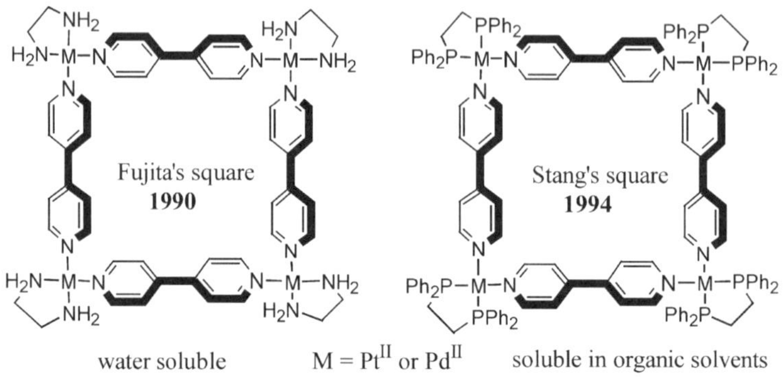

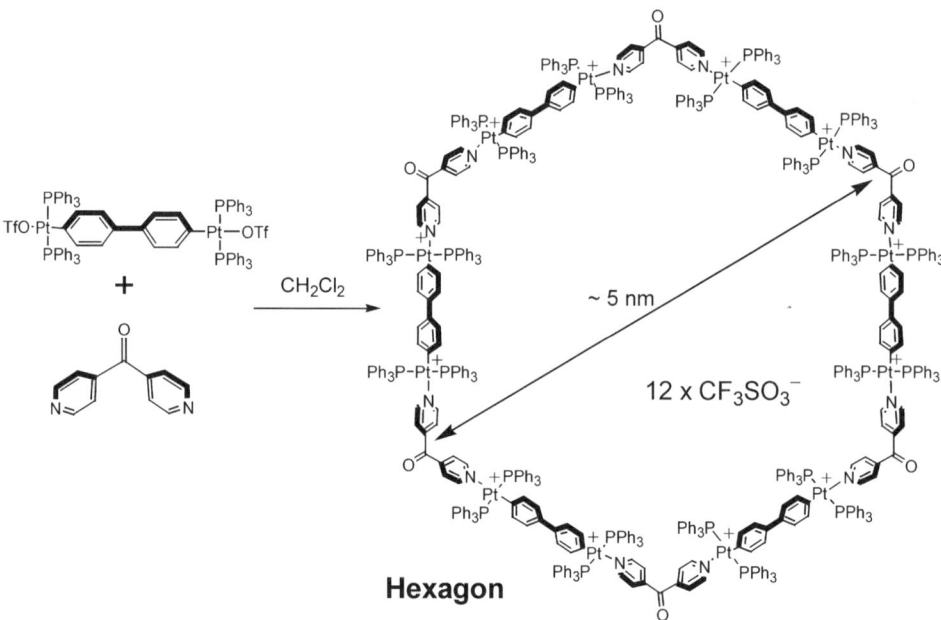

Scheme 6 Supramolecular squares as developed by Fujita et al. [40a] and Stang and Cao [43]

Scheme 7 A supramolecular heteroleptic hexagon [45c]

coordination}, such as porphyrin squares [47], ferrocene [48], 1,1′-bi-2-naphthol (BINOL) [49, 50] and anthracene [51] containing squares, and molecular necklaces [52]. Depending on the conditions and the metal centres used, these structures are dynamic. Würthner et al. reported on the dynamics of interconverting molecular triangles and squares held together by Pd/Pt(II)-pyridine interactions [53]. In a similar system the relative stabilities of the square and triangle complexes were explored by using force field methods [54]. Concentration- and temperature-dependent studies on the dynamic equilibrium of a supramolecular dimeric rhomboid and trimeric hexagon provided the thermodynamic parameters [55].

The less labile Pt-acetylene interaction and the limited conformational flexibility of bridging BINOL ligands was facilitating a rare one-pot self-assembly of molecular polygons ranging from triangle to octagon assemblies (with 1.4–4.3 nm cavity). According to Lin et al. [56] all macrocycles were separated and well-characterised by ^{1}H NMR, ^{13}C NMR, FAB-MS and circular dichroism (Scheme 8). The production of multiple self-assembled nanoscale macrocycles is of great interest in the context of dynamic, combinatorial chemistry.

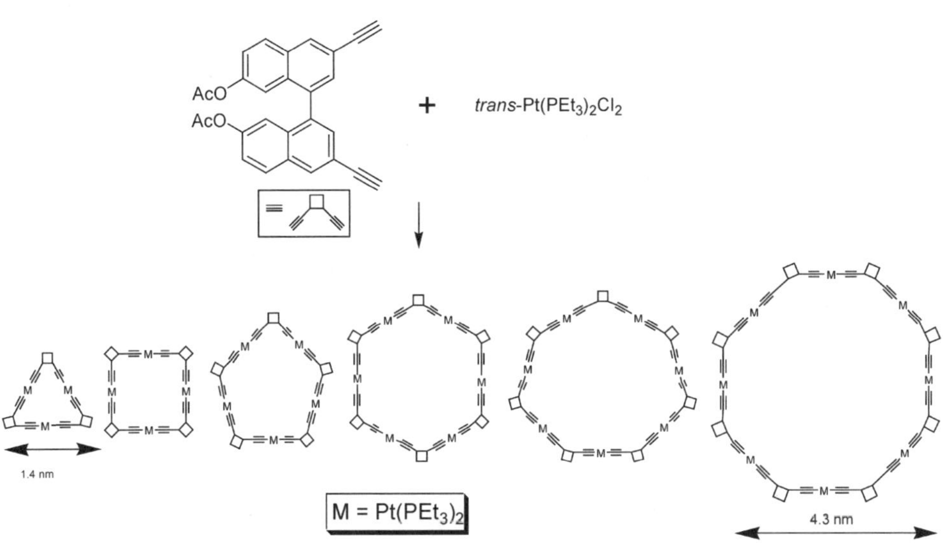

Scheme 8 One-pot self-assembly of molecular polygons [56]

4.2.2
3D Nanoscale Architectures (Polyhedra)

The expansion of metal–ligand self-assembly strategies into the third dimension leads to eye-catching 3D nanoscopic architectures. The first nanoscale cage was produced by Fujita et al., through metal-directed self-assembly of two tritopic ligands with three metal corners [57]. Interestingly, these molecular cages only formed in the presence of a suitable guest (induced-fit recognition) [58]. The same tritopic ligand, when reacted with Pd(NO$_3$)$_2$, produced a molecular sphere, M$_6$L$_8$, of octahedral geometry [59]. The utility of rigid tritopic bridging ligands in constructing nanoscale 3D structures exhibiting dimensions of 2–4 nm was also impressively demonstrated by Stang et al. [60]. Likewise, tetratopic bridging ligands have been utilised to construct nanocages [61]. For example, Shinkai et al. described the construction of homo-oxacalix[3]arene-based dimeric cages showing reversible inclusion of C$_{60}$ [62].

Induced-fit recognition was also a key motif in the construction of nanotubes with disodium-4,4'-biphenyldicarboxylate as a rod-like guest [63]. In the absence of any template or guest molecule, only uncharacterised products were formed. Several nanotubes were reported using tape-shaped oligo(3,5-pyridine) ligands and end-capped Pd(II) complexes [63a,b]. The length of these nanotubes can be controlled by simply varying the number of pyridine nuclei in the backbone of the ligand.

Fujita et al. recently demonstrated the self-assembly process of a dodecapyridine ligand where two discrete assemblies, an end-capped tube and an open-tube arrangement, are interconverted into each other on a relatively flat potential energy surface (Scheme 9) [64]. The conversion of the major to the minor component in the presence of sodium biphenylcarboxylate is slowly promoted only at high concentrations, and much more strongly by removing the minor complex through crystallisation. A similar strategy has been utilised to construct quantitatively hetero-metallic nanoscale catenanes [65].

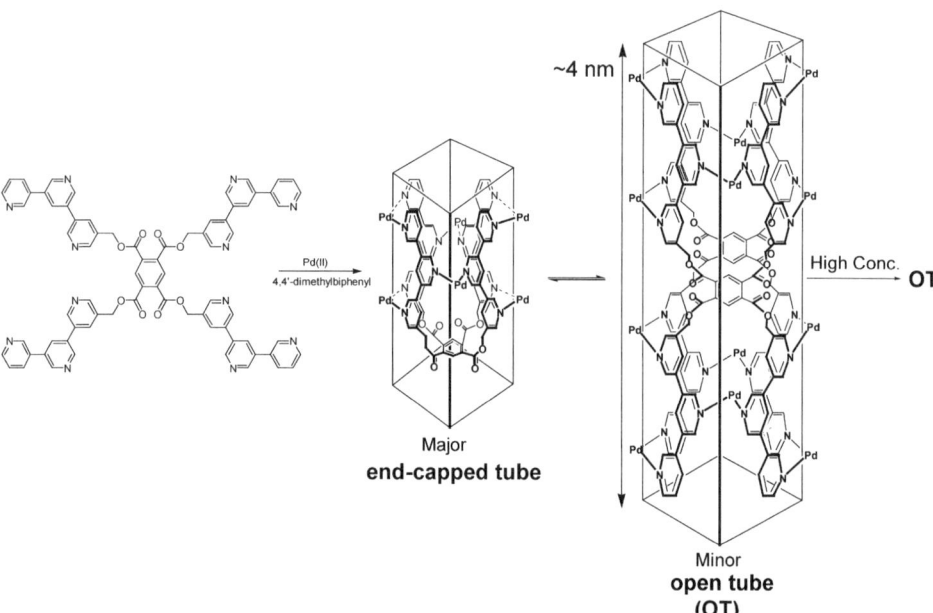

Scheme 9 A dodecapyridine ligand forming an end-capped tube and an open tube arrangement in the presence of Pd(II) [64]

Stang et al. prepared a family of nanoscale 3D arrays via non-covalent interactions using rigid tritopic bridging ligands and rectangular corner units [66]. The highly symmetrical cage compounds are described as face-directed, self-assembled truncated tetrahedra with T_d symmetry. A nanoscale adamantanoid framework was constructed using an angle-directed, coordination-driven approach [67] by combining six angular ditopic units with four angular tritopic units (Scheme 10). A family of nanoscopic prisms of approximate size 1–4 nm has also been reported, using a similar approach [68].

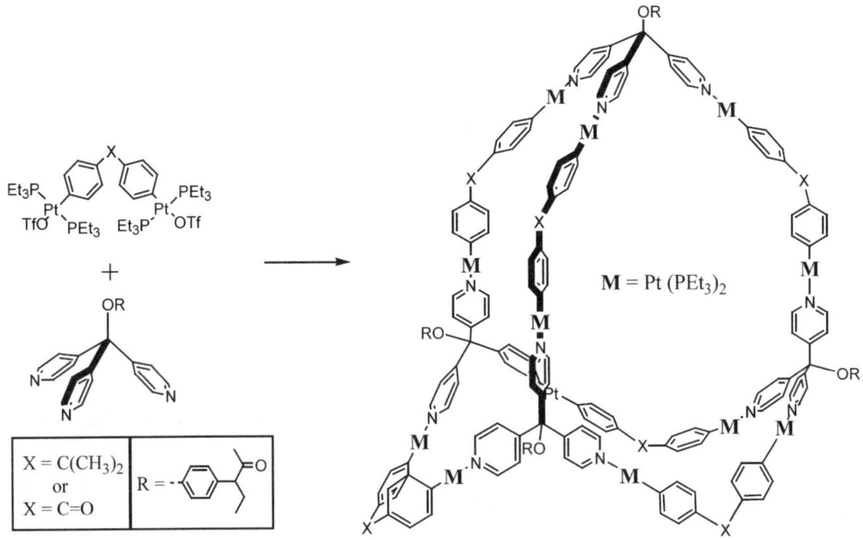

Scheme 10 A nanoscale adamantanoid framework by Stang et al. [67]

Later work by the same group produced a fascinating self-assembled dodecahedra with outer dimensions of ca. 5 and 8 nm from 50 predesigned components via 60 metal-ligand bonds [69]. The resultant assembly was characterised by pulse gradient spin-echo (PGSE) NMR techniques [70] and ESI-MS.

4.3
Nanoscale Self-Assemblies Built from Bidentate and Tridentate Ligands

2,2′-Bipyridines, [1,10]phenanthrolines, catechols and terpyridines (Scheme 11) are among the most commonly used chelating ligands for metallo-supramolecular architectures. As a ligand, phenanthroline is structurally very similar to bipyridine, but its higher rigidity leads to a better directionality in coordination processes. Similarly, catechols are rigid bidentate ligands with a high directionality, as are terpyridines, despite their conformational flexibility.

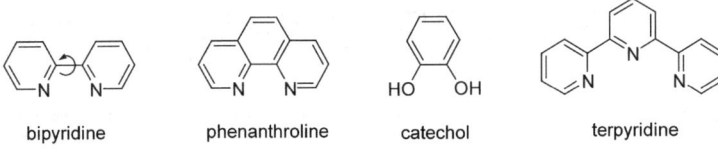

Scheme 11

Multimetallic nanoscale arrays utilising bi- or terpyridines have been prepared by means of two different approaches (Scheme 12). In approach **A**, a bi- or terpyridine-type ligand self-assembles in the presence of metal ions with itself to homoleptic nanoscopic arrays, such as grids or circular helicates depending on the rigidity of the bridge. Alternatively, in approach **B** a bi- or terpyridine type ligand is combined with another ligand, furnishing heteroleptic motifs, such as ladders, cylinders or grids.

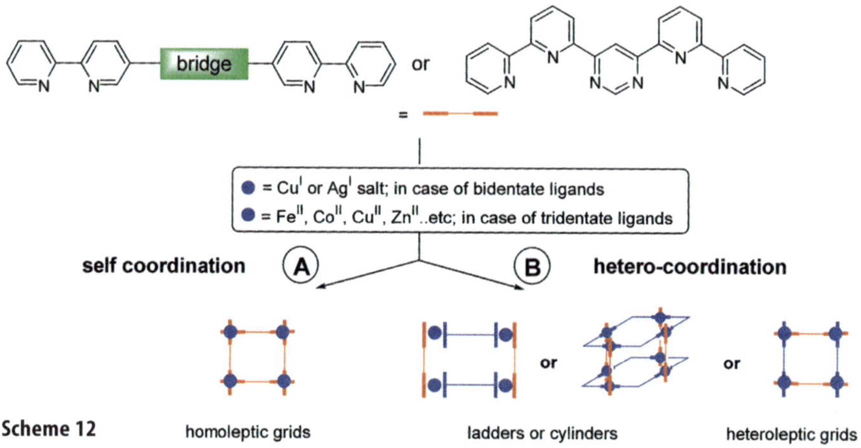

Scheme 12

A variety of supramolecular structural motifs were created using these ligands as the chelating group, such as racks [71], ladders [72], grids [73], ircular helicates [74], and cylinders [75]. In this article ladders, grids, and circular helicates of nanoscale dimensions will be discussed.

4.3.1
2D Assemblies (Homo- and Heteroleptic Aggregation)

4.3.1.1
Homoleptic Aggregates

The metal-directed self-assembly of rigid and linear oligobipyridines can be used for the programmed formation of nanoscale grids. J.-M. Lehn, the leading pioneer in this area, and co-workers, have reported numerous examples of homoleptic nanoscale grids ranging from ~2 to 5 nm in size [73a–m]. While the smaller grids had been prepared by simply mixing oligopyridines with metal ions, such as Cu^+ or Ag^+ salts, later investigations aimed at constructing higher order nanoscale grids, such as a [5×5] grid, resulted in a mixture of equilibrating [4×5] grid and a quadruple helicate [76]. Both structures could be solved by X-ray structure analysis. A similar equilibrium, now involving a double helical, a triangular and a square structure was detected in solution for another, but much shorter, bis(bipyridine) ligand [73d]. A mechanistic investigation of such self-assembly processes clearly suggests a stepwise construction of these multicomponent architectures [77].

Another level of complexity addresses the controlled introduction of a given metal ion at specific sites resulting in a heterometallic [2×2] grid of about 2 nm size [78]. A multi-grid array was built both by metal ion coordination and hydrogen-bonding [79].

A few nanogrid structures based on terpyridine-type ligands have also been reported [73k], e.g. a chiral self-assembled [2×2] grid with predetermined configuration by Zelewsky et al. [80], a rigid tetra-cobalt(III) [2×2] grid by Glass et al. [81], and a related iron(II) box by Constable et al. [82].

Nanoscale hexagons, such as terpyridine-Ru(II) and Fe(II) macrocycles, were constructed by stepwise self-assembly procedures [83]. The resulting macrocycles are expected to be ~4 nm along the diagonal.

4.3.1.2
Heteroleptic Aggregates

A higher level of structural complexity can be obtained when heteroleptic versus homoleptic interactions can be controlled in dynamic complexes, and not only in kinetically inert systems (e.g. ruthenium complexes). An interesting example solely attributed to recognition motifs [73c] of hetero-recognition versus homo-recognition can be illustrated by the reaction of **2** with **3** in the presence of silver ions (Scheme 13).

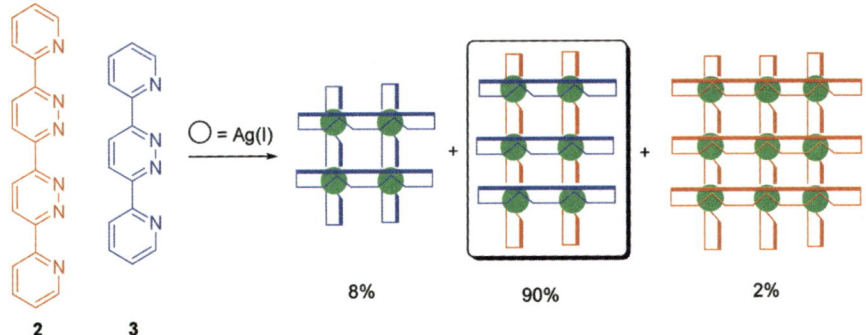

Scheme 13 An interesting case of hetero vs. homo recognition in the formation of grids [73c]

Another concept to control heteroleptic versus homoleptic interactions was introduced by Schmittel et al. [84]. The so-called HETPHEN (where HET denotes heteroleptic and PHEN, phenanthroline) concept allows the building of selectively heteroleptic bisphenanthroline complexes at Cu^+, Zn^{2+} and Ag^+ centres when using specially designed phenanthroline binding sites [85]. As a test case for the concept, heteroleptic [2×2] and [2×3] nanogrids were prepared [86]. When parent phenanthrolines were incorporated into linear bisphenanthrolines, a mixture of homoleptic and heteroleptic grids resulted. Using phenanthrolines along the HETPHEN protocol, however, afforded the heteroleptic [2×2], [2×3], and [3×3] nanoscale grids in quantitative yields [86]. The resulting grids were estimated to have a length of 6.0 nm while exhibiting remarkable square openings (>370 Å^2). These assemblies represent the largest discrete supramolecular grids to date (Scheme 14).

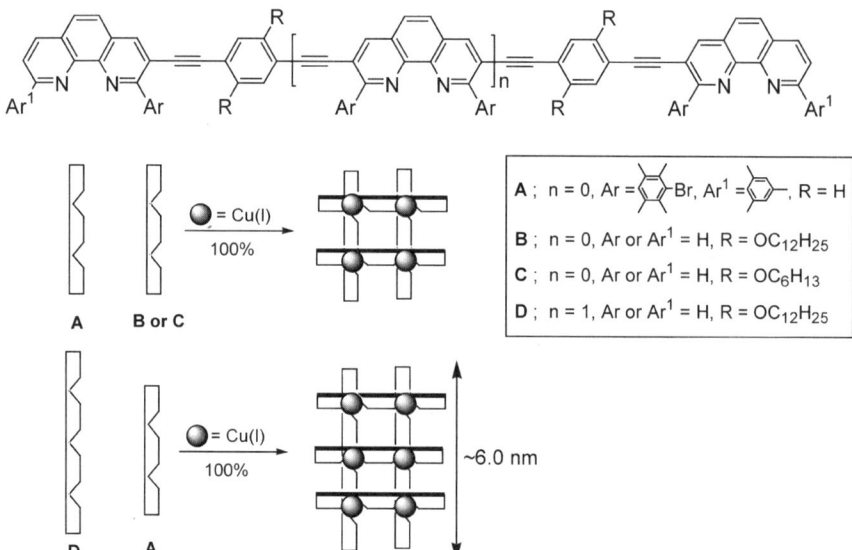

Scheme 14 Heteroleptic nanoscale grids along the HETPHEN (HET=heteroleptic, PHEN= phenanthroline) concept [86]

4.3.2
3D (Heteroleptic) Assemblies

4.3.2.1
Tubes and Cylinders

Several examples of 3D architectures using bidentate ligands have been reported. As outlined above, these motifs are heteroleptic in nature. The approach by Lehn et al. to nanoscale heteroleptic cylinders makes use of rigid linear polytopic ligands, disc-like molecules and appropriate metal ions [75]. Following this strategy several nanocylinders ranging from ~2 to 4 nm in length were prepared (Scheme 15). Again, the selective heteroleptic over homoleptic complexation is ascribed to recognition motifs and maximum site occupancy.

Functional, Discrete, Nanoscale Supramolecular Assemblies

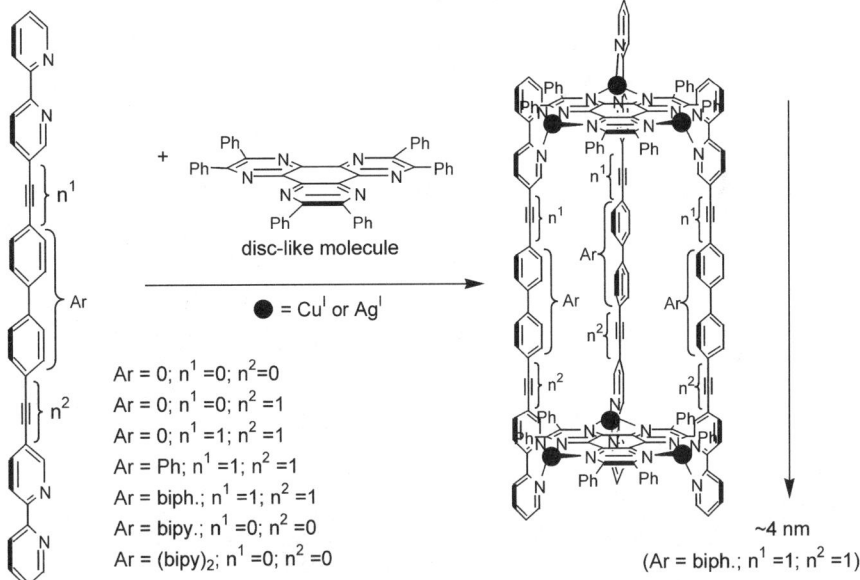

Scheme 15 Preparation of nanoscale cylinders [75]

While the first generation of cylindrical architectures was based on bipyridines and tetrahedrally coordinating metal ions, the concept was later extended to terpyridines and octahedrally coordinating metal ions by Lehn et al. [75]. A family of nanoscale cylinders was prepared by multicomponent hetero-self-assembly. As above, the process was explained by maximum site occupancy and cooperativity rules.

A more general approach to heteroleptic aggregates is again possible by using the HETPHEN concept [87]. A family of silver and copper nanoscaffolds was constructed by Schmittel et al. ranging from ~3.5 to 6 nm in height and with remarkable cavities (each cavity >5,000 Å^3). The nanoboxes and triple-deckers were constructed by reacting rigid polytopic ligands and a macrocycle [88] by means of tetrahedrally coordinating metal ions (Cu$^+$, Ag$^+$). In the case of the triple-decker, the self-assembly process involved 11 components of three different types. Copper and silver nanoboxes could readily be prepared and the silver aggregates converted into the copper(I) assemblies as monitored by UV/vis and by ESI-MS [87]. The silver box could be regenerated from the Cu$^+$ box in the presence of cyanide and excess Ag$^+$ (Scheme 16, Fig. 2).

4.3.2.2
Cyclophanes and Ring-In-Ring Structures

While cyclophanes are well-known from ditopic monodentate ligands, very few reports are available on cyclophanes using the bidentate ligands bipyridine and

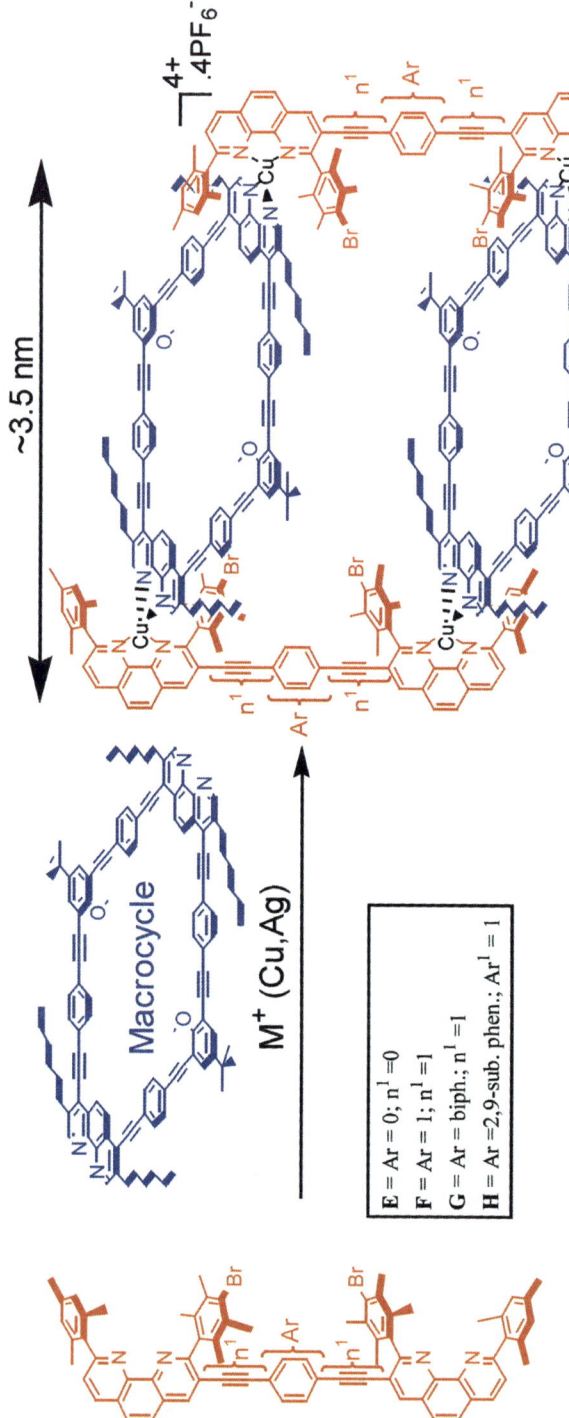

Scheme 16 Preparation of a heteroleptic nanoscaffold [87]

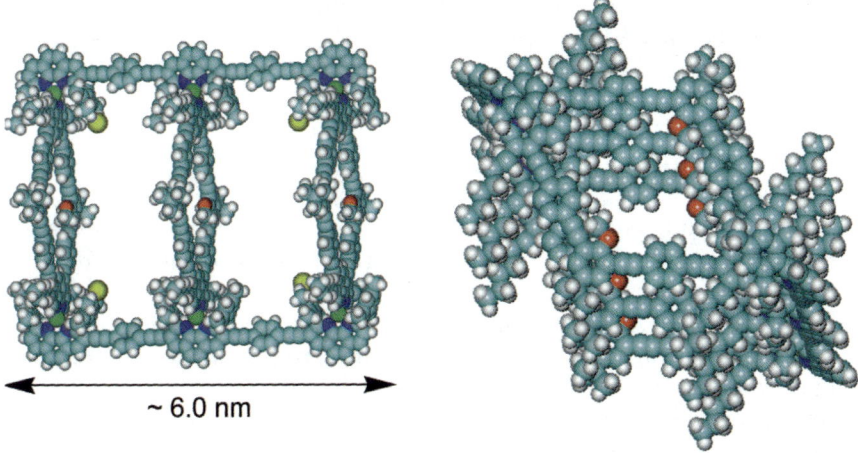

Fig. 2 Calculated structure of a triple-decker structure (view from side, left; view from top, right) [89]

phenanthroline. Lehn et al. conceived of the construction of a metallo-cyclophane by a thermodynamic self-assembly process [90]. A helical metallo-cyclophane has also been reported [91]. However, the size of these assemblies is well below nanoscale. The first nanoscale cyclophane was assembled using the HETPHEN approach by reacting a bis-bidentate macrocycle and a flexible bis-bidentate ligand in the presence of Cu(I) and Ag(I) ions [92]. A small series of nanoscale ring-in-ring structures was also constructed along those lines by Schmittel et al. [93], and later by Siegel et al. [94].

4.3.2.3
Other Structures

Saalfrank et al. [5] and others [95] have developed a predictive design strategy resulting in the synthesis of various high-symmetry coordination clusters. Structural motifs including helicates [95b, 96], mesocates, tetrahedra, cylinders, and octahedra have been reported [97]. On the basis of a bis(pyrazolyl-pyridine) ligand, Ward et al. developed a way to achieve a self-assembled molecular cube $Zn_8L_{12}^{16+}$ with S_6 symmetry [98]. By simultaneously fulfilling the symmetry requirements of both the ligand and metal centres, discrete high-symmetry clusters were generated under thermodynamic control. However, most of these structures fall well below the 2 nm scale. Analogously, Powell et al. have utilised the specific interaction of metals with small organic species (oxyhydroxy compounds) to create nanoscale aggregates [6], such as a huge polyiron cluster, in which a mineral core (metal cluster) is trapped within an organic shell, showing a lot of structural similarities with the protein ferritin.

Helical architectures have received special attention because of their structural similarity with double-stranded DNA. Several circular helicates have been reported by Lehn and others utilising bipyridine [74] and other bidentate ligands [99] as chelating groups.

5
Functional Devices

The directed and programmed construction of multicomponent nanoscale assemblies furnishes new opportunities and tremendous potential in the areas of supramolecular catalysis, molecular electronics, sensor design and optics. At the present time, however, the design of functional working models is still in the fledgling stage and lags far behind the progress made in structural engineering. We would therefore like to sketch some recent advances in the following areas: supramolecular catalysis, photoactive assemblies, molecular recognition and switches, and electroactive assemblies. Despite some recent progress, in terms of interdependence and emergence, nature and its masterfully designed functional systems outclass any known artificial systems by far.

5.1
Supramolecular Catalysis

True supramolecular catalysis has to amalgamate peculiarities and vantage grounds of supramolecular architectures and those of catalysis in order to generate a surplus value. Obviously, there is ample room for fascinating combinations of both areas that is only limited by our imagination: e.g. the nanoscale assembly can serve as a container to select molecules by size and shape, it can place reactants in adequate distance and geometry, it can use allosteric control over catalysis etc. The art and the beauty of these systems are evident from the fine interplay of weak bonding interactions and transition state control needed for the catalytic process.

Rebek et al. [100] have demonstrated the utility of hydrogen-bonded cage assemblies. When they were set up to reversibly encapsulate two guest molecules, then chemical transformations proceeded faster than in homogeneous solution [100a,b, 101, 102]. As outlined below, the substrate size is crucial for recognition by the cage assembly.

In homogeneous solution the Diels-Alder reaction of 2,5-dimethylthiophene-1,1-dioxide and benzoquinone (Scheme 17) is slow at millimolar concentrations (10% conversion after 2 days). In contrast, enclathration of diene and dienophile into the cage assembly accelerates the reaction to a sizable extent (55% conversion after 2 days). The formed Diels-Alder adduct is ejected in favour of two molecules of p-quinone. A similar enhancement is observed for other Diels-Alder reactions [103].

By the same group, a spectacular example of a 1,3-dipolar cycloaddition furnishing a triazole (Scheme 18) was established in a reversibly formed, self-as-

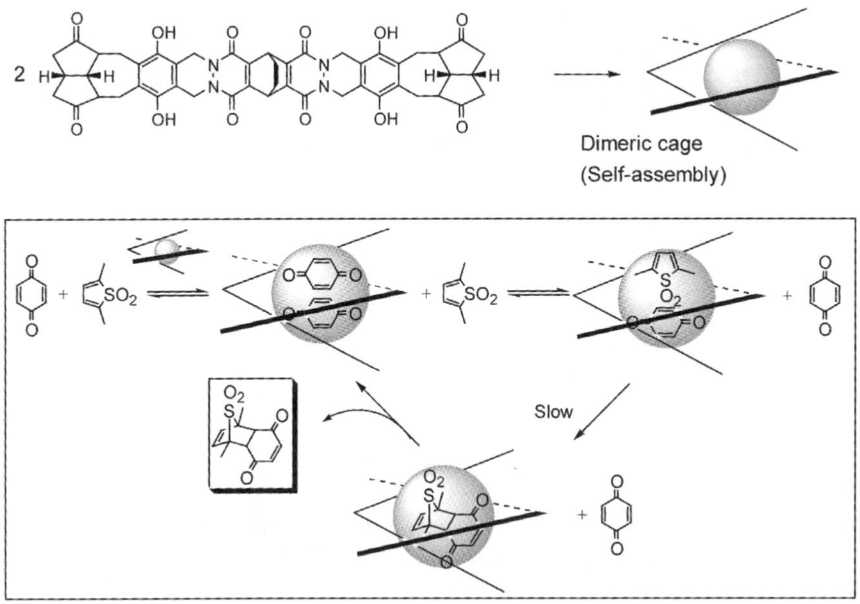

Scheme 17 A capsule-accelerated Diels-Alder cycloaddition [103]

sembled dimeric capsule [104]. This example allowed the direct observation of a Michaelis complex (unsymmetrical loaded complex). In the absence of the dimeric capsule, phenylacetylene and phenylazide reacted very slowly (4.3×10^{-9} $M^{-1}s^{-1}$) at millimolar concentrations. In contrast, in the presence of the dimeric capsule the 1,3-dipolar cycloaddition takes place smoothly. Only the 1,4-isomer is formed and this compound is liberated by the addition of N,N-dimethylformamide (DMF). The analysis indicated that for formation of the 1,4-adduct, the two substrates assumed a space-saving side-by-side arrangement, whereas the cavity space is insufficient for the 1,5-adduct. As a drawback, the system did not solve the turnover problem.

A remarkable case of autocatalysis, used to amplify amide bond formation, was noticed in the presence of dimeric capsules [105]. When **4a** or **4b** and **5** were reacted in the presence of N,N-dicyclohexylcarbodiimide (DCC) in homogeneous solution (Scheme 19), the anilides **6a** and **6b** formed at almost identical rates along with dicyclohexylurea (DCU). In the presence of the capsule, however, the initial rates were much lower due to the selective binding of DCC by the dimeric capsule. Under such conditions, the reaction of the shorter acid **4a** was much faster than that of the longer **4b**. This can be explained, since **6a** and DCU are better guests for the dimeric capsule than DCC. So, once the products are formed, they displace DCC from the capsule to the bulk solution where it can react with the remaining acid and **5**. Hence, the rate of the reaction increased with higher conversion. Since **6b** is too large to fit into the capsule, it cannot displace DCC and therefore no acceleration in kinetics was observed.

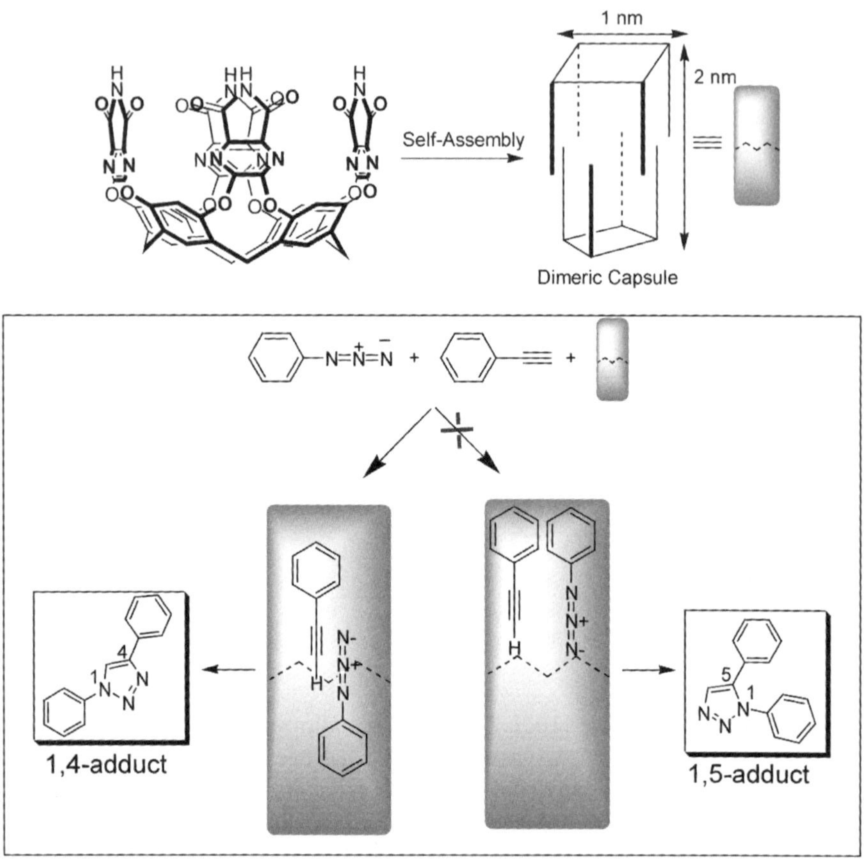

Scheme 18 A capsule-accelerated 1,3-dipolar cycloaddition [104]

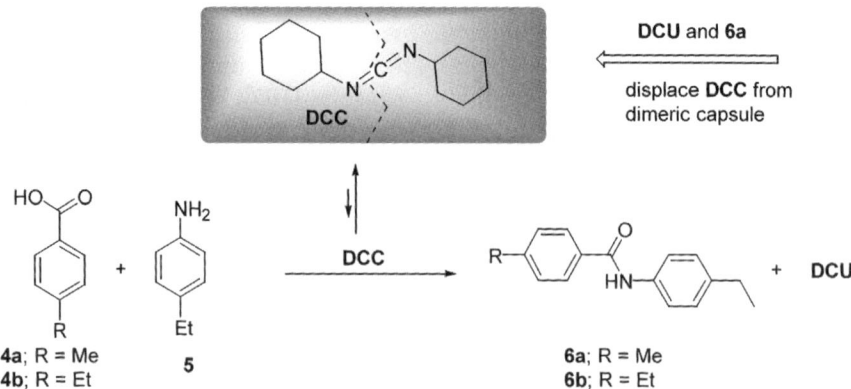

Scheme 19 Amplification by compartmentalisation [105]. *DCU* Dicyclohexylurea, *DCC* N,N-dicyclohexylcarbodiimide

Fujita et al. [106] have shown that a nanocage complex can act as a reverse phase-transfer catalyst for the Wacker oxidation of olefins by free [Pd(en)(NO$_3$)$_2$] (10%) in aqueous media.

Though the Pd(II)-N bond is dynamic, the coordinative Pd(II)-cage is not affected by the presence of [Pd(en)(NO$_3$)$_2$]. The water-soluble cage has a hydrophobic cavity that can enclathrate hydrophobic molecules such as aromatic alkenes (Scheme 20). Transport of the filled molecular container {accommodating [Pd(en)(NO$_3$)$_2$] as catalyst} from the organic to the aqueous phase leads to the oxidation of the aromatic alkenes. As the resultant ketones are less hydrophobic than aromatic alkenes, the net result is declathration of the product and enclathration of reactant. This cycle continues until the oxidation is complete. Electron-rich compounds that have strong enclathration could be readily oxidised while the electrodeficient nitrostyrene was hardly reacted.

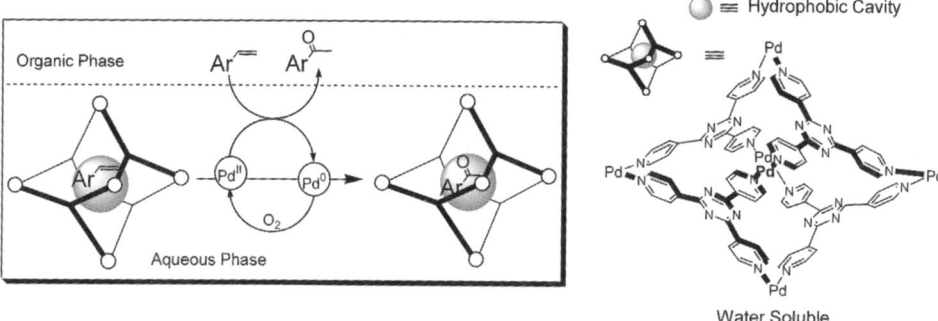

Scheme 20 Wacker oxidation of styrene with a nanocage acting as reverse phase catalyst [106]

A few other reports are also available where supramolecular assemblies were used for catalytic purposes [107] or cavity-directed synthesis [108].

Lin et al. reported on a family of Pt(II) polygons and their utility in enantioselective catalytic diethylzinc additions to aromatic aldehydes [50]. The basic idea is to incorporate chiral ligands such as 1,1′-bi-2-naphthol (BINOL) into supramolecular architectures. Triangle 7 in combination with Ti(O^{i}Pr)$_4$ showed excellent catalytic activity for the addition of diethylzinc to aromatic aldehydes, furnishing secondary alcohols in high yield and enantioselectivity (Scheme 21). A later report by the same group described the assembly of a chiral cyclophane and its utility in asymmetric catalysis [109]. Interestingly, this case showed some size selectivity, because enantioselectivity dropped significantly when smaller aromatic aldehydes were used. Analogous chiral molecular rhenium squares proved useful for enantioselective sensing, due to luminescence quenching by chiral amino alcohols [49].

To mimic cytochrome P450, Hupp et al. encapsulated a manganese porphyrin into a metallosupramolecular square [110]. The superstructure served

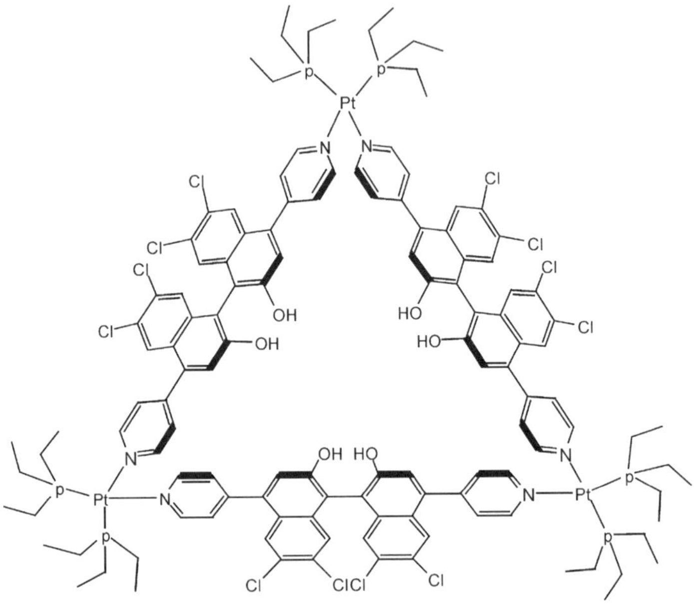

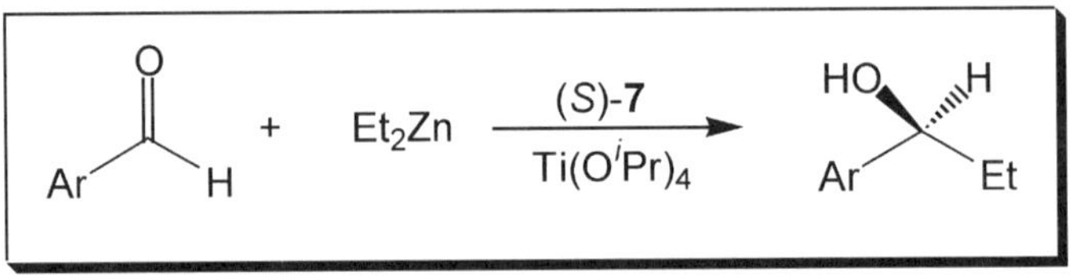

Scheme 21 Enantioselective catalytic diethylzinc additions to aromatic aldehydes [50]

to isolate the catalytic centre from other reactive centres, thus extending its functional lifetime (Scheme 22). Indeed, while the naked manganese porphyrin **9** degrades fairly rapidly under catalytic reaction conditions, the encapsulated one showed increased stability and reactivity. For example, encapsulation led to a tenfold increase in the turnover number (TON) of styrene epoxidation [111]. The half cavity defined by (**8+9**) is still large enough to permit many olefinic substrates to reach the catalytically active manganese centre. As a consequence, the reactivity of the complex is reported to be very sensitive to the olefinic substrate size.

The ring-opening of epoxides using Cr(III) and Co(III)-based salen catalysts is well-known [112]. Oligomeric versions of the Cr(III) catalysts show activities significantly greater than those of monomeric analogues [113]. Recently, Mirkin et al. reported on a supramolecular catalyst designed for allosteric control that exhibits a significant increase in the rate and selectivity of the ring-opening of cyclohexene oxide [114]. They built heteronuclear rectangles containing catalytically active Cr(III) salen sites and Rh–thioether complexes

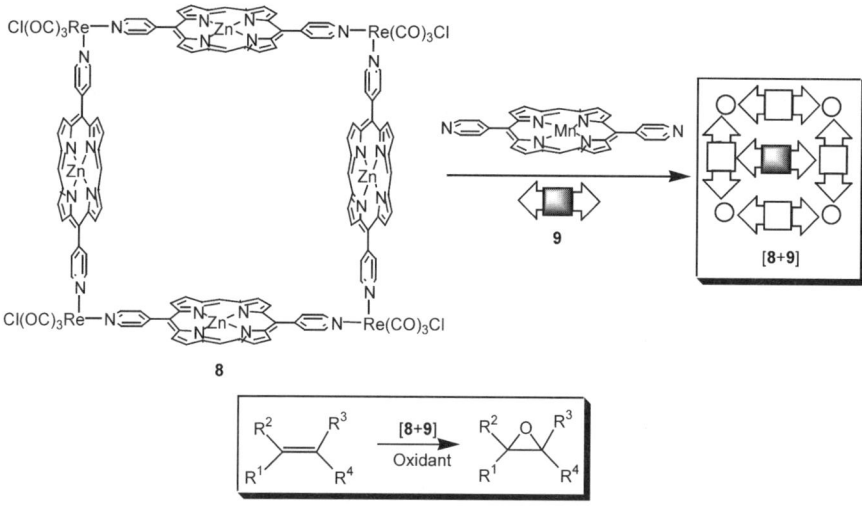

Scheme 22 Olefin epoxidation by an encapsulated manganese porphyrin [110]

for allosteric control. As depicted in Scheme 23 the weak thioether–Rh bond could be selectively cleaved by the addition of CO and chloride. As a consequence, a Rh(CO)(Cl) complex formed and the structure opened in such a way that the catalytically active Cr(III) became accessible. Upon purging with nitrogen or in high vacuum, the opened structure experienced loss of CO and Cl⁻ leading to a closed structure. Hence, the Rh(I) centres act as switches for the catalytic activity of the nanoscale capsule. For the closed structure the ring-opening of cyclohexane epoxide by trimethyl silyl azide (TMSN$_3$) resulted in product formation with 68% enantiomer excess (ee), whereas a Cr(III)-salen monomeric analogue only provided 12% ee. This was additionally accompanied by a 20-fold increase in rate compared to that observed for the monomeric analogue. The rate of formation of 1-azido-2-(trimethylsiloxy)-cyclohexane was doubled in the presence of CO and Cl⁻ compared to the closed structure.

There are currently many more efforts to apply supramolecular concepts in the field of catalysis that show wonderfully the encouraging prospects in this field. Reek et al. have prepared bidentate ligands via the assembly of bis-Zn(II)porphyrin and functionalised pyridine ligands for rhodium complexes designed for the hydroformylation of 1-octene and styrene (Scheme 24) [115]. Later this idea was extended to a hexaporphyrin assembly [116].

A porphyrin-cyclodextrin 1:2 complex proved to be a supramolecular sensitiser in water providing photo-oxygenation with high TONs, e.g. for phenol degradation. High yield oxidations were obtained for L-methionine methyl ester and uracil [117].

In the realm of smaller supramolecular aggregates (<2 nm) Fujita et al. have shown the utility of cadmium-bipyridine squares for heterogeneous catalysis [118] in the cyanosilylation of aldehydes. The catalytic activity depended very

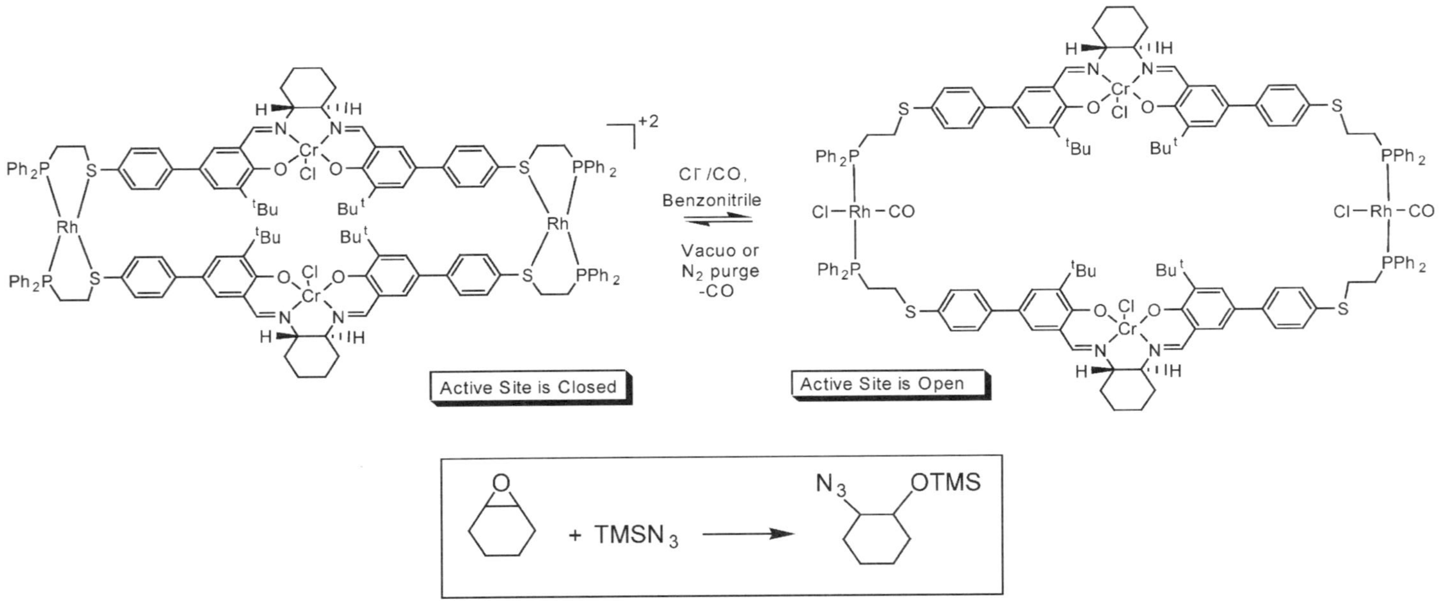

Scheme 23 Allosteric control in a heteronuclear rectangle catalyst for cyclohexene oxide opening [114]. *TMSN₃* Trimethyl silyl azide, *OTMS* trimethylsiloxy

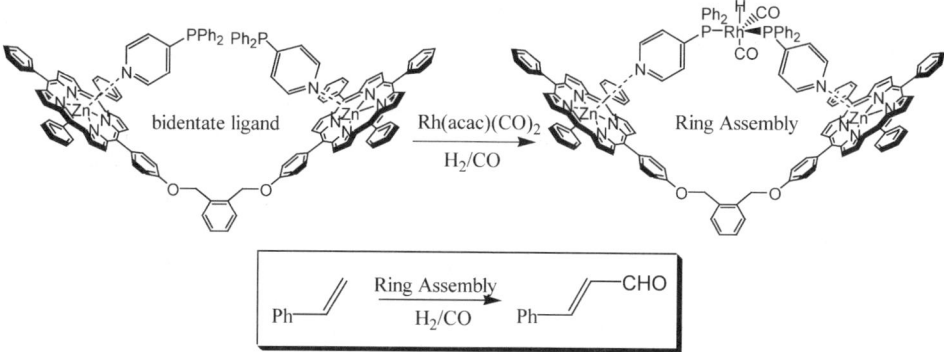

Scheme 24 Using a bidentate ligand assembled from a bis-Zn(II)porphyrin and functionalised pyridine in Rh-catalysed hydroformylation [115]

much on the size of the aldehydes. Süss-Fink et al. designed a supramolecular cluster (<2 nm) that proved to be the mildest and the most efficient catalyst for the hydrogenation of benzene [119].

Supramolecular assistance to covalent synthesis is often explored as a means of constructing ring systems such as catenanes, knots and macrocycles. A nanoscale macrocycle was constructed using a shape-persistent tricationic platinum template [120].

5.2
Photoactive Assemblies

The rich structural variety of supramolecular assemblies and photoactive groups can amalgamate profitably for a mélange of purposes: light-harvesting systems, electron or energy transfer in nanostructured systems, photochemical devices etc.

In photosynthesis the energy is collected by light-harvesting antennae systems, from which it is funnelled to the photosynthetic reaction centre. This ingenious arrangement in nature has attracted the ample attention of biomimetic chemists owing to its effective energy conversion. The structure known from the bacterial photosynthetic system revealed that the reaction centre is surrounded by many antennae pigments in cyclic arrangements, to collect and transmit light energy efficiently [121]. Subsequently, several discrete photoactive assemblies have been reported, quite often containing a porphyrin core to mimic the natural systems. Still, the construction of giant artificial photoactive aggregates using self-assembly techniques is a challenging task. It is interesting to note that almost all of the photoactive assemblies studied so far rely on metallosupramolecular building principles, in particular using Pd/Pt and Re interactions.

5.2.1
Homoleptic Devices

Würthner et al. incorporated four perylene bisimide chromophores into nanoscale Pd and Pt squares. Since the perylene fluorescence was not quenched by the metal ion coordination, these systems were able to show fluorescence quantum yields near 100% [53b,c]. Later the self-assembly of an analogous nanosize scaffold with 16 additional dyes (pyrenes) was demonstrated to result in efficient energy transfer from the outer antenna pyrenes to the inner perylene chromophore (Scheme 25) [53d]. Some other reports have also utilised Pt(II) metal corners to harvest photoactive devices [122].

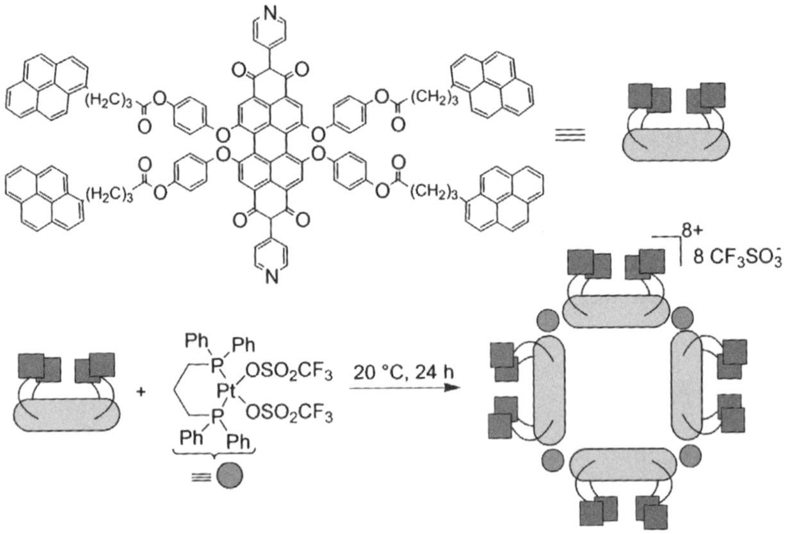

Scheme 25 A nanoscale metallosupramolecular square with 16 appended pyrenes [53d]

A recent report by Lin et al. described chiral nanoscale Pt(II) molecular squares exhibiting dual luminescence, which is of potential use as chiral sensory material [123]. Due to a stepwise, directed approach, the entropic disadvantages in building larger functional metallo-macrocycles could be overcome.

Recently, Lehn et al. reported on a [2×2] metallogrid that exhibits reversible changes in its optical properties triggered by multiple protonation and deprotonation steps [124]. Due to the metal ions in the grid architectures, the eight hydrazone N–H protons are sufficiently acidic to allow for a modulation of the protonation stage (Scheme 26). The colour of the solution changed from pale yellow at low pH to orange and finally to deep violet above neutral pH. This process was found to be reversible.

An ingenious hexameric circular assembly was recently reported by Takahashi and Kobuke (Scheme 27) [125]. Starting from the *meta*-gable bis-zinc

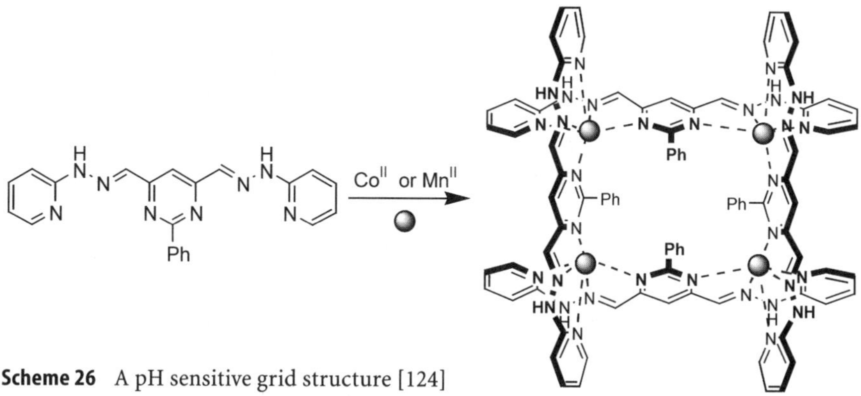

Scheme 26 A pH sensitive grid structure [124]

porphyrin **10**, the self-assembly furnished a dodecaporphyrin barrel structure with a diameter of 4.1 nm. The fluorescence quantum yield of the suprastructure relative to that of the monomeric bisporphyrin was 0.51.

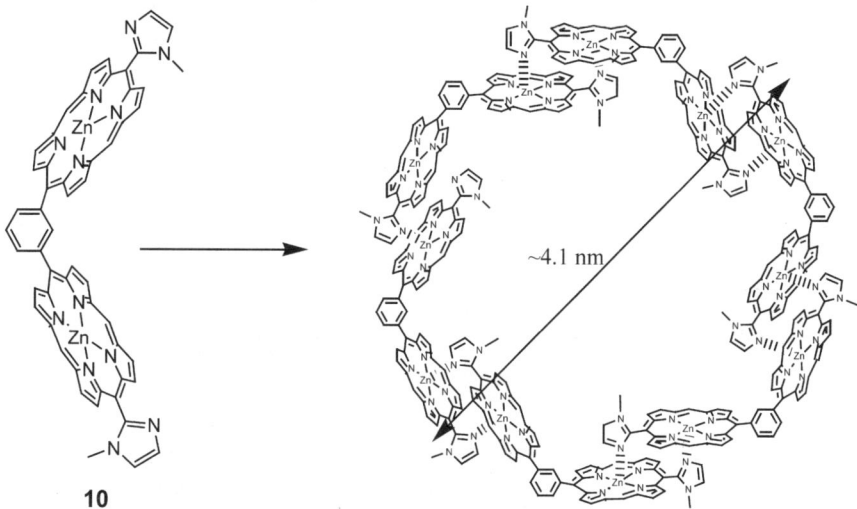

Scheme 27 A highly fluorescent dodecaporphyrin barrel structure [125]

Kuroda et al. demonstrated the construction of multiporphyrin arrays (nonameric and heptadecameric porphyrins) using a different templating strategy. The fluorescence of the central porphyrin is increased to 77 times that of the free porphyrin [126]. Several photoactive self-assembled squares were reported by Hupp et al. using rhenium metal corners [39a]. A recent example by the same group showed the utility of these devices to construct chemoresponsive photonic lattices capable of functioning as chemical sensors [127]. A smaller porphyrin dimer was constructed using rhenium metal corners and the ligand remained fluorescent even after coordination [128]. A similar-sized assembly was prepared by Lu et al. from 18 components showing on/off switching of emission by presence/absence of aqueous solvent [61b].

Aida, Tsuda et al. have reported on a 3-pyridyl substituted zinc porphyrin, which exhibits supramolecular thermochromism [129]. The compound (shown in Scheme 28) displayed a stepwise colour change from green to yellow to red on heating from 0 to 50 to 100 °C in toluene. The vivid colour change is most likely brought about by a thermal self-assembly process to a cyclic tetramer via the axial coordination of the pyridyl group to the zinc porphyrin.

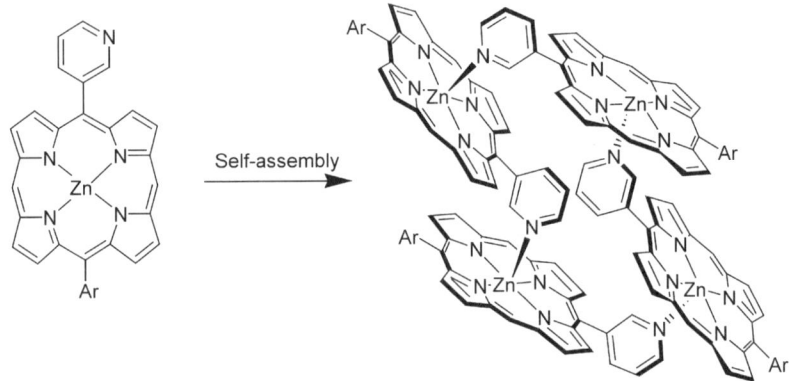

Scheme 28 A 3-pyridyl-substituted zinc porphyrin tetramer exhibiting supramolecular thermochromism [129]

5.2.2
Heteroleptic Devices

In the presence of 12 *trans* Pd(II) connectors a stunning 3+3 array was prepared by the self-assembly of 21 molecular entities of three different kinds of porphyrin core units (Scheme 29). Drain et al. demonstrated that several metal ions such as Mn(II), Ni(II), Co(II), or Zn(II) could be incorporated into the porphyrin core of this assembly [130]. As precursors to the device, nanoscale structures of the porphyrin arrays and aggregates of controlled size were deposited on surfaces. Atomic force microscopy and scanning tunnelling microscopy of these materials show that the choice of surface (gold, mica, glass, etc.) may be used to modulate the aggregate size, and thus its photophysical properties.

5.3
Molecular Recognition

Polygons and polyhedra derived from self-assembly processes often provide a defined cavity which is of good use for molecular recognition or sensing. In

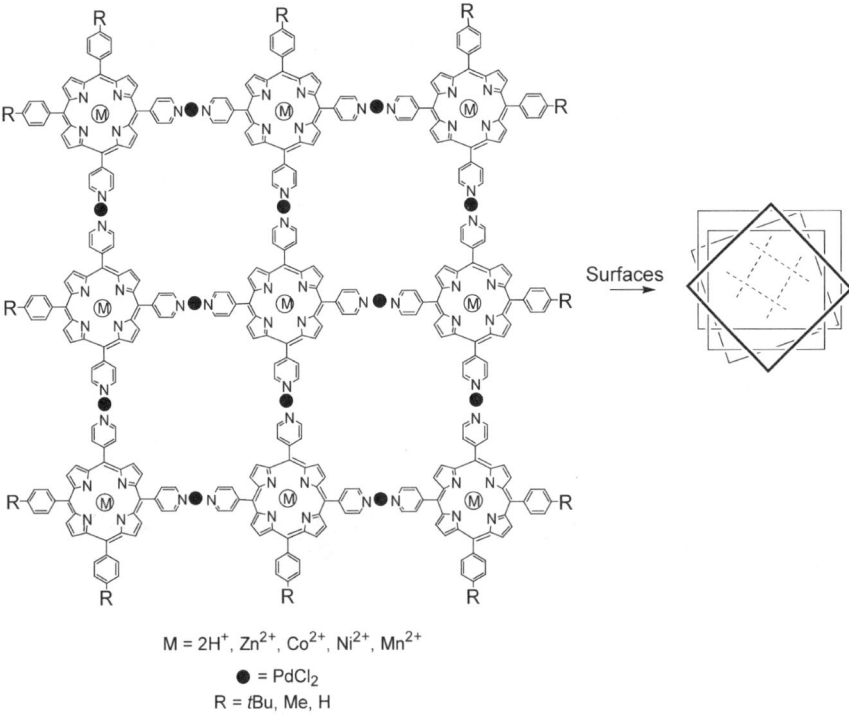

Scheme 29 3+3 array prepared by the self-assembly of 21 molecular entities [130]

some cases, the guest inclusion plays a major role in driving the equilibrium to a unique entity (induced-fit recognition). Several factors must be taken into account to tune the host-guest interactions: (1) shape and size of both the host and guest, (2) complementary binding sites and interactions, and (3) medium (solvent). The early, seminal investigations about binding of alkali cations by cryptates, cryptophanes and carcerands [1] paved the way for our understanding of molecular recognition phenomena in supramolecular assemblies. Due to space restriction, supramolecular structures with small guest molecules will not be covered, although there are several spectacular examples known [131–133]. Herein, we will discuss recent progress of host-guest chemistry of nanoscale supramolecular assemblies, with a focus on the complexation of larger molecules and of more than one guest.

5.3.1
Metallosupramolecular Systems

Metallosupramolecular nanoscale systems have been frequently analysed for their potential to accommodate larger guests. Fujita and Kusokawa [102a] were able to show that guests were enclathrated in three different manners, depending on the shape and size of the guests. As shown in Scheme 30 the presence of small guests (<8 Å) resulted in a tetrahedral 1:4 complexation, medium size guests furnished orthogonal 1:2 complexation, whereas large size guests (>8 Å) afforded simple 1:1 complexation of host and guests, respectively. Progressive addition of guest molecules to the cage revealed that the enclathration was strongly cooperative: i.e. no intermediate complexes **11**(Guest)$_n$ (n=1–3) were found. A similar guest inclusion was found in a self-assembled cage reported by Rebek et al. [101f].

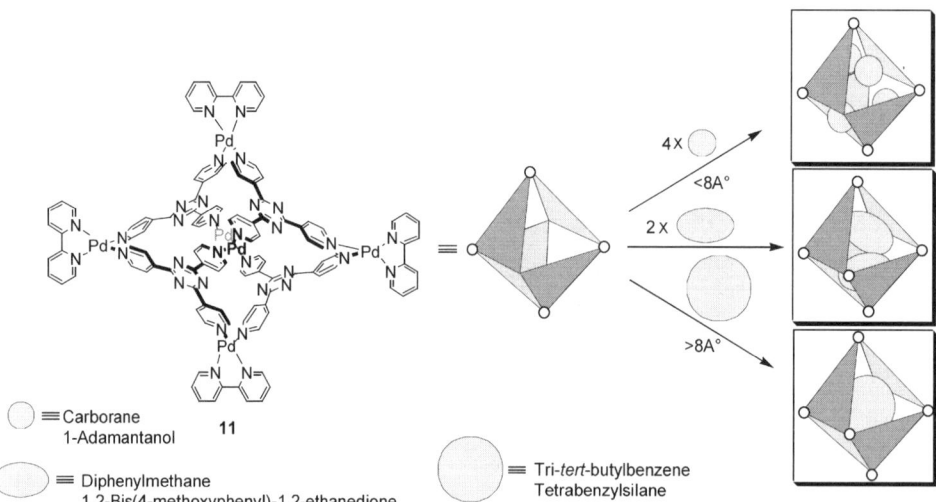

Scheme 30 1:4, 1:2 and 1:1 complexation of guests in a supramolecular cage [102a]

In a striking example, Shinkai et al. showed that a metallo-supramolecular capsule can be used to purify C_{60} very effectively [62], conceptually following a purification method for C_{60} and C_{70} that used calixarenes as capsules or sieves (Scheme 31) [134]. Shinkai constructed a Pd(II) capsule using pyridine-tethered calixarenes **12**. With smaller guest molecules such as DMF, adamantane, anthracene, pyrene, and others, no guest inclusion complexes were found, due to the large windows present in the capsule. However, ^{13}C-enriched [60] fullerene could be enclathrated efficiently, while ligand **12** itself did not exhibit enclathration of C_{60}. The enclathration could be monitored by ^{13}C and ^{1}H NMR.

In a recent report by Fujita et al. a prism-like Pt(II)-cage was used to control the equilibrium between enol and keto tautomers (Scheme 32) [135]. To avoid homoleptic assemblies in the presence of the triazine **13** and pyrazines as pillars, a template effect exerted by large aromatic molecules such as triphenylene was used. The selective formation of the heteroleptic cage was attributed to the interplane separation, which is ideal for aromatic intercalation (~3.5 Å). The resulting heteroleptic cage remained kinetically stable even after the removal of the guest molecule, giving a way to explore the cage's ability to host

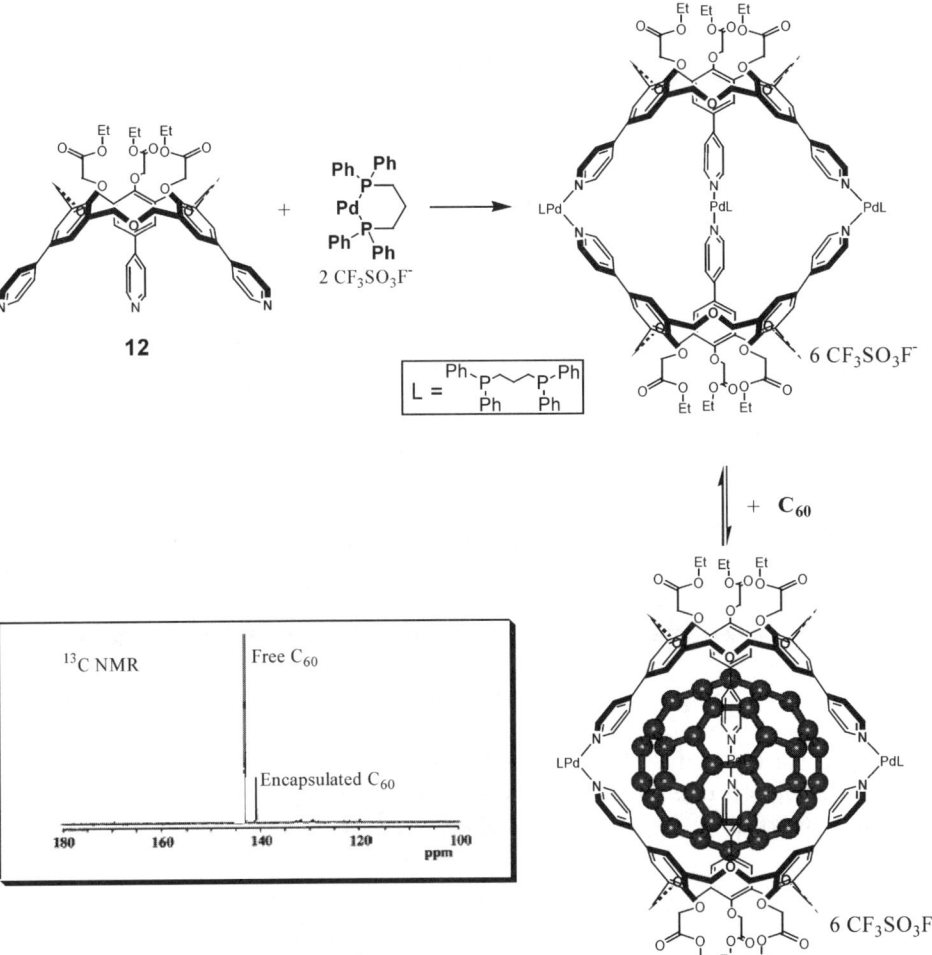

Scheme 31 A metallosupramolecular capsule entrapping C_{60} [62]. Reprinted with permission from [58]. Copyright 1999 American Chemical Society

other systems. As a consequence, it could be used to control the equilibration between the planar enol and the non-planar carbonyl tautomer of a β-diketone. In acetonitrile these exist in a 85:15 ratio that can never be separated due to fast tautomerisation. In contrast, in the presence of the heteroleptic cage molecule, the β-diketone only existed as enol due to the selective inclusion of the planar enol form.

Metallosupramolecular assemblies can also be used to stabilise labile species. The reactive, water-sensitive [$Me_2C(OH)PEt_3$]$^+$ ion was entrapped by Raymond et al. in a small metallosupramolecular capsule [136], so that it could survive for a few hours in D_2O. Fujita et al. [137] have shown that a molecular capsule can stabilise cyclic trimers of siloxanes considerably. The labile trimers

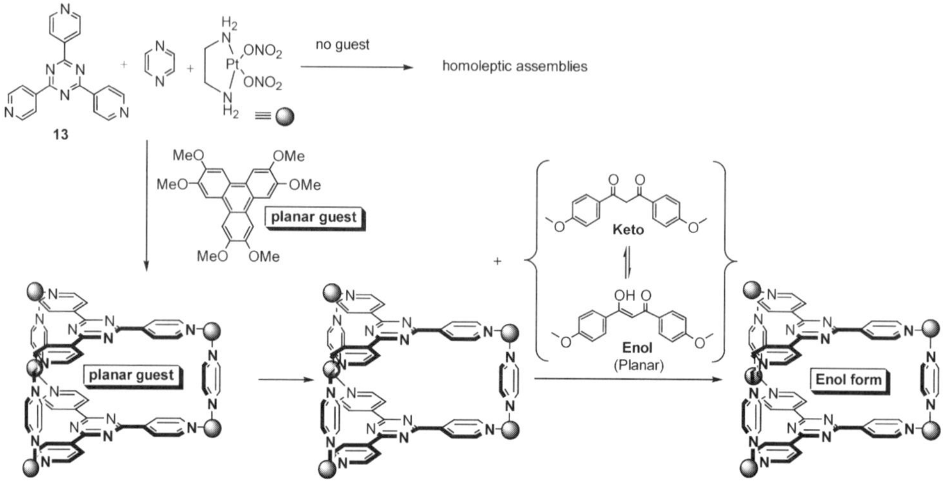

Scheme 32 Selective inclusion of a planar enol form derived from a β-diketone [135]

were effectively trapped during self-assembly of the capsules from which they could no longer escape owing to a small portal size. These clathrate structures were isolated and proved to be very stable even in acidic aqueous solutions (pH<1).

The same capsule **14** was used to stabilise the tennis-ball-shaped noncovalent dimers of *cis*-azobenzenes [138], which were selectively enclathrated from a mixture of *cis* and *trans* azobenzene (Scheme 33). The enclathrated

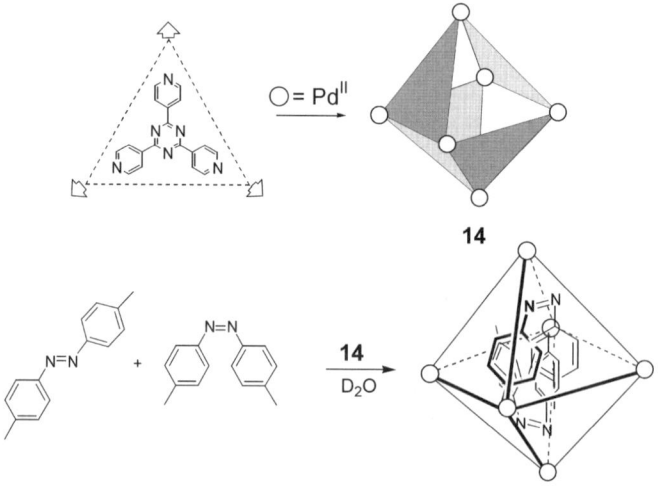

Scheme 33 Stabilisation of noncovalent dimers of *cis*-azobenzenes in a capsule [138]

hydrophobic *cis*-azobenzene dimer stability increased considerably and did not change to the *trans*-isomer even after exposure to visible light for a few weeks.

A cationic tetrahedral host with 24 positive charges can even provide shelter for a cationic guest [139] (Scheme 34). The host accommodated the tetraalkylammonium cation together with four BF_4^- ions thus creating a small sphere of inversed polarity. Competition experiments with two differently sized guest molecules (NEt_4^+ vs. NBu_4^+) indicated that the smaller guest forming a kinetically preferred complex first is slowly exchanged by the larger guest to furnish the thermodynamic product. Similar guest-exchange processes had been reported earlier [140].

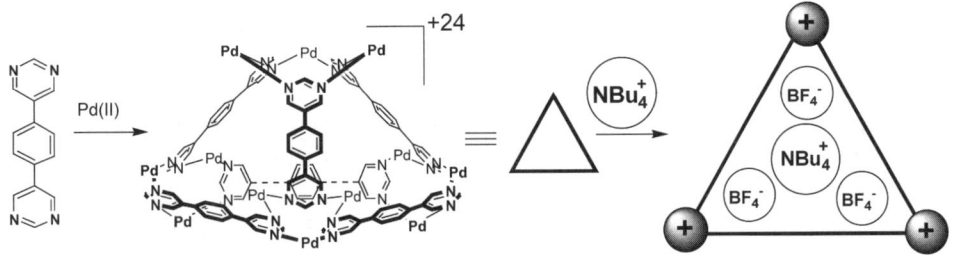

Scheme 34 Cationic guest enclathrated in a cationic host [139]

Fujita et al. have shown the electrochemically driven clathration/declathration of ferrocene in a Pd(II) capsule (Scheme 35) [141]. Because neutral ferrocene enclathrates efficiently into the cage assembly but not the ferrocenium, the encapsulated neutral molecule can be moved in and out by tuning the charge of the guest molecule. From the peak currents it was concluded that a 1:4 complex of cage and ferrocenes had formed. These observations are in line with their previous reports in which a similar-sized guest formed 1:4 complexes with cage **11**. Bulk electrolysis gave the direct evidence for the clathration and declathration process.

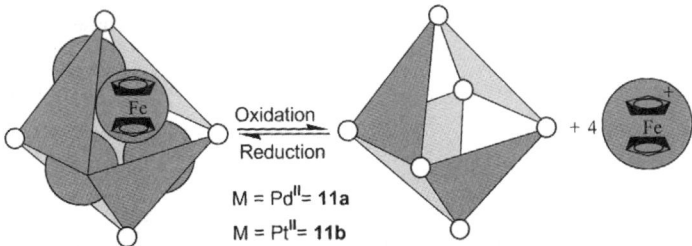

Scheme 35 Electrochemically driven clathration/declathration of ferrocene in a Pd(II) capsule [141]

Because of the high kinetic stability of the Pt(II)-cage **11b** towards acidic and basic conditions (pH<1 and pH>11), it was explored as a pH-responsive host guest system. Indeed, while dimethylaniline was enclathrated efficiently it was liberated upon protonation due to the repulsions between the generated ammonium ion and the Pt cations [142]. Several other systems were also reported that exhibit a distinct response to the pH [143].

Bosnich et al. have reported the assembly of nanoscale heteroleptic trigonal prisms and their host-guest chemistry [144]. Remarkably, 1:6 and 1:7 host-guest complexes with 9-methylanthracene formed, while another one of the prisms furnished a 1:2 host-guest complex with a very large tritopic tri-anthracene guest. Clearly, the development of such multi-site hosts may eventually lead to the construction of molecular switches and motors, which rely on guests moving between the multiple sites of a molecular host.

A dynamic metallo-supramolecular receptor library was generated by Fujita et al. [58c] in which all the receptors were in an equilibrium from which the appropriate one was selected by an optimal guest. The dynamic library was generated from the Pd(II) complexes of two different ligands. Large (spherical) or small (flat) guests stabilized the homoleptic cages $M_3(L1)_2+M_3(L2)_2$, whereas medium sized guests induced the heteroleptic receptor $M_3(L1)(L2)$.

Molecular sensors have been developed, where the binding of the guest could be detected through luminescence [145].

5.3.2
Hydrogen-bonded Aggregates

But in the field of molecular recognition, hydrogen-bonded aggregates also play an important role. Rebek et al. have reported a dimeric capsule that is able to encapsulate cryptates inside its cavity. Hence, the resulting systems are complexes within complexes and represent a second sphere of supramolecular chemistry [101a]. A similar system was set up by Raymond et al., by incorporating an alkali ion–crown ether into a metallosupramolecular cluster [102b].

A self-assembled dimeric capsule was described to accommodate two different molecules (benzene and *p*-xylene) [146]. Such a hetero-guest inclusion reflects how optimal occupancy of the capsule can be reached: two benzene molecules are too small and two *p*-xylene molecules are too crowded for the capsule. This leads to a new form of isomerism, coined as "social isomerism", which arises from the orientation of two different molecules in the container, e.g. in a cylindrical capsule. The orientation of one guest depends critically on the presence and nature of the second guest molecule.

In a separate study [147], Atwood et al. have reported the self-assembly of calix[4]arenes to yield trimeric nanoscale structures which can very efficiently entrap highly volatile gases such as methane and freon under relatively extreme conditions of temperature and pressure.

In apolar solvents six calix[4]resorcinarenes undergo self-assembly by means of 60 hydrogen bonds leading to a chiral spherical molecular capsule

[148]. Kaifer et al. have shown the efficient formation of a robust molecular capsule of six calix[4]resorcinarenes (inner volume ~1,375 Å^3) around an electrochemically generated ferrocenium ion, which behaves as the nucleus of the assembly (Scheme 36) [149]. This encapsulation process is driven by cation–π interactions with the inner surface of the hydrogen-bonded calix[4]resorcinarene capsule. Similar results were obtained with cobaltocenium ion. The encapsulation of ferrocenium or cobaltocenium ions slows down their electrochemical reduction, requiring an overpotential of 500 mV.

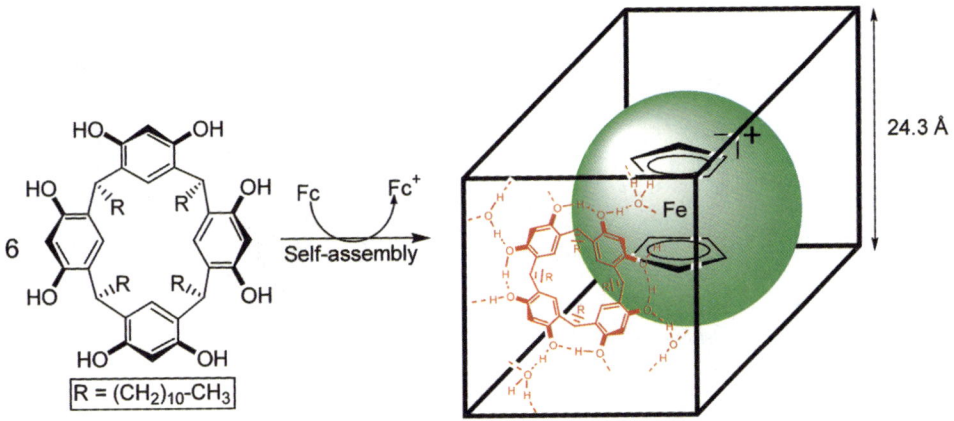

Scheme 36 Six calix[4]resorcinarenes forming a hydrogen-bonded capsule [149]

Encapsulation can be used to induce a conformational preference as demonstrated by Rebek et al. [150]. When a dimeric capsule incarcerated longer n-alkanes, the $C_{10}H_{22}$ was still extended while the $C_{14}H_{30}$ ended coiled up in a helical conformation.

5.4
Switching the Molecular Shape of Assemblies

The development of switchable nanoscale assemblies is of great importance to the development of molecular-level sensors and devices. In seminal contributions, Stoddart et al. [151], Sauvage et al. [152] and others have engineered topologically linked systems which exhibit systematic molecular motions upon photochemical, electrochemical or chemical enticement. While the whole field is still in its infancy, the number of examples coming from metallosupramolecular aggregates seem to outnumber the ones from hydrogen-bonded assemblies. This is readily understandable, since the stronger driving force per binding contact in metallosupramolecular stuctures needs a smaller number of binding sites, thus allowing for higher structural flexibility.

5.4.1
Metallosupramolecular Structures

A recent report by Lehn et al. demonstrated a spectacular metal-mediated interconversion of single helices to a multi-metallic nanoscale grid structure (Scheme 37) [73m]. Upon addition of metal ions the helical strand **15** starts to unwrap to give the fully extended strand incorporated in a [4×4] Pb_{16}^{32+} grid assembly. Demetallation is expected to bring back the helical structure. Similar results were found by Shionoya et al., in which the self-assembly structure depends on the metal ion concentration [153].

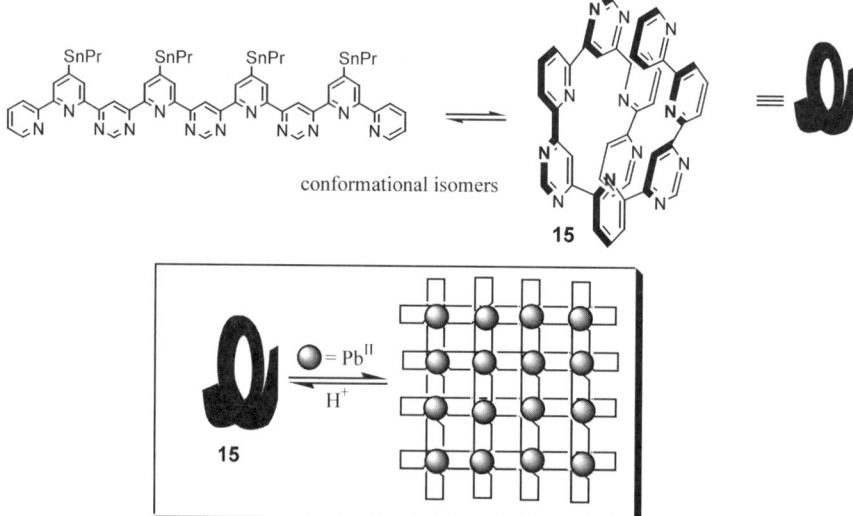

Scheme 37 Metal-mediated interconversion of a single helix to a nanoscale grid [73m]. Reprinted with permission from [70m]. Copyright 2003 American Chemical Society

Hexanuclear hexagons were formed after spontaneous reorganisation of osmate-ester-based molecular squares in the presence of a naphthalene-2,6-diol as a guest as described by Jeong et al. (Scheme 38) [154]. A solid-state structure illustrated the guest's role in driving the tetragon to a hexagon motif. The guest molecule is not filling the void but ends up bound to each side of the cyclic osmate ester by two hydrogen bonds as well as by face-to-face aromatic stacking forces. These observations (guest-induced organisation or induced fit organisation) are very similar to the earlier reports by Fujita et al. [63].

Solvent coordination can also play a major role in governing these multicomponent self-assembly processes (Scheme 39) [155]. The 1:1 stoichiometric reaction between Cu(II) and **16** in acetonitrile afforded an equilibrating mixture of a tetranuclear grid and a hexameric cyclophane. However, when acetonitrile was replaced by nitromethane only the hexameric arrangement could

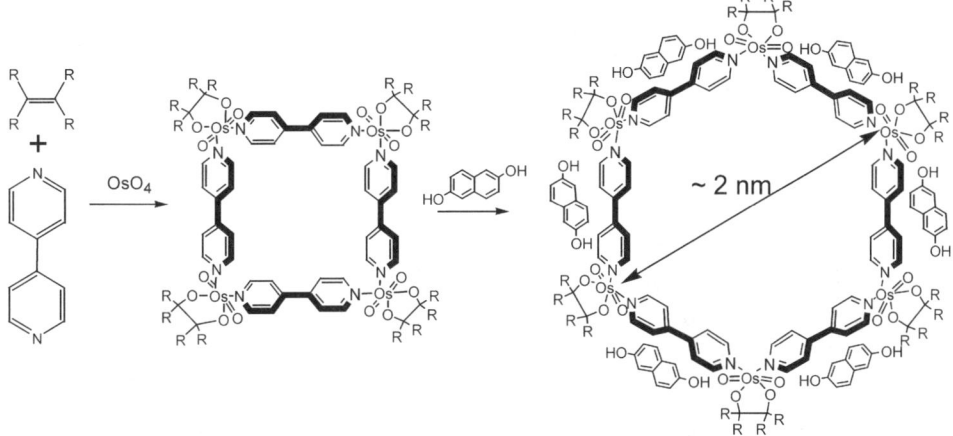

Scheme 38 Hexanuclear hexagons formed after spontaneous reorganisation of molecular squares [154]

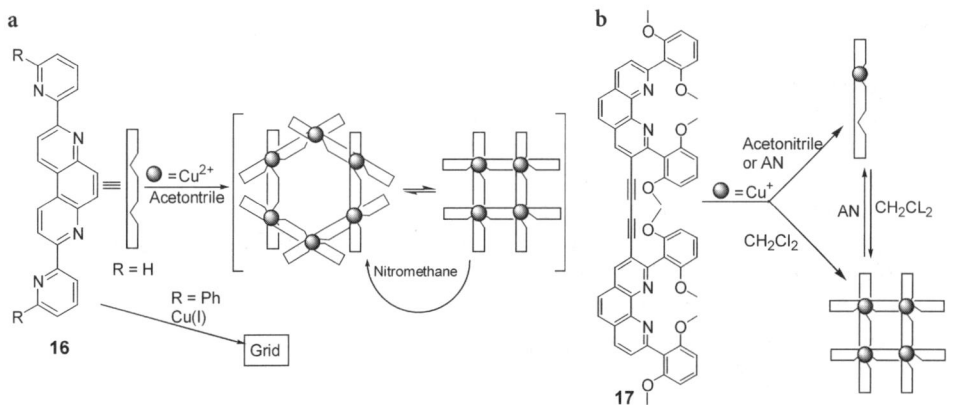

Scheme 39a,b Switching the supramolecular architecture by solvent effects [155, 156]

be observed. Similar results were obtained when **17** was reacted with Cu(I) salt in a 1:1 ratio [156]. In acetonitrile the reaction mixture was dominated by an open structure, whereas in methylene chloride it self-assembled quantitatively to a tetrameric grid.

A remarkable example of redox/metal ion controlled generation of different self-assembled structures has been conceived by Lehn et al. [74b]. The two different types of complexation subunits in ligand **18** convey the source code for the assembly of two divergent structures, such as a nanoscale grid and a circular helicate, depending on the metal ion being used (Scheme 40). A linear combination assembly resulted when eight Cu(I) and four Cu(II) were used as metal templates. Cu(I) is well-known to prefer tetra-coordination, whereas

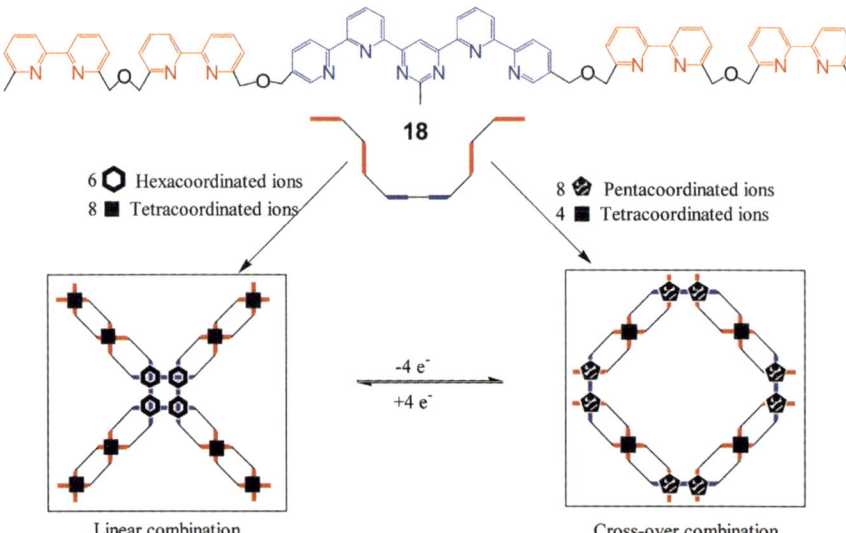

Scheme 40 Redox/metal ion controlled generation of different self-assemblies [74b]

Cu(II) favours penta-coordination. The same ligand in the presence of fourCu(I) and eight Cu(II) formed an intertwined cross-over assembly. Moreover, it was shown that these two discrete assemblies can be interconverted by means of oxidation/reduction.

An intramolecular coordination behaviour was set up in a similar molecule bringing together three different binding subunits [74d]. The reaction of the polytopic ligand with Cu(II) or Zn(II) ions generated double helical structures at the expense of grid systems.

An interesting conversion of a triple-helicate and a tetrahedral cluster by means of host-guest interactions has been reported by Raymond et al. (Scheme 41)

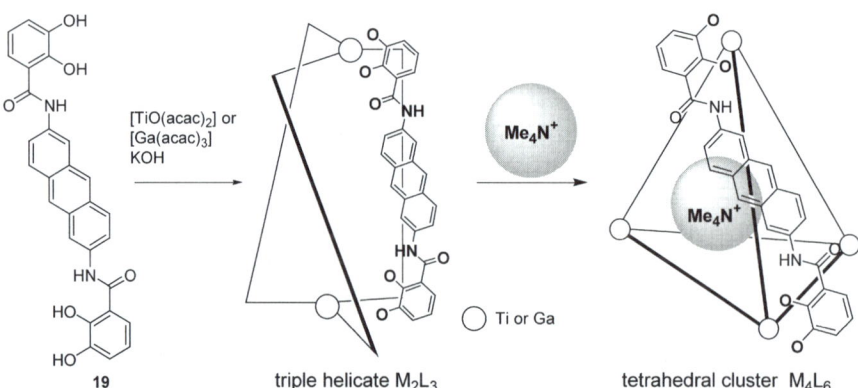

Scheme 41 Conversion of a triple-helicate and a tetrahedral cluster [157]

[157]. The bis(bidentate) catecholamide ligand **19** self-assembled in the presence of Ti or Ga ions to a triple helicate M_2L_3. However, upon addition of Me_4N^+ the triple helicate quantitatively converted to the tetrahedral cluster M_4L_6 by encapsulating the guest molecule inside the cavity.

Lees et al. described a series of self-assembled molecular squares containing a photo-isomerisable azobenzene unit (Scheme 42) [158]. The photophysical and photochemical properties of the squares were investigated, and all of them showed a lack of luminescence at room temperature. However, on irradiation, the Pd(II) and Pd(II)-Re(I) tetranuclear squares could be converted to dinuclear squares which returned thermally to the tetranuclear squares.

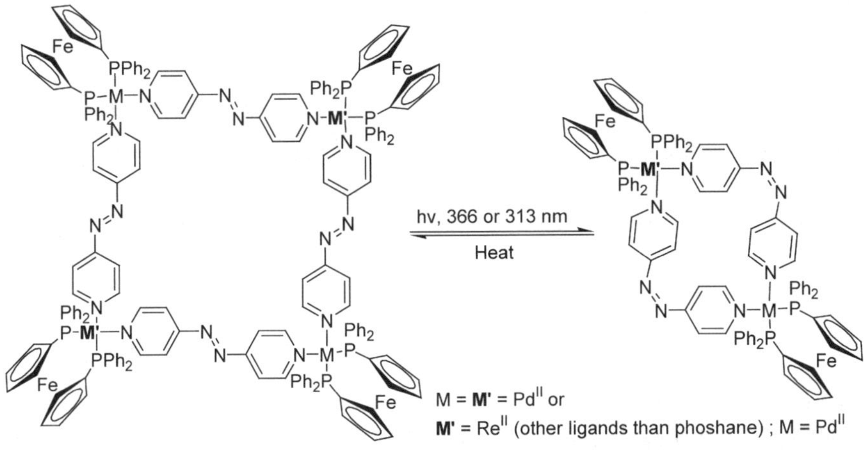

Scheme 42 Molecular squares containing a photo-isomerisable azobenzene unit [158]

5.4.2
Hydrogen-Bonded Structures

In a recent example by Reinhoudt et al. [159], the symmetry of a self-assembled rosette was demonstrated to change in the presence of a self-assembled guest molecule (Scheme 43). Notably, the guest molecule only self-assembled in the presence of the host rosette. Upon guest inclusion, an allosteric regulation of the rosette was observed and in order to accommodate the trimeric guest, the rosette capsule changed its symmetry from D_3 to C_{3h}. This process led to an increase in the intermolecular separation between the two rosette layers from ~3.5 to ~6.7 Å. Guest expulsion was achieved by the addition of butyl cyanurate, which forms much stronger hydrogen bonds with melamines than barbiturates, forming a new rosette.

Supramolecular polymers are interesting examples of self-assembling materials, in which monomeric units are held together by reversible secondary forces – hydrogen bonds, metal-ligand interactions, van der Waals attractions,

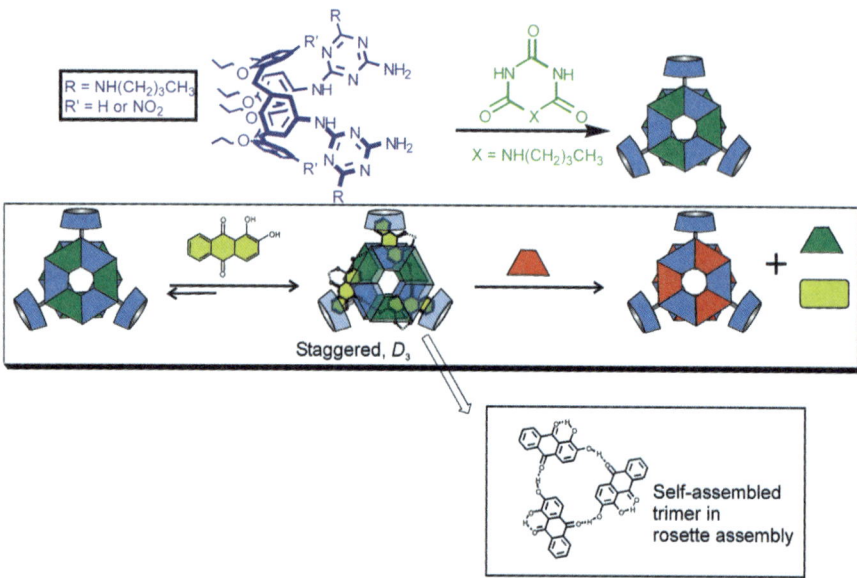

Scheme 43 Guest-induced allosteric regulation of a hydrogen-bonded rosette [159]

etc. [160] When two calix[4]arene tetraureas were covalently linked at their lower rims, hydrogen-bonding yielded a polymeric chain of capsules [161]. In continuation of their earlier work, Rudkevich et al. applied reversible carbamate linkages and hydrogen-bonding to link calix[4]arene tetraurea dimers [162]. Thereby, a novel supramolecular polymer is constructed through chemical fixation of CO_2 and hydrogen-bonding.

5.5
Electroactive and Magnetic Assemblies

The search for high-density information storage devices has stimulated the search for self-assembled electroactive and magnetic structures having multi-stable behaviour that can be precisely controlled by external stimuli. Though many of the supramolecular architectures discussed above incorporate several metal centres with distinct stable redox stages, often their electrochemistry and their magnetic properties in paramagnetic redox stages are not thoroughly investigated.

One approach is to study the oxidation/reduction behavior of the coordinating metal centres in metallosupramolecular assemblies. Lehn et al. reported on a unique Co(II) grid assembly possessing exceptional electrochemical properties, in which the metal centres undergo ten well-resolved, reversible reduction steps involving 11 electrons [163]. A later report by the same group extended these studies to several other grid architectures varying transition metal centres and ligands [164]. A reversible multielectron redox behaviour was also found in Pt [165] (Stang et al.) and rhenium molecular squares by Kaim et al. [166].

Another approach relies on incorporating known reversible redox systems into the metal ligand sphere, such as the ferrocenephosphanes, as described by Stang et al. [167] and Lees et al. [158] for several molecular squares.

A large number of ferrocenes could be assembled about a supramolecular square using the bridging ligand for attaching the electroactive units [168]. Würthner et al. compared the electrochemical properties of these assemblies and the monomeric analogues. The redox response of ferrocene was greatly influenced by the nano environment of the squares. Upon oxidation of all ferrocene units, decomposition of the assembly was observed (Scheme 44).

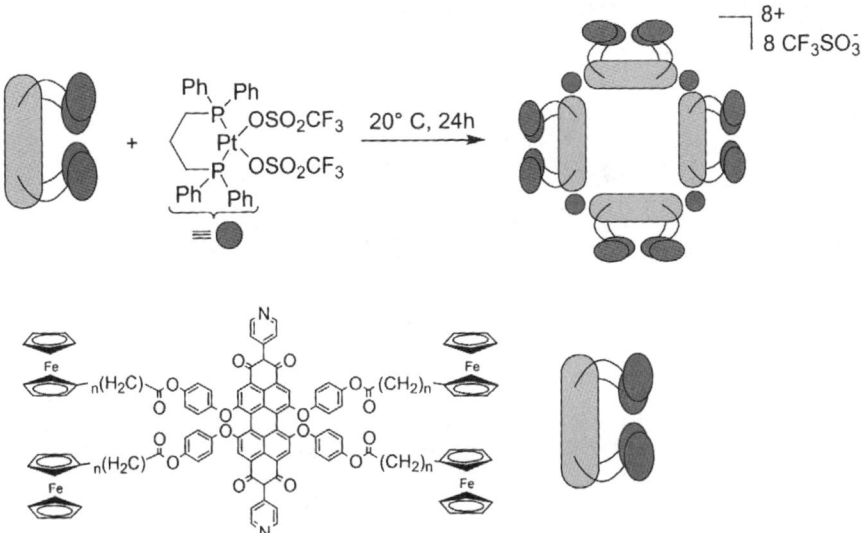

Scheme 44 Electroactive (ferrocene based) supramolecular square [168]

Bucher et al. reported on a self-assembled, ferrocene-substituted porphyrin, which can sense electrochemically neutral molecules via a "tail on-tail off" process [169]. A nitrogen tail was introduced at the ferrocene moiety in order to facilitate the tail on-tail off process. The ferrocene-substituted porphyrin underwent self-assembly to furnish a dimeric structure **20**. Addition of neutral molecules, such as pyridine, imidazole or 2-methylimidazole, triggered the cleavage of the dimeric structure and thereby shifted the potential of the ferrocene moiety (Scheme 45).

Sensing interfaces have been constructed by Willner et al. through the noncovalent crosslinking of 12 nm Au colloids with Pd(II) molecular squares [170]. The molecular squares act as π-receptors for the complexation of π-donor substrates, such as dialkoxybenzene derivatives. Binding of p-hydroquinone increased its local concentration at the electrode surface thus allowing for its

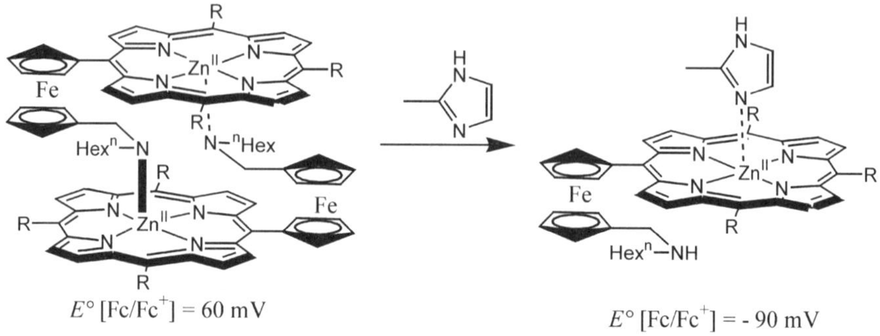

E° [Fc/Fc$^+$] = 60 mV

20

E° [Fc/Fc$^+$] = – 90 mV

Scheme 45 Electroactive sensor for neutral coordinating molecules [169]

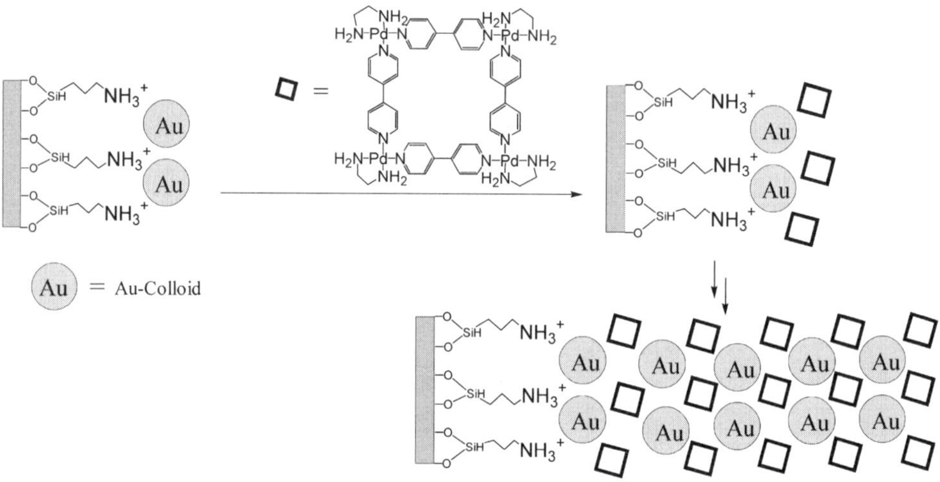

Au = Au-Colloid

Scheme 46 Electrochemical sensing by a 3D conductive Au array [170]

electrochemical sensing by the 3D conductive Au array (Scheme 46). The electrochemical response could be increased by augmenting the number of layers.

In a similar study, Hupp et al. have analysed the size-selective transport of guests through nanocrystalline and/or amorphous thin films made from supramolecular Re-squares by deposition [171]. A major advantage of these systems is the absence of counter ions inside the cavity, making them suitable for membrane-like permeation processes. Electrochemical transport experiments with [Ru(NH$_3$)$_6$]$^{3+}$ (diameter~5.5 Å) and [Co(phen)$_3$]$^{2+}$ revealed that the material is exceptionally porous to small molecules and that large molecules were blocked.

A Fe(II) molecular [2×2] grid exhibits a spin-crossover that is triggered by temperature, pressure and light [172]. The four iron atoms in the assembly can be present in either high-spin (HS) or low-spin (LS) state. The availability of three distinct magnetic levels (3HS/1LS, 2HS/2LS and 1HS/3LS) in this assembly makes this system a "multiply switchable multilevel device" which is driven by three different triggers (temperature, pressure and light).

6
Conclusions

In recent years, metal coordination and hydrogen-bond-driven self-assembly has produced an impressive variety of startling supramolecular architectures. While a truly biomimetic approach would typically employ a balanced mixture of weak interactions (hydrogen-bonding, π-π interactions, etc.), stronger non-covalent interactions (such as the metal ligand coordinative bond) have equally proven their high utility in the preparation of nanoscale assemblies. The time has come now to install functional elements on nanoscale aggregates in order to build nanoscale devices that exhibit non-linearity, interdependence and emergence, i.e. typical characteristics of higher complexity systems. Needless to say, artificial systems, in their size and functional complexity, are still lagging far behind biological molecules/entities (Fig. 3).

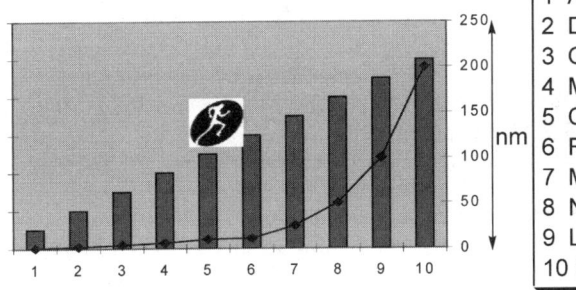

Fig. 3 Size of selected biological molecules/entities

Acknowledgements The authors are indebted to the Deutsche Forschungsgemeinschaft and the Fonds der Chemischen Industrie for continued support. This review is dedicated to Prof. Dr. Christoph Rüchardt (Universität Freiburg) on the occasion of his 75th birthday.

References

1. Lehn J-M (1995) Supramolecular chemistry. VCH, New York
2. Gianni S, Guydosh NR, Khan F, Caldas TD, Mayor U, White GWN, DeMarco ML, Daggett V, Fersht AR (2003) Proc Natl Acad Sci USA 100:13286–13291
3. Due to the limitation to multicomponent assemblies a large amount of supramolecular, albeit spectacular systems composed of a small number of components will not be covered: (a) Raymo FM, Stoddart JF (1999) Chem Rev 99:1643–1663; (b) Stoddart JF (2001) Acc Chem Res 34:410–411; (c) Vögtle F, Safarowsky O, Heim C, Affeld A, Braun O, Mohry A (1999) Pure Appl Chem 71:247–251; (d) Schalley CA, Beizai K, Vögtle F (2001) Acc Chem Res 34:465–476; (e) Brouwer AM, Frochot C, Gatti FG, Leigh DA, Mottier L, Paolucci F, Roffia S, Wurpel GWH (2001) Science 291:2124–2128
4. see for example; (a) Schmid G, Pugin R, Malm J, Bovin J-O (1998) EurJ Inorg Chem 813–818; (b) Müller A, Serain C (2000) Acc Chem Res 33:2–10; (c) Gabriel J-CP, Boubekeur K, Uriel S, Batail P (2001) Chem Rev 101:2037–2066; (d) Albrecht M (2001) Chem Rev 101:3457–3497; (e) Selby HD, Roland BK, Zheng Z (2003) Acc Chem Res 36:933–944, and refs. therein
5. (a) Saalfrank RW, Demleitner B, Glaser H, Maid H, Reihs S, Bauer W, Maluenga M, Hampel F, Teichert M, Krautscheid H (2003) Eur J Inorg Chem 6:822–829; (b) Glaser H, Puchta R, Hommes NJRVE, Leusser D, Murso A, Stalke D, Bauer W, Saalfrank RW (2002) Helv Chim Acta 85:3828–3841; (c) Saalfrank RW, Reimann U, Göritz M, Hampel F, Scheurer A, Heinemann FW, Büschel M, Daub J, Schünemann V, Trautwein AX (2002) Chem Eur J 8:3614–3619; (d) Saalfrank RW, Demleitner B, Glaser H, Maid H, Bathelt D, Hampel F, Bauer W, Teichert M (2002) Chem Eur J 8:2679–2683; (e) Saalfrank RW, Maid H, Mooren N, Hampel F (2002) Angew Chem Int Ed 41:304–307; (f) Saalfrank RW, Glaser H, Demleitner B, Hampel F, Chowdhry MM, Schünemann V, Trautwein AX, Vaughan GBM, Yeh R, Davis AV, Raymond KN (2002) Chem Eur J 8:493–497; (g) Waldmann O, Koch R, Schromm S, Schülein J, Müller P, Bernt I, Saalfrank RW, Hampel F, Balthes E (2001) Inorg Chem 40:2986–2995; (h) Saalfrank RW, Bernt I, Hampel F (2001) Chem Eur J 7:2770–2774; (i) Saalfrank RW, Bernt I, Chowdhry MM, Hampel F, Vaughan GBM (2001) Chem Eur J 7:2765–2769, and refs therein
6. Mandel A, Schmitt W, Womack TG, Bhalla R, Henderson RK, Heath SL, Powell AK (1999) Coord Chem Rev 190–192:1067–1083, and refs therein
7. Prins LJ, Reinhoudt DN, Timmerman P (2001) Angew Chem Int Ed 40:2383–2426
8. (a) Fenniri H, Deng B-L, Ribbe AE, Hallenga K, Jacob J, Thiyagarajan O (2002) Proc Natl Acad Sci USA 99:6487–6492; (b) Fenniri H, Packiarajan M, Vidale KL, Sherman DM, Hallenga K, Wood KV, Stowell JG (2001) J Am Chem Soc 123:3854–3855; (c) Fenniri H, Deng B-L, Ribbe AE (2002) J Am Chem Soc 124:11064–11072
9. For reviews see: (a) Rebek J (1999) Acc Chem Res 32:278–286; (b) Rebek J (1996) Chem Soc Rev 25:255–264; (c) Conn MM, Rebek J (1997) Chem Rev 97:1647–1668; (d) Bohmer V, Vysotsky MO (2001) Aust J Chem 54:671–677; (e) Rudkevich DA (2002) Bull Chem Soc Jpn 75:393–413; (f) MacGillivray LR, Atwood JL (2000) Spherical molecular assemblies: a class of hosts for the next millennium. In: Keinan E, Schechter I (eds) Chemistry for the 21st Century. Wiley-VCH, New York, pp 130–149
10. (a) Berl V, Huc I, Khoury RG, Krische MJ, Lehn J-M (2000) Nature 407:720–723; (b) Berl V, Huc I, Khoury RG, Lehn J-M (2001) Chem Eur J 7:2810–2820; (c) Berl V, Huc I, Khoury RG, Lehn J-M (2001) Chem Eur J 7:2798–2809; (d) Suarez M, Lehn J-M, Zimmerman SC, Skoulios A, Heinrich B (1998) J Am Chem Soc 120:9526–9532; (e) Drain CM, Russell KC, Lehn J-M (1996) Chem Commun 337–338
11. Sessler JL, Magda D, Furuta H (1992) J Org Chem 57:818–826

12. (a) Sessler JL, Wang R (1996) J Am Chem Soc 118:9808–9809; (b) Sessler JL, Wang R (1998) Angew Chem Int Ed 37:1726–1729
13. Ranganathan A, Pedireddi VR, Rao CNR (1999) J Am Chem Soc 121:1752–1753; Lehn J-M, Mascal M, Decian A, Fischer J (1990) J Chem Soc Chem Commun 479–481; Zerkowski JA, Seto CT, Wierda DA, Whitesides GM (1990) J Am Chem Soc 112:9025–9026; Seto CT, Whitesides GM (1990) J Am Chem Soc 112:6409–6411; Zerkowski JA, Mathias JP, Whitesides GM (1994) J Am Chem Soc 116:4305–4315
14. Lipkowski P, Bielejewska A, Kooijman H, Spek AL, Timmerman P, Reinhoudt DN (1999) Chem Commun 1311–1312
15. (a) Zerkowski JA, Seto CT, Whitesides GM (1992) J Am Chem Soc 114:5473–5475; (b) Mathias JP, Simanek EE, Zerkowski JA, Seto CT, Whitesides GM (1994) J Am Chem Soc 116:4316–4325
16. (a) Mammen M, Simanek EE, Whitesides GM (1996) J Am Chem Soc 118:12614–12623; (b) Seto CT, Whitesides GM (1993) J Am Chem Soc 115:905–916; (c) Seto CT, Whitesides GM (1993) J Am Chem Soc 115:1330–1340
17. Isaacs L, Chin DN, Bowden N, Xia Y, Whitesides GM (1999) Supramolecular materials and technologies In: Reinhoudt DN (ed) Perspectives in supramolecular chemistry, vol. 4. Wiley, New York
18. (a) Mathias JP, Seto CT, Simanek EE, Whitesides GM (1994) J Am Chem Soc 116:1725–1736; (b) Seto CT, Whitesides GM (1991) J Am Chem Soc 113:712–713
19. Vreekamp RH, Van Duynhoven JPM, Hubert M, Verboom W, Reinhoudt DN (1996) Angew Chem Int Ed Engl 35:1215–1218
20. (a) Klok HA, Jolliffe KA, Schauer CL, Prins LJ, Spatz JP, Möller M, Timmerman P, Reinhoudt DN (1999) J Am Chem Soc 121:7154–7155; (b) Schönherr H, Paraschiv V, Zapotoczny S, Crego-Calama M, Timmerman P, Frank CW, Vancso GJ, Reinhoudt DN (2002) Proc Natl Acad Sci USA 99:5024–5027; (c) Garcia-Lopez JJ, Zapotoczny S, Timmerman P, Van Veggel FCJM, Vansco GJ, Crego-Calama M, Reinhoudt DN (2003) Chem Commun 352–353
21. Jolliffe KA, Timmerman P, Reinhoudt DN (1999) Angew Chem Int Ed 38:933–937
22. (a) Kramer R, Lehn J-M, Marquis-Rigault A (1993) Proc Natl Acad Sci USA 90:5394–5398; (b) Caulder DL, Raymond KN (1997) Angew Chem Int Ed Engl 36:1440–1442; (c) Albrecht M, Blau O, Wegelius E, Rissanen K (1997) New J Chem 23:667–668; (d) Albrecht M, Blau O, Fröhlich R (2002) Proc Natl Acad Sci USA 99:4867–4872
23. Prins LJ, Neuteboom EE, Paraschiv V, Crego-Calama M, Timmerman P, Reinhoudt DN (2002) J Org Chem 67:4808–4820
24. Paraschiv V, Crego-Calama M, Fokkens RH, Padberg CJ, Timmerman P, Reinhoudt DN (2001) J Org Chem 66:8297–8301
25. Porath D, Bezryadin A, De Vries S, Dekker C (2000) Nature (London) 403:635–638
26. (a) Cram DJ (1983) Science 219:1177–1183; (b) Cram DJ, Karbach S, Kim YH, Baczynskyj L, Kalleymeyn GW (1985) J Am Chem Soc 107:2575–2576; (c) Conceill J, Lacombe L, Collet A (1985) J Am Chem Soc 107:6993–6996; (d) Cram DJ, Tanner ME, Knobler CB (1991) J Am Chem Soc 113:7717–7727; (e) Timmerman P, Verboom W, Van Veggel FCJM, Wan Hoorn WP, Reinhoudt DN (1994) Angew Chem Int Ed Engl 33:1292–1295
27. (a) Rebek J (1996) Pure Appl Chem 68:1261–1266; (b) De Mendoza J (1998) Chem Eur J 4:1373–1377; (c) Hof F, Rebek J (2002) Proc Natl Acad Sci USA 99:4775–4777
28. (a) Shimizu KD, Rebek J (1995) Proc Natl Acad Sci USA 92:12403–12407; (b) Hamann BC, Shimizu KD, Rebek J (1996) Angew Chem Int Ed Engl 35:1326–1329
29. Bonar-Law RP, Sanders JKM (1993) Tetrahedron Lett 34:1677
30. Lee SB, Hong J-I (1996) Tetrahedron Lett 37:8501

31. (a) Zhao S, Luong JHT (1994) J Chem Soc Chem Commun 2307–2308; (b) Zhao S, Luong JHT (1995) J Chem Soc Chem Commun 663–664
32. (a) Wyler R, De Mendoza J, Rebek J (1993) Angew Chem Int Ed Engl 32:1699–1701; (b) Martin T, Obst U, Rebek J (1998) Science 281:1842–1845; (c) O'Leary BM, Grotzfeld RM, Rebek J (1997) J Am Chem Soc 119:11701–11702
33. (a) Higler I, Grave L, Breuning E, Verboom W, de Jong F, Fyles TM, Reinhoudt DN (2000) Eur J Org Chem 1727–1734; (b) Vreekamp R, Verboom W, Reinhoudt DN (1996) J Org Chem 61:4282–4288; (c) Corbellini F, Fiammengo R, Timmerman P, Crego-Calama M, Versluis K, Heck AJR, Luyten I, Reinhoudt DN (2002) J Am Chem Soc 124:6569–6575
34. (a) Castellano RK, Rebek J (1998) J Am Chem Soc 120:3657–3663; (b) Castellano RK, Craig SL, Nuckolls C, Rebek J (2000) J Am Chem Soc 122:7876–7882
35. (a) Shivanyuk A, Paulus EF, Böhmer V (1999) Angew Chem Int Ed 38:2906–2909; (b) Koh K, Araki K, Shinkai S (1994) Tetrahedron Lett 35:8255–8258; (c) Vysotsky MO, Thondorf I, Böhmer V (2000) Angew Chem Int Ed 39:1264–1267
36. Kobayashi K, Ishii K, Sakamoto S, Shirasaka T, Yamaguchi K (2003) J Am Chem Soc 125:10615–10624
37. a) Leininger S, Olenyuk B, Stang PJ (2000) Chem Rev 100:853–907; b) Seidel SR, Stang PJ (2002) Acc Chem Res 35:972–983
38. (a) Fujita M, Ogura K (1996) Coord Chem Rev 148:249–264; (b) Fujita M (1998) Chem Soc Rev 27:417–425; (c) Fujita M (1999) Acc Chem Res 32:53–61
39. (a) Slone RV, Benkstein KD, Belanger S, Hupp JT, Guzei IA, Rheingold AL (1998) Coord Chem Rev 171:221–243; (b) Keefe MH, Benkstein KD, Hupp JT (2000) Coord Chem Rev 205:201–228
40. (a) Fujita M, Yazaki J, Ogura K (1990) J Am Chem Soc 112:5645–5647; (b) Fujita M, Yazaki J, Ogura K (1991) Tetrahedron Lett 32:5589–5592; (c) Fujita M, Yazaki J, Ogura K (1991) Chem Lett 1031–1032; (d) Fujita M, Sasaki O, Mitsuhashi T, Fujita T, Yazaki J, Yamaguchi K, Ogura K (1996) Chem Commun 1535–1536
41. Lee SB, Hwang S, Chung DS, Yun H, Hong JI (1998) Tetrahedron Lett 39:873–876
42. Schalley CA, Müller T, Linnartz P, Witt M, Schäfer M, Lützen A (2003) Chem Eur J 8:3538–3551
43. Stang PJ, Cao DH (1994) J Am Chem Soc 116:4981–4982
44. Stang PJ, Olenyuk B (1997) Acc Chem Res 30:502–518
45. (a) Manna J, Whiteford JA, Stang PJ, Muddiman DC, Smith RD (1996) J Am Chem Soc 118:8731–8732; (b) Manna J, Whiteford JA, Kuehl C, Stang PJ (1997) Organometallics 16:1897–1905; (c) Stang PJ, Persky NE, Manna J (1997) J Am Chem Soc 119:4777–4778
46. Sun S-S, Lees AJ (2001) Inorg Chem 40:3154–3160
47. (a) Drain CM, Lehn J-M (1994) J Chem Soc Chem Commun 2313–2315; Fan J, Whiteford JA, Olenyuk B, Levin MD, Stang PJ, Fleischer EB (1999) J Am Chem Soc 121:2741–2752; (b) Iengo E, Zangrando E, Alessio E (2003) Eur J Inorg Chem 13:2371–2384, and refs therein; (c) Fukushima K, Funatsu K, Ichimura A, Sasaki Y, Suzuki M, Fujihara T, Tsuge K, Imamura T (2003) Inorg Chem 42:3187–3193
48. Li G, Song Y, Hou H, Li L, Fan Y, Zhu Y, Meng X, Mi L (2003) Inorg Chem 42:913–920, and refs therein
49. Lee SJ, Lin W (2002) J Am Chem Soc 124:4554–4555
50. (a) Lee SJ, Hu A, Lin W (2002) J Am Chem Soc 124:12948–12949; (b) Hua J, Lin W (2004) Org Lett 6:861–864
51. Kuehl CJ, Huang SD, Stang PJ (2001) J Am Chem Soc 123:9634–9641
52. Park KM, Kim S-Y, Heo J, Whang D, Sakamoto S, Yamaguchi K, Kim K (2002) J Am Chem Soc 124:2140–2147, and refs therein

53. (a) Sautter A, Schmid D, Jung G, Würthner F (2001) J Am Chem Soc 123:5424–5430; (b) Würthner F, Sautter A (2000) Chem Commun 445–446; (c) Würthner F, Sautter A, Schmid D, Weber PJA (2001) Chem Eur J 7:894–902; (d) Würthner F, Sautter A (2003) Org Biomol Chem 1:240–243
54. Ferrer M, Mounir M, Rossell O, Ruiz E, Maestro MA (2003) Inorg Chem 42:5890–5899
55. Yamamoto T, Arif AM, Stang PJ (2003) J Am Chem Soc 125:2309–12317
56. Jiang H, Lin W (2003) J Am Chem Soc 125:8084–8085
57. Fujita M, Yu S-Y, Kusukawa T, Funaki H, Ogura K, Yamaguchi K (1998) Angew Chem Int Ed 37:2082–2085
58. (a) Fujita M, Nagao S, Ogura K (1995) J Am Chem Soc 117:1649–1650; (b) Hiroaka S, Fujita M (1999) J Am Chem Soc 121:10239–10240; (c) Hiraoka S, Kubota Y, Fujita M (2000) Chem Commun 1509–1510
59. Chand DK, Biradha K, Fujita M, Sakamoto S, Yamaguchi K (2002) Chem Commun 2486–2487
60. (a) Kuehl CJ, Kryschenko YK, Radhakrishnan U, Seidel SR, Huang SD, Stang PJ (2002) Proc Natl Acad Sci USA 99:4932–4936; (b) Hartshorn CM, Steel PJ (1997) Chem Commun 541–542; (c) Kuehl CJ, Yamamoto T, Seidel SR, Stang PJ (2002) Org Lett 6:913–915
61. (a) Pirondini L, Bertolini F, Cantadori B, Ugozzoli F, Massera C, Dalcanale E (2002) Proc Natl Acad Sci USA 99:4911–4915; (b) Manimaran B, Thanasekaran P, Rajendran T, Liao RT, Liu YH, Lee GH, Peng SM, Rajagopal S, Lu KL (2003) Inorg Chem 42:4795–4797 and refs therein
62. Ikeda A, Yoshimura M, Udzu H, Fukuhara C, Shinkai S (1999) J Am Chem Soc 121:4296–4297
63. (a) Aoyagi M, Biradha K, Fujita M (1999) J Am Chem Soc 121:7457–7458; (b) Aoyagi M, Tashiro S, Tominaga M, Biradha K, Fujita M (2002) Chem Commun 2036–2037; (c) Tominaga M, Tashiro S, Aoyagi M, Fujita M (2002) Chem Commun 2038–2039
64. Tashiro S, Tominaga M, Kusukawa T, Kawano M, Sakamoto S, Yamaguchi K, Fujita M (2003) Angew Chem Int Ed 42:3267–3270
65. (a) Hori A, Kataoka H, Okano T, Sakamoto S, Yamaguchi K, Fujita M (2003) Chem Commun 182–183; (b) Dietrich-Buchecker C, Colasson B, Fujita M, Hori A, Geum N, Sakamoto S, Yamaguchi K, Sauvage J-P (2003) J Am Chem Soc 125:5717–5725
66. Leininger S, Fan J, Schmitz M, Stang PJ (2000) Proc Natl Acad Sci USA 97:1380–1384
67. Schweiger M, Seidel SR, Schmitz M, Stang PJ (2000) Org Lett 2:1255–1257
68. Kryschenko YK, Seidel SR, Muddiman DC, Nepomuceno AI, Stang PJ (2003) J Am Chem Soc 125:9647–9652
69. Olenyuk B, Levin MD, Whiteford JA, Shield JE, Stang PJ (1999) J Am Chem Soc 121:10434–10435
70. (a) Stejskal EO, Tanner JE (1965) J Chem Phys 42:288; (b) Tanner J (1970) J Chem Phys 52:2523–2526; (c) Mayzel O, Cohen Y (1994) J Chem Soc Chem Commun 1901–1902
71. Hanan GS, Arana CR, Lehn J-M, Baum G, Fenske D (1996) Chem Eur J 10:1292–1302
72. Hanan GS, Lehn J-M (1996) Chem Commun 2019–2020
73. Bidentate based grids; (a) Baxter PNW, Lehn J-M, Fischer J, Youinou MT (1994) Angew Chem Int Ed Engl 33:2284–2287; (b) Baxter PNW, Lehn J-M, Kneisel BO, Fenske D (1997) Chem Commun 2231–2232; (c) Baxter PNW, Lehn J-M, Kneisel BO, Fenske D (1997) Angew Chem, Int Ed Engl 36:1978–1981; (d) Baxter PNW, Lehn J-M, Rissanen K (1997) Chem Commun 1323–1324; (e) Weissbuch I, Baxter PNW, Kuzmenko I, Cohen H, Cohen S, Kjaer K, Howes PB, Nielsen J, Lehn J-M, Leiserowitz L, Lahav M (2000) Chem Eur J 6:725–734; Tridentate based grids; (f) Hanan GS, Volkmer D, Schubert US, Lehn J-M, Baum G, Fenske D (1997) Angew Chem Int Ed Engl 36:1842–1844; (g) Rojo J, Lehn J-M, Baum G, Fenske D, Waldmann O, Müller P (1999) Eur J Inorg Chem 517–522;

(h) Waldmann O, Hassmann J, Müller P, Volkmer D, Schubert US, Lehn J-M (1998) Phys Rev B Cond Matter 58:3277–3285; (i) Weissbuch I, Baxter PNW, Cohen S, Cohen H, Kjaer K, Howes PB, Nielsen JA, Hanan GS, Schubert US, Lehn J-M, Leiserowitz L, Lahav M (1998) J Am Chem Soc 120:4850–4860; (j) Garcia AM, Romero-Salguero FJ, Bassani DM, Lehn J-M, Baum G, Fenske D (1999) Chem Eur J 6:1803–1808; (k) Nierengarten H, Leize E, Breuning E, Garcia A, Romero-Salguero F, Rojo J, Lehn J-M, Van Dorsselaer A (2002) J Mass Spectrom 37:56–62; (l) Nitschke JR, Lehn J-M (2003) Proc Natl Acad Sci USA 100:11970–11974; (m) Barboiu M, Vaughan G, Graff R, Lehn J-M (2003) J Am Chem Soc 125:10257–10265 and refs therein; (n) Youinou MT, Rahmouni N, Fischer J, Osborn JA (1992) Angew Chem Int Ed Engl 31:733–735
74. (a) Funeriu DP, Lehn J-M, Baum G, Fenske D (1997) Chem Eur J 3:99–104; (b) Funeriu DP, Lehn J-M, Fromm KM, Fenske D (2000) Chem Eur J 6:2103–2111; (c) Lehn J-M (2000) Chem Eur J 6:2097–2102; (d) Funeriu DP, Rissanen K, Lehn J-M (2001) Proc Natl Acad Sci USA 98:10546–10551
75. Garcia AM, Bassani DM, Lehn JM, Baum G, Fenske D (1999) Chem Eur J 5:1234–1238; Baxter PNW, Lehn J-M, Kneisel OB, Baum G, Fenske D (1999) Chem Eur J 5:113–120; Baxter PNW, Lehn J-M, Baum G, Fenske D (1999) Chem Eur J 5:102–112; Marquis Rigault A, Gervais AD, Baxter PNW, Van Dorsselaer A, Lehn J-M (1996) Inorg Chem 35:2307–2310; Baxter PNW, Lehn J-M, De Cian A, Fischer J (1993) Angew Chem Int Ed Engl 32:69–72; Lehn J-M (1995) Supramolecular Chemistry. VCH, Weinheim, p 155
76. Baxter PNW, Lehn J-M, Baum G, Fenske D (2000) Chem Eur J 6:4510–4517
77. Marquis A, Kintzinger J-P, Graff R, Baxter PNW, Lehn J-M (2002) Angew Chem Int Ed 41:2760–2764
78. Bassani DM, Lehn J-M, Fromm K, Fenske D (1998) Angew Chem Int Ed 37:2364–2367
79. Breuning E, Ziener U, Lehn J-M, Wegelius E, Rissanen K (2001) Eur J Inorg Chem 1515–1521
80. Bark T, Düggeli M, Stoeckli-Evans H, von Zelewsky A (2001) Angew Chem Int Ed 15:2848–2851
81. Plante JP, Jones PD, Powell DR, Glass TE (2003) Chem Commun 336–337
82. Constable EC, Schofield E (1998) Chem Commun 403–404
83. Newkome GR, Cho TJ, Moorefield CN, Cush R, Russo PS, Godinez LA, Saunders MJ, Mohapatra P (2002) Chem Eur J 8:2946–2954
84. (a) Schmittel M, Ganz A (1997) Chem Commun 999–1000; (b) Schmittel M, Lüning U, Meder M, Ganz A, Michel C, Herderich M (1997) Heterocycl Commun 3:493–498
85. (a) Schmittel M, Ganz A, Fenske D, Herderich M (2000) J Chem Soc, Dalton Trans 353–359; (b) Schmittel M, Michel C, Liu S-X, Schildbach D, Fenske D (2001) Eur J Inorg Chem 1155–1166
86. Schmittel M, Kalsani V, Fenske D, Wiegrefe A (2004) Chem Commun 490–491
87. Schmittel M, Ammon H, Kalsani V, Wiegrefe A (2002) Chem Commun 2566–2567
88. (a) Schmittel M, Ammon H (1999) Synlett 750–752; (b) see review on shape-persistent macrocycles: Grave C, Schlüter AD (2002) Eur J Org Chem 3075–3098
89. Michel C (2001) PhD Thesis, Würzburg, Germany. Schmittel M, Michel C, Wiegrefe A (private communication)
90. Leize E, Van Dorsselaer A, Krämer R, Lehn J-M (1993) J Chem Soc Chem Commun 990–993
91. Bark T, Weyhermüller T, Heirtzler F (1998) Chem Commun 1475–1476
92. Kalsani V, Ammon H, Jäckel F, Rabe JP, Schmittel M (2004) Chem Eur J 10:5481–5492
93. Schmittel M, Ganz A, Fenske D (2002) Org Lett 14:2289–2292
94. Loren JC, Yoshizawa M, Haldimann RF, Linden A, Siegel JS (2003) Angew Chem Int Ed 42:5702–5705

95. (a) Enemark EJ, Stack TDP (1998) Angew Chem Int Ed 37:932–935; (b) Albrecht M (1998) Chem Soc Rev 27:281–287
96. Albrecht M (2001) Chem Rev 101:3457–3498
97. Caulder DL, Raymond KN (1999) Acc Chem Res 32:975–982
98. Bell ZR, Harding LP, Ward MD (2003) Chem Commun 2432–2433
99. (a) De Wolf P, Heath SL, Thomas JA (2002) Chem Commun 2540–2541; (b) Bonnefous C, Bellec N, Thummel RP (1999) Chem Commun 1243–1244; (c) Hall JR, Loeb SL, Shimizu GKH, Yap GPA (1998) Angew Chem Int Ed 37:121–123
100. (a) Kang J, Rebek J (1997) Nature 385:50–52; (b) Kang J, Hilmersson G, Santamaria J, Rebek J (1998) J Am Chem Soc 120:3650–3656; (c) Kang J, Santamaría J, Hilmersson G, Rebek J (1998) J Am Chem Soc 120:7389–7390
101. (a) Lützen AR, Renslo CA, Schalley BM, O'Leary, Rebek J (1999) J Am Chem Soc 121:7455–7456; (b) Shivanyuk A, Rebek J (2002) J Am Chem Soc 124:12074–12075; (c) Ballester P, Shivanyuk A, Rafai Far A, Rebek J (2002) J Am Chem Soc 124:14014–14016; (d) Craig SL, Lin S, Chen J, Rebek J (2002) J Am Chem Soc 124:8780–8781; (e) Hayashida O, Shivanyuk A, Rebek J (2002) Angew Chem Int Ed 41:3423–3426; (f) Johnson DW, Hof F, Iovine PM, Nuckolls C, Rebek J (2002) Angew Chem Int Ed 41:3793–3796
102. (a) Kusukawa T, Fujita M (2002) J Am Chem Soc 124:13576–13582; (b) Parac TN, Scherer M, Raymond K (2000) Angew Chem Int Ed 39:1239–1242; (c) Kusukawa T, Fujita M (1998) Angew Chem Int Ed 37:3142–3144; (d) Ziessel R, Charbonniere L, Cesario M, Prange T, Nierengarten H (2002) Angew Chem Int Ed 6:975–979
103. Kang J, Rebek J (1996) Nature 382:239–241
104. Chen J, Rebek J (2002) Org Lett 3:327–329
105. Chen J, Craig SL, Rebek J (2002) Nature 415:385–386
106. Ito H, Kusukawa T, Fujita M (2000) Chem Lett 598–599
107. (a) Walter CJ, Anderson HL, Sanders JKM (1993) J Chem Soc Chem Commun 458–460; (b) Leung DH, Fiedler D, Bergman RG, Raymond KN (2004) Angew Chem Int Ed 43:963–933
108. (a) Yoshizawa M, Kusukawa T, Fujita M, Sakamoto S, Yamaguchi K (2001) J Am Chem Soc 123:10454–10459; (b) Yoshizawa M, Takeyama Y, Kusukawa T, Fujita M (2002) Angew Chem Int Ed 41:1347–1349
109. Jiang H, Hu A, Lin W (2003) Chem Commun 96–97
110. Merlau ML, Mejia MDP, Nguyen ST, Hupp JT (2001) Angew Chem Int Ed 22:4239–4242
111. The dynamic equilibrium between bound and unbound porphyrin catalyst could account for the observed eventual destruction of all of the catalyst after 3 h
112. Jacobsen EN (2000) Acc Chem Res 33:421–431
113. (a) Konsler RG, Karl J, Jacobson EN (1998) J Am Chem Soc 120:10780–10781; (b) Ready JM, Jacobson EN (2001) J Am Chem Soc 123:2687–2688; (c) Ready JM, Jacobsen EN (2002) Angew Chem Int Ed 41:1374–1377
114. Gianneschi NC, Bertin PA, Nguyen ST, Mirkin CA, Zakharov LN, Rheingold AL (2003) J Am Chem Soc 125:10508–10509
115. Slagt VF, Van Leeuwen PWNM, Reek JNH (2003) Chem Commun 2474–2475
116. Slagt VF, Van Leeuwen PWNM, Reek JNH (2003) Angew Chem Int Ed 42:5619–5623
117. Bonchio M, Carofiglio T, Carraro M, Fornasier R, Tonellato U (2002) Org Lett 26:4635–4637
118. Fujita M, Kwon YJ, Washizu S, Ogura K (1994) J Am Chem Soc 116:1151–1152
119. Süss-Fink G, Faure M, Ward TR (2002) Angew Chem Int Ed 41:99–101
120. Chuchuryukin AV, Dijkstra HP, Suijkerbuijk BMJM, Gebbink RJMK, Klink GPMV, Mills AM, Spek AL, Koten GV (2003) Angew Chem Int Ed 42:228–230

121. Ogoshi H, Mizutani T, Hayashi T, Kuroda Y (2000) The porphyrin handbook. Academic, San Diego
122. Hui CK, Chu BWK, Zhu NY, Yam VWW (2002) Inorg Chem 41:6178–6180
123. Lee SJ, Luman CR, Castellano FN, Lin W (2003) Chem Commun 2124–2125
124. Ruben M, Lehn J-M, Vaughan G (2003) Chem Commun 1338–1339
125. Takahashi R, Kobuke Y (2003) J Am Chem Soc 125:2372–2373
126. (a) Kuroda Y, Sugou K, Sasaki K (2000) J Am Chem Soc 122:7833–7834; (b) Sugou K, Sasaki K, Kitajima K, Iwaki T, Kuroda Y (2002) J Am Chem Soc 124:1182–1183
127. Mines GA, Tzeng BC, Stevenson KJ, Li J, Hupp JT (2002) Angew Chem Int Ed 41:154–157
128. Splan KE, Keefe MH, Assari AM, Walters KA, Hupp JT (2002) Inorg Chem 41:619–621
129. Tsuda A, Sakamoto S, Yamaguchi K, Aida T (2003) J Am Chem Soc 125:15722–15723
130. Drain CM, Batteas JD, Flynn GW, Milic T, Chi N, Yablon DG, Sommers H (2002) Proc Nat Acad Sci USA 99:6498–6502
131. Hasenknopf B, Lehn J-M, Boumediene N, Gervais AD, Dorsselaer AV, Kniesel B, Fenske D (1997) J Am Chem Soc 119:10956–10962
132. (a) Baum G, Constable EC, Fenske D, Housecroft CE, Kulke T (1999) Chem Commun 195–196; (b) Mamula O, Von Zelewsky A, Bernardinelli G (1998) Angew Chem Int Ed 37:290–293; (c) Childs LJ, Alcock NW, Hannon MJ (2002) Angew Chem Int Ed 22:4244–4247; (d) Bell ZR, Jeffery JC, McCleverty JA, Ward MD (2002) Angew Chem Int Ed 14:2515–2518
133. Fujita N, Biradha K, Fujita M, Sakamoto A, Yamaguchi K (2001) Angew Chem Int Ed 9:1718–1721
134. Atwood JL, Koutsantonis GA, Raston CL (1994) Nature 368:229–231
135. Kumazawa K, Biradha K, Kusukawa T, Okano T, Fujita M (2003) Angew Chem Int Ed 42:3909–3913
136. Ziegler M, Brumaghim JL, Raymond KN (2000) Angew Chem Int Ed 39:4119–4121
137. Yoshizawa M, Kusukawa T, Fujita M, Yamaguchi K (2000) J Am Chem Soc 122:6311–6312
138. Kusukawa T, Fujita M (1999) J Am Chem Soc 121:1397–1398
139. Bourgeois J-P, Fujita M, Kawano M, Sakamoto S, Yamaguchi K (2003) J Am Chem Soc 125:9260–9261
140. Claulder DL, Powers RE, Parac TN, Raymond KN (1998) Angew Chem Int Ed 37:1840–1843
141. Sun W-Y, Kusukawa T, Fujita M (2002) J Am Chem Soc 124:11570–11571
142. Ibukuro F, Kusukawa T, Fujita M (1998) J Am Chem Soc 120:8561–8562
143. (a) Matsumoto N, Motoda Y, Matsuao T, Nakashima T, Re N, Dahan F, Tuchagues J-P (1999) Inorg Chem 38:1165–1173 and refs therein; (b) Du M, Bu X-H, Ribas J, Diaz C (2002) Chem Commun 2550–2551
144. Crowley JD, Goshe AJ, Bosnich B (2003) Chem Commun 2824–2825
145. Holliday BJ, Farrell JR, Mirkin CA, Lam K-C, Rheingold AL (1999) J Am Chem Soc 121:6316–6317
146. Heinz T, Rudkevich DM, Rebek J (1998) Nature 394:764–766
147. Atwood JL, Barbour LJ, Jerga A (2002) Science 296:2367–2369
148. MacGillivray LR, Atwood JL (1997) Nature 389:469–472
149. Philip IE, Kaifer AE (2002) J Am Chem Soc 124:12678–12679
150. Scarso A, Trembleau L, Rebek J (2003) Angew Chem Int Ed 42:5499–5502
151. Balzani V, Credi A, Raymo FM, Stoddart JF (2000) Angew Chem Int Ed 39:3349–3391
152. (a) Collin JP, Dietrich-Buchecker C, Gavina P, Jimenez-Molero MC, Sauvage JP (2001) Acc Chem Res 34:477–487; (b) Blanco MJ, Jimenez MC, Chambron JC, Heitz V, Linke M, Sauvage JP (1999) Chem Soc Rev 293–305
153. Hiraoka S, Yi T, Shiro M, Shionoya M (2002) J Am Chem Soc 124:14510–14511

154. Cho YL, Uh H, Chang S-Y, Chang H-Y, Choi M-G, Shin I, Jeong K-S (2001) J Am Chem Soc 123:1258–1259, and ref. therein
155. Baxter PNW, Khoury RG, Lehn J-M, Baum G, Fenske D (2000) Chem Eur J 22:4140–4148
156. Schmittel M, Kalsani V (private communication)
157. Scherer M, Caulder DL, Johnson DW, Raymond KR (1999) Angew Chem Int Ed 38: 1588–1592
158. Sun S-S, Anspach JA, Lees AJ (2002) Inorg Chem 41:1862–1869
159. Kerckhoffs JMCA, van Leeuwen FWB, Spek AL, Kooijman H, Crego-Calama M, Reinhoudt DN (2003) Angew Chem Int Ed 42:5717–5722
160. Brunsveld L, Folmer BJB, Meijer EW, Sijbesma RP (2001) Chem Rev 101:4071–4097
161. Castellano RK, Rudkevich DM, Rebek J (1997) Proc Natl Acad Sci USA 94:7132–7137
162. Xu H, Hampe EM, Rudkevich DM (2003) Chem Commun 2828–2829
163. Ruben M, Breuning E, Gisselbrecht J-P, Lehn J-M (2000) Angew Chem Int Ed 39: 4139–4142
164. Ruben M, Breuning E, Barboiu M, Gisselbrecht J-P, Lehn J-M (2003) Chem Eur J 9:291–299
165. Kaim W, Schwederski B, Dogan A, Fiedler J, Kuehl CJ, Stang PJ (2002) Inorg Chem 41:4025–4028
166. Hartmann H, Berger S, Winter R, Fiedler J, Kaim W (2000) Inorg Chem 39:4977–4980
167. Stang PJ, Olenyuk B, Fan J, Arif AM (1996) Organometallics 15:904–908
168. You C-C, Würthner F (2003) J Am Chem Soc 125:9716–9725
169. Bucher C, Devillers CH, Moutet J-C, Royal G, Saint-Aman E (2003) Chem Commun 888–889
170. Lahav M, Gabai R, Shipway AN, Willner I (1999) Chem Commun 1937–1938
171. Bélanger S, Hupp JT, Stern CL, Slone RV, Watson DF, Carrell TG (1999) J Am Chem Soc 121:557–563
172. (a) Breuning E, Ruben M, Lehn J-M, Renz F, Garcia Y, Ksenofontov V, Gütlich P, Wegelius E, Rissanen K (2000) Angew Chem Int Ed 39:2504–2507; (b) Bassani DM, Lehn J-M, Serroni S, Puntoriero F, Campagna S (2003) Chem Eur J 9:4422–4429

Synthesis and Applications of Supramolecular Porphyrinic Materials

Charles Michael Drain[1] (✉) · Israel Goldberg[2] · Isabelle Sylvain[1] · Alexander Falber[1]

[1] Department of Chemistry, Hunter College and Graduate School of The City University of New York, 695 Park Avenue, New York, NY 10021, USA
cdrain@hunter.cuny.edu, isylvain33@hotmail.com, afalber@hunter.cuny.edu
[2] School of Chemistry, Sackler Faculty of Exact Sciences, Tel Aviv University, 69978 Tel Aviv, Israel
goldberg@chemsg7.tau.ac.il

1	Introduction	56
1.1	Supramolecular Chemistry	56
1.2	Barriers to Commercial Applications of Supramolecular Chemistry	58
1.3	Porphyrinoids	59
1.4	Potential Applications of Supramolecular Porphyrinoid Materials	62
2	Self-Assembled Porphyrinic Arrays	63
2.1	Hydrogen-Bonding	63
2.2	Coordination Bonding	66
2.3	Electrostatics	69
3	Hierarchical Assemblies: Aggregates	72
4	Crystal Engineering of Porphyrins	73
4.1	Porphyrin Tectons	73
4.2	Hydrogen-Bonding	75
4.3	Coordination Chemistry	77
4.4	Hybrid Approaches	82
4.5	Electrostatic	84
5	Concluding Remarks	84
	References	85

Abstract The supramolecular chemistry of porphyrinic systems is extraordinarily diverse because the rigid macrocycles offer a variety of topologies that can be matched to the topologies of hydrogen-bonding moieties and metal ion coordination geometries. Other modes to self-assemble multi porphyrin arrays and crystals exploit electrostatic interactions, combinations of hydrogen-bonding, and metal ion coordination. Porphyrinoids are ideal building blocks because they impart a high degree of functionality to the materials due to their rich electrochemical and photophysical properties, and their extraordinary stability. These properties will continue to be incorporated into supramolecular porphyrinic arrays and crystalline materials that can serve as the active components in sensors, molecular sieves, photonics, molecular electronics, and catalysts. Herein we summarize the various approaches to the synthetic supramolecular chemistry of porphyrins that results in discrete systems, and as a mode of crystal engineering. The functionality of many of these materials is discussed in relation to the supramolecular structure.

Keywords Supramolecular · Porphyrin · Crystal engineering · Nanotechnology · Molecular electronics

There are few human beings who receive the truth, complete and staggering, by instant illumination. Most of them acquire it fragment by fragment, on a small scale, by successive developments, cellularly, like a laborious mosaic – Anais Nin.

1
Introduction

1.1
Supramolecular Chemistry

Supramolecular chemistry is chemistry beyond the covalent bond [1–7]. That molecules and ions associate via non-covalent interactions has been known for more than a century (e.g. the work of van der Waals in the late 1800s) and is experimentally exemplified by day-to-day phenomena such as freezing point depression and boiling point elevation. As our knowledge of chemical structure and molecular properties has increased, so too has our understanding of intermolecular forces. Based on the types of atoms and/or interactions observed, these forces are heuristically classified as generally falling into several categories – electrostatic, coordination, van der Waals, and hydrogen-bonding to these aid in conceptualization and understanding. However, since all intermolecular, interionic, and molecule-ion interactions arise from the electronic distributions in the molecules and/or ions, there are substantial overlaps among the above categories and a wide distribution of interaction energies [8, 9].

In addition, there are geometric or structural factors that oftentimes come into play. The propensity for a given metal ion to bind certain types of ligand atoms in certain geometries is exploited often in supramolecular chemistry. The molecular shape and conformational dynamics can also play crucial roles in binding to ions and/or other molecules. The importance of shape complementarity was realized by Emil Fischer in 1894, who proposed the lock and key concept for enzyme-substrate interactions [10, 11]. Though all of these intermolecular interactions have been studied for nearly a century in terms of inorganic salts, coordination chemistry, and for many decades in terms of hydrogen bonds, the specific design of host molecules for specific guests can be traced to a classic paper by Pedersen in 1967 on crown ethers [12]. Work on similar compounds by Cram and on the aza derivatives by Lehn broadened our understanding of the roles of specific intermolecular interactions, shape, and dynamics in the molecular recognition of a guest by a host [1–3]. Naturally occurring macrocyclic ligands such as the cyclic peptide antibiotic cyclosporin and the porphyrins can be considered effective hosts for many metal ion guests; moreover, the extended aromatic system of the latter adds a degree of functionality unobtainable with many other natural or synthetic macrocycles, *vide infra*.

The spontaneous association of matter into organized systems is the foundation on which life builds complex systems, and can be categorized as self-assembly and self-organization. Self-assembly [1, 2, 8, 13, 14] results in the formation of a discrete supramolecular entity wherein all copies are the same (the yields may vary, but the desired structure is intolerant of defects such as missing a component atom or molecule). On the other hand, self-organization of ions and molecules results in non-discrete systems wherein at least one dimension is variable, such as a linear polymer or a two-dimensional polymeric array; such systems tend to be more dynamic and tolerant of defects. Though it is not necessarily the case, self-assembled systems are usually topologically saturated in that there are no free (or unused) specific intermolecular bonding sites. Non-specific dispersion forces are, of course, always present. With the exception of simple dimers, the most common self-assembled systems are also closed in that they are rings, triangles, squares, rosettes, etc. Notable exceptions to this latter trend are the inorganic helicates [3] and those systems that use secondary steps to obtain the target supermolecule. Biological inspiration includes multimeric enzymes, for example the spontaneous formation of hexameric protein complexes that serve specific functions such as ion channels, helicases [15], and ion pumps are examples of self-assembled systems because they are discrete and each copy is the same. Defects in primary, secondary, tertiary, or quaternary structure of these protein complexes oftentimes result in a malfunctioning enzyme. On the other hand, the self-organization of lipids into bilayer membranes results in a two-dimensional system wherein only the thickness is defined.

In analogy to the structure of biological systems, the structure of supramolecular materials can be generally defined in terms of increasing complexity [8, 13]. The primary structure is that of the molecule and the coordination chemistry of the ions. The breadth of organic transformations and knowledge of inorganic chemistry allows us to construct a staggering variety of molecules. The secondary structure is that of the self-assembled entity, and nearly 40 years of supramolecular chemistry has defined many of the principles that govern the self-assembly and self-organization processes, *vide supra*. The tertiary structure is how the self-assembled array self-organizes into solids, which may or may not be crystalline. Routine prediction of how a self-assembled system will pack into a crystal lattice remains an elusive goal and an area of vigorous research. The quaternary structure pertains to how the supramolecular material is incorporated into a device. The quaternary structure must then also depend on interactions with surfaces and on surface chemistry. Though many quite beautiful crystal structures continue to appear frequently in the literature, the formation of functional supramolecular materials requires the examination of the structure and the function of self-assembled and/or self-organized systems on surfaces.

1.2
Barriers to Commercial Applications of Supramolecular Chemistry

For the foreseeable future, commercially viable devices that exploit self-assembled and self-organized materials will have them supported on surfaces (Fig. 1). The co-mingled roles of surface composition, surface energetics, and surface structure are of paramount importance to the structure and function of self-organized and self-assembled systems for real-world applications. There

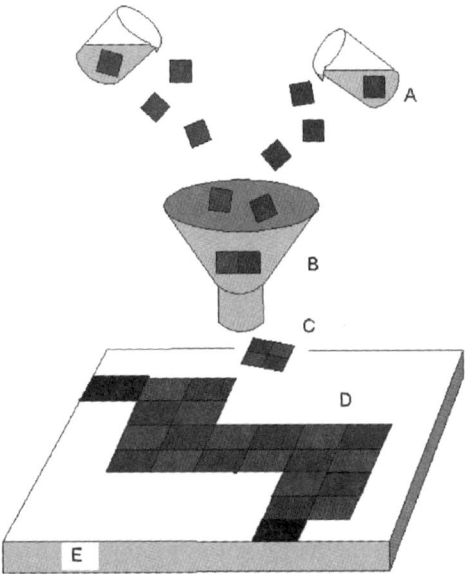

Fig. 1 The ambition of using supramolecular chemistry in the fabrication of device components involves a structural hierarchy. *A* Primary structure – the design and synthesis of molecules with both the desired properties for the material and the recognition motifs for self-assembly and/or self-organization. *B* Secondary structure – the molecules are then mixed in the appropriate stoichiometries under conditions that maximize the self-assembly process to form supramolecular entities. *C* Tertiary structure – the self-assembled supermolecules then self-organize into higher ordered structures such as crystalline materials. *D* Quaternary structure – the last step is the organization of the material into a functional device on some appropriate support. *E* There are a variety of stamps and soft lithographic techniques that are capable of patterning a surface such that self-assembled and/or self-organized systems can be deposited onto the surface with desired patterns. The *black voids* represent parts of the surface that have been modified to be attractive to the supramolecular system, while the *white part* of the surface has been modified to be repulsive. Though accomplishing hierarchical assembly of functional molecules into working devices is in its infancy, recent developments in the self-assembly of hard, micro scaled devices into working circuitry have been reported [55]

is no *a priori* reason that the structure and function of self-assembled systems in solution or as solids can be deposited on surfaces with high structural fidelity [13, 14, 16]. Since the interactions of molecules and ions with a given surface can be stronger than the relatively weak intermolecular and/or coordination bonds that hold together materials formed via self- processes, it should not be surprising that these structures can be disassembled or significantly altered upon surface deposition. The reorganization of these materials oftentimes results in significantly altered functions or loss of function. From a different perspective, knowledge of the surface properties can be used as another tool in the design and assembly of supramolecular systems on surfaces.

Supramolecular chemistry generally results in thermodynamic products [1–3, 5, 9]. The advantage of this is that the reversibility of the intermolecular interactions allows for "error checking", to afford greater yields of complex products that are likely unobtainable using covalent chemistry. Therein lies the fundamental barrier to using supramolecular systems in commercially viable products. All but the simplest self-assembled molecules are formed in solution. The inherent concentration increases as these supermolecules are deposited on surfaces (via drop casting, spin casting, airbrush, etc.) can significantly alter the equilibrium, and therefore the structure of the deposited product. Secondly, variations in temperature, humidity, and substrate morphology can change the structure and function of the deposited supramolecular devices. These obstacles can be overcome by designing robust systems wherein there is substantial cooperativity mediated by multiple intermolecular interactions per molecule and/or ion, or by using surface interactions and concentration gradients as additional tools in the formation of surface-bound structures.

At present, one of the shortcomings of supramolecular chemistry is the inability to design discrete self-assembled arrays that are in the range ~10–100 nm, though these sizes are routinely observed for inorganic systems that are made either by bottom-up or top-down approaches [9]. One exception can be nanocrystalline organic materials, but these are unusual, and difficult to obtain predictably in high yields. Though large supramolecular systems with many copies of a few molecules/ions can self-organize (non-discrete and of varying dispersity), the vast majority of self-assembled and self-organized systems are constituted from ≤three kinds of molecules/ions. Thus, another area that warrants attention is the issue of molecular complexity in terms of expanding the number of different molecular/ionic entities that are incorporated into supramolecular systems with defined geometries.

1.3
Porphyrinoids

Life as we know it rests on the manifold functions of the porphyrinoids [17]. A few examples of our dependence on these tetrapyrrolic macrocycles are given (Fig. 2) [18]:

1. Oxygen transport via iron containing hemoglobin and myoglobin [19].
2. Oxygen is activated toward oxidation of xenobiotics by the iron porphyrins in cytochromes P-450.
3. Energy and electron transport mediated by various magnesium and free-base chlorophyll pigments in photosynthesis to ultimately produce both dioxygen and free energy in terms of carbohydrates.

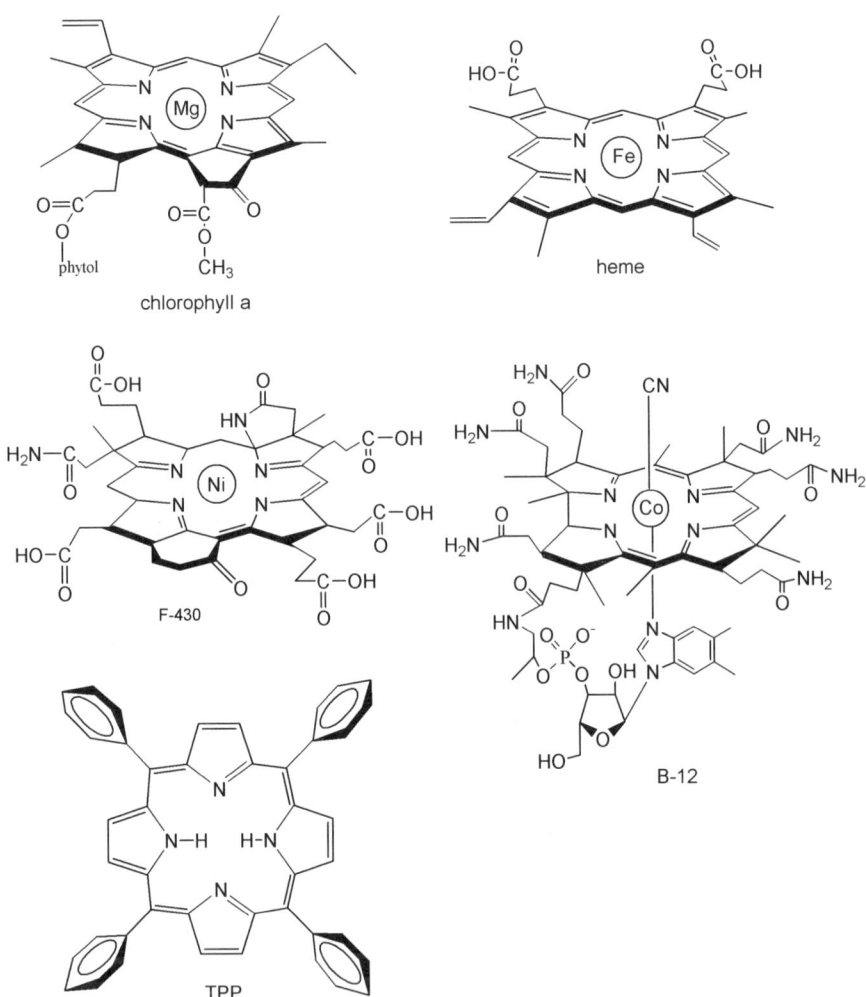

Fig. 2 Some of the porphyrinic co-factors found in nature harvest light and convert it to chemical potential (magnesium chlorophylls), transport or activate dioxygen (iron, hemes), use organometallic chemistry to form carbon-carbon bonds (cobalt, B_{12}), and reduce a methylthioether to form methane (nickel, F-430). Many synthetic multi-porphyrinic systems use derivatives of tetraphenylporphyrin (TPP). No stereochemistry is implied

4. A thioether is reduced in methanogenic bacteria by a nickel porphyrinoid cofactor in coenzyme M reductase [20].
5. The cobalt porphyrinoid of vitamin B_{12} serves as a rare example of organometallic chemistry in biological systems.

Though the external organic substituents have varying degrees of stability, the porphyrin core is remarkably robust, as evidenced by its presence in various geochemical deposits such as shale oil [17]. This stability has led some to propose that the macrocycle is prebiotic, but little evidence so far supports this hypothesis. Porphyrins are also used as therapeutics [21, 22]. The obvious question then is, "How do all these functions arise from porphyrinoids?" In broad strokes, these functions arise from three factors [18]: (1) The coordination chemistry and redox chemistry of the metal ion complexed by the porphyrin act in concert with the electronics of the macrocycle; (2) the number of reduced double bonds in the macrocycle and the exocyclic substituents affect the redox chemistry of both the porphyrin and the metal ion, thus concomitantly the coordination chemistry of the metal; and (3) the environment, i.e. protein, modulates the properties of both the porphyrin and the metal center. It is becoming increasingly clear that nature fine-tunes the redox properties of both the macrocycle and/or the chelated metal by sometimes quite subtle changes in the conformation and the conformation flexibility of the porphyrinoids [23–25].

Scientific interest in blood goes back millennia, and to be "born to the purple," meant to be the legitimate child of the emperor of Byzantium [26]. However the isolation and characterization of the heme of hemoglobin and myoglobin and the chlorophylls of plants began in the late nineteenth century [17]. Landmarks include the work by Hans Fischer et al. on the synthesis and characterization of heme and chlorophyll compounds, which resulted in the proposal that these pigments were related macrocycles [27, 28].

Though less studied than the porphyrins, porphyrazines have physiochemical properties that are in many ways complementary to the porphyrins due to the four nitrogen atoms in the bridging (or *meso*) positions. Phthalocyanine dyes, which played an integral role in the early synthetic dye industry in Europe, are also related macrocycles with nitrogen atoms at the four bridging positions and fused benzene rings on the four pyrroles. These latter two pigments and dyes posses the same richness in inorganic chemistry, redox chemistry, and photophysical properties as the porphyrins – and thus are also intensely studied [18].

In addition to their usefulness as dyes, in the last 40 years or so many researchers have demonstrated that the porphyrinoids can be used as catalysts and as components of materials and devices [18]. Just over 20 years of research on metalloporphyrin oxidation catalysts, cytochromes P450 mimics, have revealed that there are remarkably diverse ways to fine-tune the reactivity and substrate selectivity via using different metals, halogenation of the macrocycle and its substituents, and bulky peripheral groups [29]. The greatest limitation of these catalysts is deactivation by auto- and self-oxidation processes. Per-halogenation of the catalyst partially ameliorates the self-destructiveness. However, in what will be a classic paper, Hupp and coworkers describe a self-assembled porphyrinic catalyst [30, 31] that lasts "as long as the attention span of a graduate student." [1] Given their role in photosynthesis and the interplay between photophysical and electronic properties (photonics), the porphyrinoids have been studied as one of the light-harvesting components of solar energy conversion cells. Photocells containing porphyrins as solar energy collectors have been reported, for example [32, 33]. Long-term stability of the macrocycle under these conditions remains a limitation in these applications, but perhaps a well-designed self-assembled or self-organized porphyrinic system will overcome the degradation of the solar energy antenna, enhance efficiency, and lengthen the device usability. Again, owing to their various roles in nature such as in cytochromes c, the photonic properties of these three chromophores have also been exploited in terms of molecular electronics [8, 34–40]. The construction of multiporphyrinic arrays by covalent bonds has yielded a variety of molecules with suitable photonic properties for diverse applications [41–45]. The photophysical and electrochemical properties of these molecular constructs have yielded important information on the design principles and properties of multichromophoric systems [39, 46–49], the advantage being that they are single molecules, readily purified, and with clearly defined molecular properties. However, the great disadvantage of all but the simplest system is that the overall yields of these molecules are too poor for any viable commercial application.

1.4
Potential Applications of Supramolecular Porphyrinoid Materials

The mantra that the top-down – subtractive – approach to the miniaturization of electronic devices will reach the limits of practicality due to quantum effects in the small structures and/or due to production costs in the next decade or so has been widely cited as a motivation for the development of bottom-up – additive – approaches [9, 40, 50–52]. There are however, a few methods in the pipeline that may well significantly extend this timeline, such as multilayered devices [53] and changes in the fundamental mode of operation from bits to quadbits [54]. Nonetheless, self-assembled and self-organized systems are poised to make important contributions in some electronic applications [8, 34–40, 55]. Other applications include molecular sieves that are designed to be specific for a given molecule or set of molecules [31, 56–59]. Non-linear optical materials have been made wherein the crystal is designed to contain certain chromophores in non-centrosymmetric space groups [60, 61]. Nano-crystalline and nano-aggregated materials of organic chromophores may well have properties that are obtainable neither from the single molecule nor from macroscale

---

[1] J. Hupp; though the unit of graduate attention span likely varies, this certainly is much better than the hours to days for virtually all other porphyrin catalysts.

solids [9]. Self-assembled systems can provide significant advantages over both the homogeneous and supported catalysts in terms of catalytic turnover, substrate specificity, and the selective transformation of a given site on the substrate. Self-assembled chromophoric arrays can be designed to recognize a target analyte with increased specificity over simple chromophoric systems. The aforementioned applications rely on the redox or photo-physical properties of the materials, whereas the interconversion of electronic and photo signals – photonics – are also envisioned for applications in sensors, photo-gated electronic devices, and electroluminescent materials. More recently, supramolecular chemistry has been used in the formation of high-spin materials [62] with designed spin properties – spintronics – that may be used in a variety of devices. The magnetic properties of these oftentimes-crystalline materials can be used not only for magnetic storage devices, but for other applications, including sensors and actuators.

In summary, self-assembly and self-organization may offer some alternative methods for constructing multi-chromophore arrays, but should not be considered the solution to all the challenges confronting molecular electronics and nanotechnology, because there are significant difficulties with this molecule-up approach.

2
Self-Assembled Porphyrinic Arrays

2.1
Hydrogen-Bonding

There are many variations on the nature-inspired motifs for hydrogen-bonding that result in complementary pairs held together by a 10–30 kJ/mol bond. Building upon these, there are a variety of complementary pairs that use more than three hydrogen bonds that are much stronger, >40 kJ/mol [63–66]. One advantage of this mode of self-assembly is that the direction of the interactions and the energetics are predictable, so that for the most part the challenge lies in the synthesis of the building blocks. There have been many interesting self-assembled structures created by these means. One key motivation for this work is to develop systems to study electron and energy transfer between chromophores brought together by hydrogen bonds (rather than covalent bonds), as is found for the chlorophyll molecules in both photosynthetic reaction centers and in photosynthetic antenna complexes [67, 68]. The lability of hydrogen bonds likely precludes commercial applications of supramolecular materials assembled solely by these interactions. The first discrete porphyrinic array was a simple tetrameric square species assembled from two bis-uracyl-appended porphyrins and two barbituric acid molecules with complementary hydrogen bonds [69]. The combination of three triaminotriazines, each bearing two porphyrins, and three barbituric acids results in a rosette (Fig. 3, left) wherein the

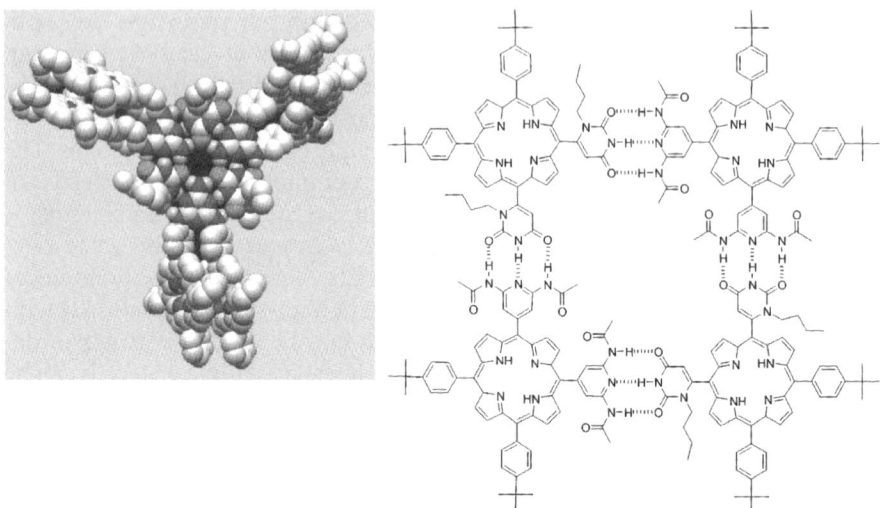

Fig. 3 Complementary hydrogen-bonding moieties are used in the formation of discrete supramolecular arrays of porphyrins by to form a rosette (*left*) [70] and a square (*right*) [74]

porphyrins are in a slipped cofacial arrangement [70] similar to that found in the special pair in the photosynthetic reaction center [71] and in the rings of chlorophylls in the photosynthetic antenna complex [68, 72]. The cofacial porphyrins in the rosette are perpendicular to the ring, and in antenna complexes the ring of cofacially interacting porphyrins is parallel to the protein cylinder. The three pairs of slipped pairs in the rosette energetically communicate, as the fluorescence of this array is completely depolarized. Also, rosettes formed from the statistical mixture of two triazines bearing two Zn(II) porphyrins and one triazine bearing two free-base porphyrins emit primarily from the free-base chromophores upon irradiation in the Q band of the metalloporphyrin derivatives. This latter observation is strong evidence that there is substantial energy transfer from the metalloporphyrin slipped pairs to the free-base slipped pair mediated by the hydrogen bonds of the molecular recognition moieties. To further address the complex role of cooperativity in the self-assembly of closed arrays, a series of hydrogen-bonding porphyrins were made that can be assembled into square tetrameric (see Fig. 3, right) or linear systems [73, 74]. Comparison of the equilibria and thermodynamics of the closed versus the open systems yields insight into the energetic contribution of cooperativity to the closed systems [64]. In these particular systems the cooperativity accounts for about a third of the interaction energy.

A noteworthy observation is that a porphyrin bearing two diacetamidopyridyl groups at 90° forms a self-complementary tetrameric square in solution, but crystallizes into a square helical structure – where one of the porphyrins in the square cants upward to bind to a porphyrin in the tetramer above it, and its neighbor cants down to connect to the tetramer below [74]. This is one of many confirmations that the most efficient way to pack these types of supramolecular systems into solids is in a helical arrangement. Some of the hydrogen bonds to the two acetamide groups in this crystal are so strong that the dihedral angle of the pyridyl ring to the porphyrin is ~30°, which causes a significant distortion in the otherwise planar macrocycle. As one might expect, the interaction energy between the self-complementary hydrogen bond pairs is less than that of the complementary pairs by ~15 kJ/mol, and there is a concomitant reduction in cooperativity. In addition to rendering a more stable system, the stronger intermolecular interactions in the complementary squares is manifested by a greater degree of energy transfer among the porphyrins than in the self-complementary squares, as observed by energy transfer from Zn(II) porphyrins to free-bases macrocycles in a square.

These and many other results indicate that the formation of supramolecular squares (and other closed systems) is an almost unavoidable consequence of assembling component molecules that are rigid and geometrically poised to do so – as long as assembly is done in the solvent and concentration range appropriate for the intermolecular forces used. Thus the design and synthesis of closed supramolecular systems can be routinely accomplished (see section on self-assembly by coordination chemistry).

Is it possible to both increase the size of self-assembled systems held together with hydrogen bonds and increase the number of different kinds of component molecules? The next generation of the 2×2 arrangement of the simple one- or two-component squares above is a tessellation of porphyrins in a 3×3 grid (Fig. 4). Unlike the 3×3 grids made from six equivalents of a single linear, tritopic ligand and nine equivalents of a metal ion [1–4], this arrangement of porphyrins requires four 'L'-topic, four 'T'-topic and one 'X'-topic porphyrin – each with complementary hydrogen-bonding groups. Not only are three topologically different molecules needed, but two different molecular recognition moieties (the self-complementary bonds being too weak to hope for an efficient assembly). The formation of the ~5×5 nm, 3×3 porphyrinic array is observed by NMR (via global analysis of all hydrogen-bonding hydrogens), resonant light-scattering, and dynamic light-scattering in tetrahydrofuran and chloroform under a narrow concentration range at 20 °C. After mixing and allowing the solution to equilibrate for at least 1 h at 30 °C, unassembled porphyrins and small multimers are observed below ~1 mM, whereas above ~10 mM large polymers/aggregates are observed. This is not unexpected given the hydrogen-bond strength in these solvents, the formation of closed systems, and the design of the molecular tiles. Not surprisingly, attempts to crystallize the hydrogen-bond nonamer have yet to be successful. These results indicate that there is a careful balance between the reversibility of the intermolecular interactions and the need to lock the correct tile into place (see section on coordination chemistry). These results also indicate that the formation of closed, discrete systems larger than tetramers by hydrogen bonds may require secondary cooperative interactions – such as π-stacking, or coordination chemistry.

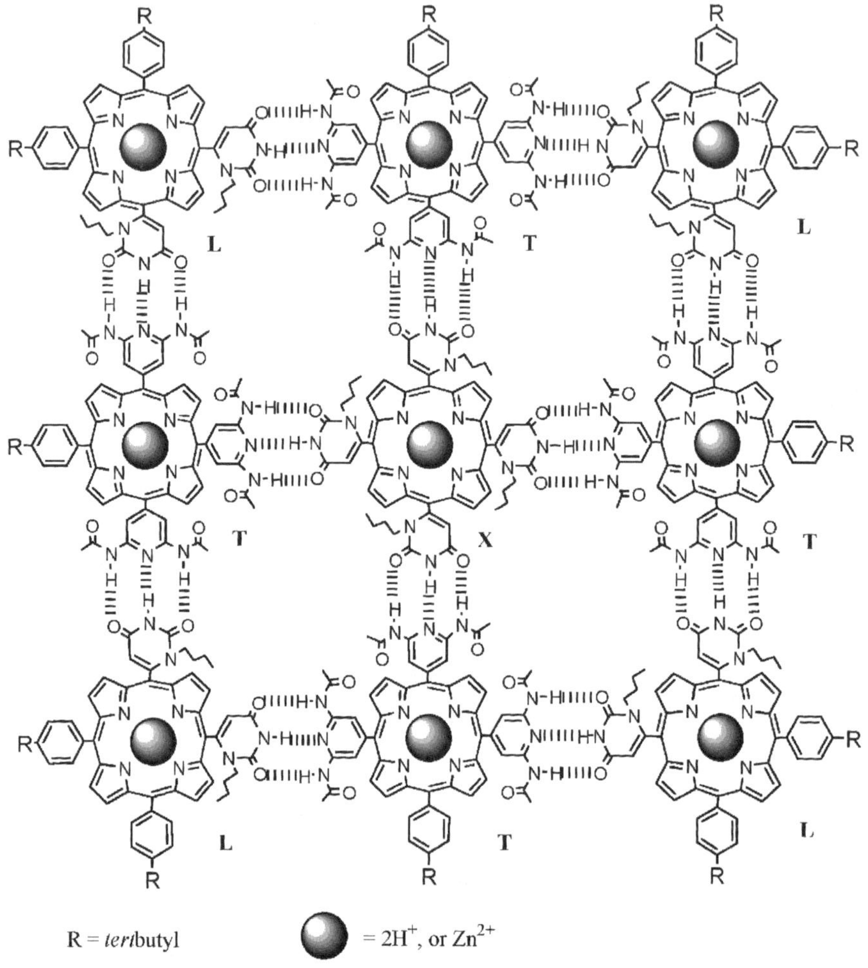

Fig. 4 Dynamic and resonance light scattering, mass spectroscopy, and ^{1}H NMR evidence for the formation of a nonameric array held together by complementary hydrogen bonds indicate that it exists only at limited concentration ranges and with a maximum yield of ~50% in CHCl$_3$ and tetrahydrofuran at 20 °C. As expected, below this concentration window monomers and undefined small oligomers are present, and above it large undefined aggregates are found

2.2
Coordination Bonding

The rich coordination chemistry of the metals in the periodic table, along with the very large number of possible organic ligands, allows the design of discrete and open supramolecular systems that are only limited by our imagination. In addition to serving as structural tectons, many transition metals add functionality, such as paramagnetism, multiple redox states, catalysis, luminescence, and the ability to change coordination geometry upon changes in oxidation state. Coordination polymers have been known for nearly a century and are still a fruitful area of much research [75]. The formation of discrete multimers of inorganic coordination complexes has also been known for many decades, e.g., oxo-bridged metalloporphyrins and metallophthalocyanines [18], and some metal carbonyl clusters are both multimeric and closed [75]. One of the first reports on the deliberate assembly of discrete arrays by coordination chemistry – squares – was made by Fujita in the early 1990s using 4,4'-bipyridine ligands [76]. Our report on the formation of two types of squares using porphyrins bearing exocyclic pyridyl groups as either corners or sides soon followed [77]. The significant result in the latter report is that the rigid porphyrin ligands can serve as corners thereby facilitating the self-assembly of the squares even when the metal ion linker has some propensity for isomerization (in this case from *trans* Pd(II) to *cis* Pd(II). Unlike simple linear ligands, this paper demonstrated that the known coordination geometry of various metal ions could be combined with the designed topology of the ligands to result in predictable self-assembled architectures. Additionally, the use of porphyrins as components added a significant functionality to the system that was not present in systems using simple polypyridyl ligands. The porphyrin chromophores also serve as highly sensitive reporters of the assembly process using UV-visible and fluorescence spectroscopy. These photophysical properties allow study of the self-assembly/self-organization processes into the less than micromolar regime. The formation of closed molecular architectures using coordination chemistry is now routinely accomplished. After these reports the number of papers on supramolecular squares and some larger supramolecular macrocycles increased rapidly. Subsequently, several reviews appeared on the concept of matching ligand architecture with metal ion coordination geometry [42, 78, 79].

Again, given that the formation of squares is almost inevitable with the correct geometries of the ligands and metals, can the next generation, the 3×3 array, be self-assembled in reasonable yields? This also requires four L-topic corners, four T-topic sides, and one X-topic central porphyrin, but in this case all the intermolecular interactions, and pyridyl coordination to Pd(II), are the same. This latter point is important because it significantly reduces the complexity of the self-assembly process as the molecules reversibly rearrange to find the most stable supramolecular conformation. The answer is yes and the products are robust and can be processed onto surfaces (Fig. 5) [80]. Since there are three different porphyrins in the nonameric array and each of these can contain a different metal ion, there are many possible arrangements of metalloporphyrins in the nonamer. For example, with two metals there are 8 different arrangements (Fig. 5) and with three there are 21 possible ways to design the arrangement of the molecular tiles. This variety of tessellation can be used to design materials with a variety of photonic properties, examine electron/energy transfer, or create other novel materials.

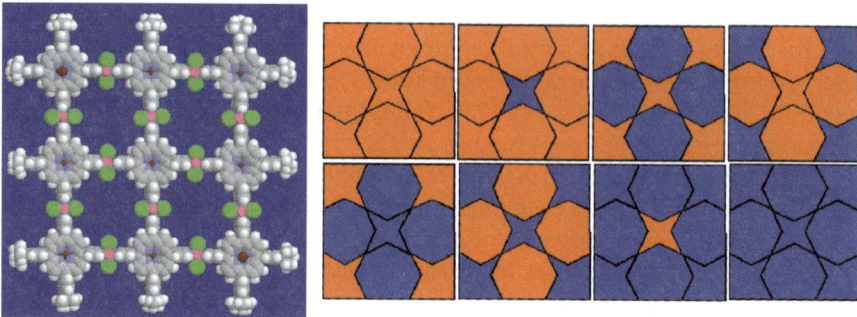

Fig. 5 The tessellation by self-assembly of a 21-component nonamer using pyridyl coordination to Pd(II)Cl$_2$ proceeds with high yields, and the products are robust enough to deposit on surfaces with high structural fidelity. Before tessellation, each of the three porphyrins can be metalated with a different ion or remains the free base (*left*). On the *right* is a representation of the eight possible arrangements in which the component porphyrins with two chelated metals can be predictably and reliable assembled. For example, inserting Fe(III) (*red*) into the 'L'-shaped porphyrins that will later be assembled as the corners of the nonameric array, and Fe(III) into X shaped porphyrin that will constitute the center, and Mg(II) (*blue*) into the four T shaped sides, will result in an arrangement depicted in the *top row, third from the left*

At room temperature in toluene in <10 µM concentrations, the formation of the nonamer proceeds at >80% yield. Later it was found that nine equivalents of one of several first row transition metals can be added to the milieu to metalate the nine porphyrins in situ [13]. This latter process represents the self-assembly of a 30-component system with five different kinds of building blocks! As expected, UV-visible titrations, temperature studies, and kinetics all point to a possible cooperativity in the nonamer assembly process [14]. The formation of the free-base nonamer with 21 particles proceeds within a time constant of ~1.4 min., while the corresponding 8-particle tetrameric square forms in ~1.2 min, and the simple 3-particle Pd(II)-linked dimer forms in ~0.8 min. That the formation of an array with seven times more particles is completed in only half as much more time is an indicator of cooperativity. Each particle in the nonamer assembles ~fourfold faster than each particle in the trimer. A secondary self-organization process occurs with a time constant of ~30 min for the free-base nonamer, which represents the formation of columnar stacks of nonamers, *vide infra*.

The tessellation of three topologically different porphyrins into a nonameric array (Fig. 5, left) and the eight possible arrangements of the nonamer when two different metals are inserted into the macrocycles (Fig. 5, right) represents an extraordinarily diverse system that allows for the design of a variety of functional materials.

Attempts to isolate and characterize the next generation of porphyrinic arrays self-assembled by coordination chemistry – tessellation of a 4×4 grid – have been unsuccessful. For example, refluxing a 1 µM mixture of four L-shaped corners, eight T-shaped sides, four X-shaped central porphyrins, and 24 Pd(II) linkers for 2 days in toluene results in some indications that the 4×4 array may be present in the milieu but not in significant yields.

2.3
Electrostatics

The formation of discrete supramolecular systems using only electrostatics is difficult because these interactions are not directional and at least two charges per molecule are needed to align them in a somewhat predictable manner. However, electrostatics can be used to create nanoscaled aggregates both in solution [81–83] and incorporated into MCM-41 cavities [84].

In the late 1980s it was shown that a discrete linear array of porphyrins could be transiently formed inside lipid bilayers via photo-formation of metalloporphyrin cations (by interfacial electron transfer to aqueous acceptors), which subsequently formed an ion chain with lipophilic anions such as tetraphenylboride [34–37]. This 3-porphyrin-3-anion chain was proposed to be the salient mechanism for the photo-gated ionic conductance across the ~8 nm lipid bilayer under an applied voltage of 40 mV. This system represents an early example of a working molecular electronic device – in this case a photo-gated transistor that used ionic rather than electronic conductance. Illumination of this system in the porphyrin absorption region (or with white light) gates 10^2–10^3 times the dark ionic currents of the boride. Temperature studies revealed that the formation of the ion chain is highly dependent on the fluidity of the lipid bilayer. The response time of this system (~0.5 ms) is limited largely by the need for the transiently formed porphyrin cations to assemble with the lipophilic anions into the functional ion chain. Later studies demonstrated that C_{60} can be used in lieu of porphyrins, and other studies revealed that the photogenerated currents could be exploited to pump protons across the bilayer via a protonophore [85–88]. The obvious solution to decreasing the onset response time is to pre-organize the conductive molecules before the photo-gating event (Fig. 6). Thus, linear arrays assembled by hydrogen bonds and coordination chemistry were later used, and indeed the response (gating) time of the system was reduced into the microsecond time regime. A further modification placed an electron donor on the opposite side of the membrane as the acceptor, such that the system is driven by the electrochemical potential difference rather than an applied voltage [8].

The above systems function because of the electrostatics of ions and dipoles inside the self-organized lipid bilayer. The tapes assembled by hydrogen bonds self-terminate at the bilayer–water interfaces because of the competition for the hydrogen-bonding moieties on the porphyrins by water and the lipid head groups. The tapes assembled by coordination to metal ions terminate because the ions, carbonyls, and water in the lipid head groups also compete for the coordination sites on the metal [8]. The solubility of all types of tapes in the low-dielectric membrane center versus the aqueous phase is also an essential design

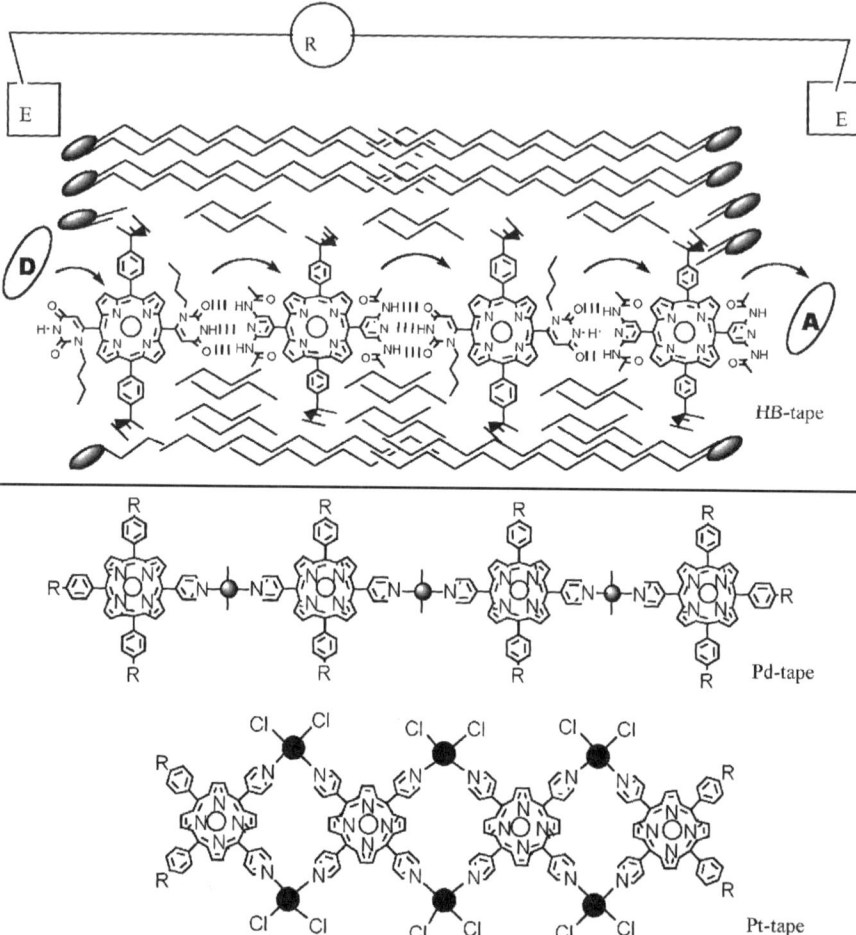

Fig. 6 The self-organization of a function molecular electronic device, in this case a photogated transistor, is accomplished by designing appropriate molecules (primary structure) that self-assemble into linear tapes (secondary structure) [8]. These tapes are then self-organized in lipid bilayers (tertiary structure) supported on a <1 mm hole in a partition separating ~5 mL solutions of 0.1 M KCl, with saturated calomel electrodes, E, in each solution connected to an operation amplifier, R [37]. A ~200 mV chemical potential is generated by placing an electron acceptor, A, such as anthroquinone sulfate in one side, and an electron donor, D, such as $K_4Fe(CN)_6$ on the opposite side. In the dark, with applied voltages within 0–60 mV, there is no measurable current (<200 fA). Using only the chemical potential between donor and acceptor, when the device is gated on by the illumination of the membrane device with white light, the current increases 100- to 1,000-fold over the dark current and 100- to 300-fold over any of the individual (unassembled) porphyrins. Illumination of the transistor causes interfacial electron transfer from the porphyrin nearest the acceptor, leaving a hole that then hops along the porphyrin tape until it is quenched by the electron donor on the opposite side of the membrane. The magnitude of the current of these self-organized photo gated transistors (quaternary structure) depends on the number, geometry, and stability of the self-assembled porphyrinic tapes. The structural/geometric factors dictating electron transfer are well known [see 37 and references therein]. The rise time of the current in these devices is ~100 times faster than the electrostatically assembled ion chains reported in the late 1980s [37] because the conducting porphyrin arrays are pre-organized

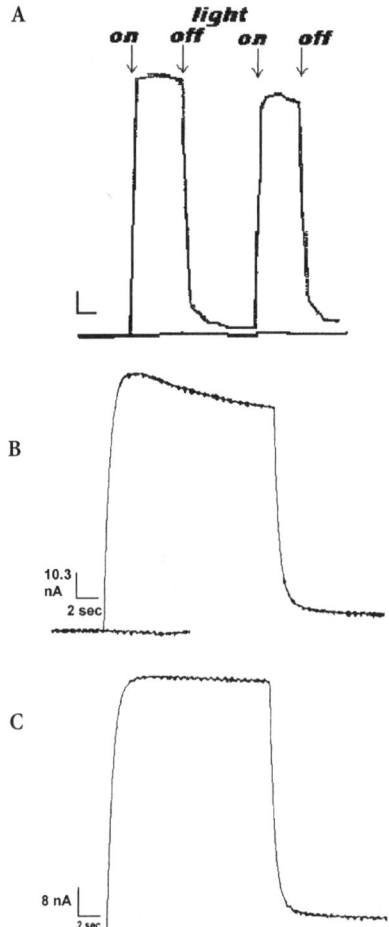

Fig. 7 Strip chart recordings of the photocurrent of the phototransistors are shown [8]. (**A**) the H-bond assembly self-organized into the bilayer where the horizontal scale bar represents 1 minute, and the vertical scale bar 6.9 nA. The lower, baseline trace is the photocurrent of $(Zn^{2+})5,15$-uracilporphyrin only. (**B**) the photo induced current due to the Pd-tape assemblies where the baseline is only the two types of pyridylporphyrins and no Pd complex is present. (**C**) the Pt-tape assembly where the baseline represents the photocurrent of the tetrapyridyl porphyrin without the Pt complex. Note that as the stability of the self-assembled porphyrinic tape increases the stability of the photocurrent increases

feature of the molecular electronic systems. Obviously, the thermodynamics of the intermolecular forces determine the distribution, or dispersity, of tape lengths. The dynamics of the self-organized device and the complexity of the membrane properties make a precise assessment of the yield of the arrays difficult; the approximate fourth power dependence of the photocurrent on the porphyrin concentration confirms the presence and functionality of the self-assembled tapes. In addition to the reduced photo-gating time, because of the pre-organized conducting elements, one would expect that the stability of the current should be proportional to the stability of the self-assembled porphyrin chain, and indeed that is what is observed (Fig. 7) [8].

3
Hierarchical Assemblies: Aggregates

After the self-assembly of the Pd(II)- linked nonameric array there is a secondary self-organization step in which the nonamers stack upon each other to form columnar stacks that are approximately the same height as the nonamer is wide [13, 14]. Though there is only one nonamer per layer in the final nanoscaled aggregate, they cannot be in exact register because the mesopyridyl groups are nominally orthogonal to the porphyrin macrocycles. The heights of the columnar stacks can be controlled by four means: (1) kinetically, where large (10–30 nm) aggregates can be trapped from the equilibrating solution; (2) chemically, by varying the size of the alkane decorating the outside of the nonamer – the larger the alkane the shorter the stacks; (3) by metalation, since the π-stacking energetics are influenced by the metalation state of the porphyrin, the height of the stacks are ~20% greater for the Zn(II) and the Ni(II) nonamers than for the free-base nonamer; and (4) by surface chemistry, since the interaction of the nonamer aggregates with surfaces also dictates the size of the nano particle: the stronger the nonamer–surface interaction, the smaller the columnar heights. This last observation illustrates the importance of surface energetics in the deposition of self-assembled species, in that it can cause structural rearrangements and/or can be used as part of the assembly process.

The exocyclic metal ions can add to or modify the functionality of the supramolecular porphyrinic system, in addition to serving as the glue that binds the material in a specific geometric arrangement. For example, transition metals tend to quench fluorescence by the heavy atom effect, but room temperature solution phase phosphorescence is not observed in the above self-assembled porphyrin squares and nonamers. However, the fluorescence intensity of columnar stacks of the free-base nonamer is more than threefold what would be expected based on the quenching of the singlet state due to binding of the exocyclic pyridyl groups by Pd(II)Cl. Some phosphorescence is also observed. This luminescence is likely due to the organization and rigidity of the nonameric arrays in the stacks, as well as the nanoscale dimensions of the aggregate [31]. Larger, amorphous particles made by reducing the stoichiometry of the corner porphyrins neither fluoresce at all, nor phosphoresce at room temperature.

Most importantly for applications, these nanoscaled aggregates can be drop cast onto a variety of solid supports with high structural fidelity, and without loss of luminescence or agglomeration of the nanoparticles. Secondly, these systems on solid supports are stable for more than 3 years at room temperature in air, again without any observable loss in structure or activity [13, 14]. The wavelength of the luminescence peak maximum of the free-base nonamer nanoparticle on glass changes by ~10 nm depending on the environment of the systems (air, hexane, acetonitrile, water), thereby demonstrating a sensor function.

The oftentimes-unavoidable aggregation of porphyrinoids into small oligomers has confounded many experiments' attempts to characterize these types of macrocycles; thus it is ingrained into porphyrinoid chemists to check for and avoid aggregates. However, it has been realized for decades that these aggregates have properties that are not observed in the monomers [84, 89], and that aggregates that are too large precipitate into amorphous powders. In between these extremes are nanoscaled particles that remain in solution. It is possible to form what are essentially colloidal dispersions of a variety of porphyrins and their metalo derivatives using procedures that are similar for the formation of inorganic colloids [16]. The addition of water to a concentrated solution of a hydrophobic porphyrin in a water-miscible solvent such as tetrahydrofuran or dimethylformamide with ~20 equivalents of a short polyethylene glycol results in nanoparticles of the porphyrin. As with the inorganic nanoparticle formation, the size of the porphyrinic nanoparticles is somewhat controllable by varying the solvent ratios and the amount of the polyethylene glycol stabilizer. In this case, the size and polydispersity of the nanoscaled aggregate products are also dependent on the chemical nature of the porphyrin.

Many of these porphyrin nanoparticles are stable for months, and can be deposited onto glass surfaces without significant change in the particle size. Similar procedures should be applicable to other planar dyes such as the phthalocyanines and porphyrazines. The applications of these porphyrinic nanoparticles include more robust catalysts and sensors. Nanoparticles of both the free-base and the Zn(II) porphyrin are luminescent. Fluorescence quenching of Zn(II) porphyrins nanoparticles by Fe(III) porphyrin dopants indicate that there are domains of ~ten Zn(II) porphyrins in the aggregate that serve as an antenna that funnels electronic energy to the Fe(III) porphyrin trap.

4
Crystal Engineering of Porphyrins

4.1
Porphyrin Tectons

The use of porphyrins as supramolecular tectons in the formation of designed crystalline materials has also been reported, and stems from the rich variety of ways that suitably functionalized porphyrin derivatives can pack into ordered solids. Simple tetraphenylporphyrin (TPP) molecules cannot pack effectively in three-dimensional space, and when crystallized from different solvents they form a myriad of clathrates and solvates [90–93]. The arrangement of the chromophores in these materials is maintained by weak dispersion forces. Many of the TPP clathrates could be characterized by a strongly conserved intercalation-type pattern, in which corrugated sheets of offset-stacked porphyrin molecules are interspaced by sheets of the included guest species [90–93]. It is evident from the large database of such structures (amounting to more than 500) that an offset layered arrangement of the roughly flat TPP species at a narrow distance range (within 4–5 Å) is a fundamental property of the porphyrin-porphyrin interaction required to optimize van der Waals' stabilization in these solids [90–96].

As stated above, the spatial order of the porphyrin chromophores in supramolecular clusters dictates the photonic properties. In another mode, associated with aggregation of these dyes and pigments into ordered crystalline solids, these materials can act as molecular sieves, sensors, actuators and other devices based on the geometric disposition of the interporphyrin voids in the solid. They can be useful also in heterogeneous catalysis, and as nano-sized reactors with a confined environment. For the free bases, exocyclic molecular recognition motifs positioned parallel to the plane of the macrocycle can be assembled into one-dimensional and two-dimensional arrangements (tetratopic X-shaped and tritopic T-shaped chromophores for two-dimensional layers; ditopic L-shaped, and ditopic I-shaped molecules for one-dimensional chains). Three-dimensional topologies are achieved by positioning the intermolecular binding sites at some angle to the porphyrin plane (usually 60 or 120°), using external bridging auxiliaries, and/or by using metalloporphyrins that can bind through their metal centers one or two axial ligands. Supramolecular design of open architectures that sustain large void space requires multiple and convergent activation of these interaction synthons to impart robustness to the assembled lattice. This is line with the observation that molecular recognition features tend to be topologically saturated in crystals, in that most of the intermolecular binding sites are used (*vide supra*) [97].

As in the aforementioned discrete porphyrin arrays, intermolecular forces to develop two-dimensional and sometimes three-dimensional order in crystals of these molecules (by specific and directional interactions rather than by virtue of van der Waals' forces only) can be roughly divided into multiple and complementary hydrogen-bonding, coordination chemistry, and electrostatics. In some cases the π-π interactions between porphyrins [95, 96] from different self-assembled arrays predictably assist in the self-organization of the supermolecules into a three-dimensional material, but in most cases this remains only a qualitative design feature. There are several reports that use a concerted mechanism of hydrogen-bonding and coordination chemistry to execute structures with predictable three-dimensional order in terms of the orientation and disposition of the chromophores [5, 98]. Understanding and predicting the subtle interplay between the designed, specific intermolecular interactions and the known, but non-specific interactions between molecules and supermolecules in the solid state remains the great challenge in the supramolecular chemistry of crystals.

4.2
Hydrogen-Bonding

Though the use of hydrogen bonds to self-assemble discrete supramolecular arrays results in systems of limited stability in solution, the cooperative interactions found in the solid state oftentimes yield remarkably stable architectures [99–101]. In the TPP-based porphyrinoids, noteworthy examples reported thus far use one of the hydroxy, amide, carboxy or diaminotriazine groups. Symmetrically substituted on the periphery of the porphyrin macrocycle; asymmetric chromophores have been less successful in the designs of extended arrays. These molecular recognition functions are self-complementary as they can act as both proton donors and proton acceptors at the same time. Functional groups placed at the 4-position of the aryl substituents may allow the horizontal (layered) organization of the chromophores parallel to the macrocyclic core framework, while their location at the 3- and 5-positions allows a vertical (columnar) organization in a ladder-type fashion (the latter has been realized only with the hydroxy derivatives to date). The preferred binding through the peripheral functions introduces voids between the porphyrin units, which are sizeable enough to accommodate other guest species. Correspondingly, when different binding patterns are possible, the guests acting as templates may play an important role in directing formation of the energetically preferred supramolecular aggregate.

The basic motif of self-assembly with the tetra(4-hydroxyphenyl)porphyrin in its free-base or metalated form, is represented by one-dimensional chain arrangement of the chromophores which involves hydrogen-bonding through the *cis*-related residues of neighboring blocks. In some cases the linear arrays are further interlinked sideways by hydrogen bridges to afford a two-dimensional grids with typically ~3.5 Å-wide interporphyrin voids (much smaller than the size of the chromophore unit itself) [102–104]. In columnar assemblies sustained by vertical hydrogen bonds involving the 3,5-dihydroxyphenyl groups, the van der Waals' voids between successive porphyrin cores can be nearly twice as large [105, 106]. These hydrogen-bonding functions form relatively weak bonds and reveal a high affinity for polar and protic solvents, and their application in supramolecular design is therefore quite limited.

The tetra(4-amidophenyl) and tetra(4-carboxyphenyl) chromophores are much better building blocks for the construction of multiply hydrogen-bonded arrays. Either cyclic-dimeric [$(COOH)_2$, $(CONH_2)_2$] or chain-polymeric aggregation modes of the carboxy and amido groups are possible with similar enthalpic stabilization (depending on the total number of hydrogen bonds every unit is associated with). The cyclic pattern requires a strict coplanar head-to-head orientation of these groups, while for the less geometrically constrained chain arrangement, their mutually perpendicular orientation is preferred. Given these degrees of freedom, the preferred mode of the intermolecular aggregation is often shaped by crystallization templates present in the reaction mixture. To this end suitable examples are provided by two-dimensional net-

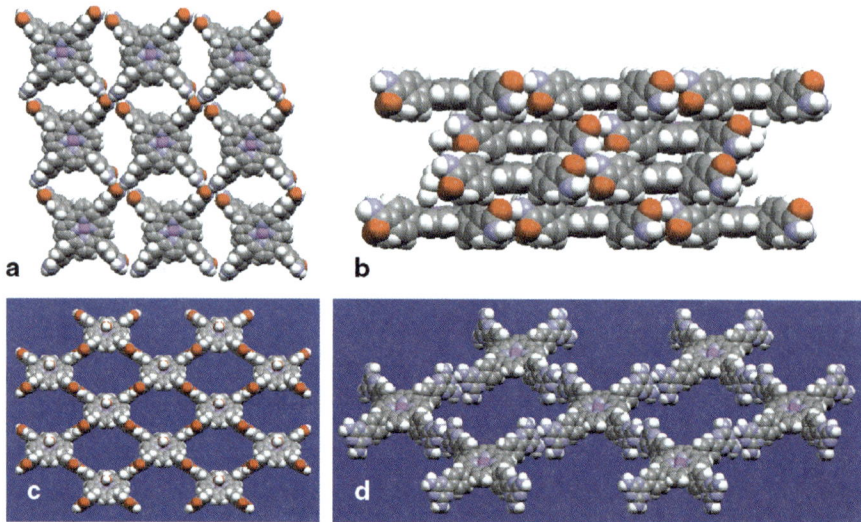

Fig. 8a–d Structures composed of hydrogen-bonding arrays. **a** Flat networks of the tetra(amidophenyl) metalloporphyrin with 0.65 nm-wide voids [79]. **b** Effective van der Waals stacking of the flat multiporphyrin hydrogen-bonded layers in crystals [58, 79, 110]. **c** Open arrays of the aquazinc-tetra(carboxyphenyl)porphyrin with 1.6 nm-wide voids [58, 109, 110]. **d** Hydrogen-bonded two-dimensional networks of the tetra(diaminotriazinophenyl) derivative with 2.2 nm-wide interporphyrin cavities [111]

works of the amido-chromophore organized by the dimethylsulphoxide template (Fig. 8a) [79], and layered arrangements of the carboxy-chromophore templated by a smaller ethyl benzoate [107] and a larger 18-crown-6 [108] guest entities. The observed widths of the interporphyrin cavities in these structures are 0.65, 0.70 and 0.85 nm, respectively. They are still smaller than the size of the chromophore building block, which makes interpenetration of the two-dimensional networks into one another during nucleation and crystal growth impossible. Correspondingly, the two-dimensional layers are regularly stacked at close distance (Fig. 8b), and in an offset parallel manner within the condensed crystalline phase (as in the TPP clathrates), to optimize the interlayer dispersive interactions which stabilize the solid structure (*vide supra*). Rarely, these layered crystals can sustain extraction of the guest template from the lattice without a significant deterioration.

Supramolecular networks with interporphyrin voids larger than the size of a single chromophore have been synthesized as well. Their preparation is more difficult, as it requires optimal utilization of multiple hydrogen bonds and the presence of suitable templates. Moreover, it is often associated with interweaving of these networks into one another, which slows down the assembly process of these species. With the tetracarboxy chromophore, the open arrays can form by means of the cyclic dimeric interaction synthon of all four functional groups to four different neighboring units along the equatorial molecular axes. Each building block is linked to its neighbors by eight O–H···O=C hydrogen bonds. The resulting layered multiporphyrin network is characterized by ~1.6×1.6 nm^2 voids (Fig. 8c) [58, 109, 110]. Crystals consisting of both interpenetrating [109] as well as layered networks [58, 110] were described. The latter were synthesized by applying sizeable templating agents of low tendency to interact with the carboxy metalloporphyrin used in this case. In this solid the hydrogen-bonded open layers stack in an almost overlapping manner along the normal direction at an average distance of 4.7 Å. The layered structure is thus stabilized by effective van der Waals' interactions, by maintaining close contact between the large, flat multiporphyrin surfaces. The interporphyrin voids combine into about 1.5 nm wide channels, which propagate through the crystal. Although, the metalloporphyrin lattice occupies less than 40% of the crystal volume, it remains stable up to 80 °C, resembling a molecular sieve material with wide channels that are readily accessible to other species. Replacement of the carboxylic functions by larger diaminotriazine substituents yields porphyrin chromophores with longer side arms, while preserving their self-complementarity and multiple functionality in hydrogen-bonding. It allows the formulation of open networks of nearly square symmetry with ~2.2×2.2 nm^2 interporphyrin voids (Fig. 8d) [111]. Each chromophore takes part in eight hydrogen bonds to its neighbors, by head-to-head interactions between the diaminotriazine functions. Stabilization of the entire crystal structure is achieved in this case by concatenation of these networks into one another, and further hydrogen-bonding between them. Straight 0.6 nm-wide channels propagate through the interweaved networks, imparting to the structure the sieving properties of small guest-species [111].

The above examples illustrate the possibilities of obtaining porphyrin-based two-dimensional hydrogen-bonded polymers in crystals, which can conveniently host other molecules of nanometric dimensions. They also demonstrate the tunability of the pore size (from less than 0.5 nm to more than 2.0 nm) by corresponding modification of the functional groups positioned on the porphyrin framework, as well as by variation of the templating agents.

4.3
Coordination Chemistry

Some of the earliest examples of using known coordination chemistry to order porphyrins in the solid state into polymeric arrays used metalo derivatives of the readily available tetrapyridylporphyrin [112–114]. More recent work has introduced the tetra(carboxyphenyl)porphyrin entity as an extremely versatile building block for effective supramolecular assembly of new polymeric materials [115]. In addition to the variety of exocyclic ligands, the topological points of designed interaction can now include one or both of the axial sites of a metalloporphyrin. The enforced square planar geometry of the porphyrin ligand results in sometimes-unusual axial binding properties of the bound metal ion. For example, Zn(II) exhibits high propensity to coordinate one axial ligand but can also appear as 4- or 6-coordinate, Ni(II) is either 4- or 6-coordinate, Mn(III) is mostly 6-coordinate and examples of 4-, 5-, and 6-coordinate Cr(III) porphyrins are known. A further degree of control is afforded by the oxidation state of the metal, e.g. the coordination chemistry of Co(II) is different than that of Co(III). Lastly, the propensity for hard metal ions such as Al(III) to bind hard ligands such as oxygen, and soft metal ions to bind soft ligands such as sulfur and nitrogen can be exploited. The stability of coordination polymers in solution is considerably greater than that of the hydrogen-bonded arrays, and their formation can be monitored reliably in solution by spectroscopic methods, as well as in the solid state by crystallographic techniques.

The coordination chemistry of porphyrins and metalloporphyrins in crystals (i.e., formation of extended coordination polymers) can be carried out with or without external auxiliaries. Thus, a given metalated chromophore can coordinate directly to neighboring porphyrin entities through its metal center as well as peripheral functions. Initial reports referred to the syntheses of one-dimensional homogeneous coordination polymers of the pyridyl type chromophores [112–114, 116]. This was followed by the synthesis of a three-dimensional single-framework coordination polymer with a honeycomb architecture, which revealed unprecedentedly high thermal stability up to about 400 °C [117, 118]. In the latter, each chromophore coordinates to four other species by attracting the pyridyl ligands of two neighboring molecules to its metal center [either Zn(II), Co(II), or Mn(II)], while donating two of its own pyridyl functions to the metal cores of other molecules. These polymers act as molecular sieves, being perforated by 0.6 nm-wide channels, and revealing molecular sorption and desorption features (Fig. 9a) [117, 118]. Ladder-type polymers in which 5- and 6-coordinate zinc tetrapyridylporphyrin coexist have also been reported [119]. Metallo-tetraarylporphyrins with iron ions in their core also yield homogeneous coordination polymers, which are characterized by a layered structure with a two-dimensional paddle-wheel like pattern [120, 121]. The coordination geometry around each iron center is octahedral, with two nitrogen sites (of the pyridyl or aminoaryl substituents) from adjacent porphyrin entities occupying the axial sites. These networks, wherein all the neighboring chromophores are mutually perpendicular to each other, exhibit a remarkable thermal stability [120]. Mixed free-base and metalloporphyrin-containing architectures have also been designed by reacting the tetrapyridyl chromophore with a manganese-TPP unit. This combination yields flat networks of nearly square symmetry with alternating metalated and non-metalated entities along the two axes [122]. Their stacking in the crystal also yields a porous material (Fig. 9b).

It is possible to widely diversify the coordination chemistry of the porphyrinoids by the use of external metal ion auxiliaries of bridging capacity between peripherally functionalized chromophores. Such external linkers add electrostatic (ion-pairing) attractions and topological versatility to the polymeric assembly. The tetrapyridylporphyrin species have been widely used to this end, being capable of forming relatively strong bonds (through its pyridyl

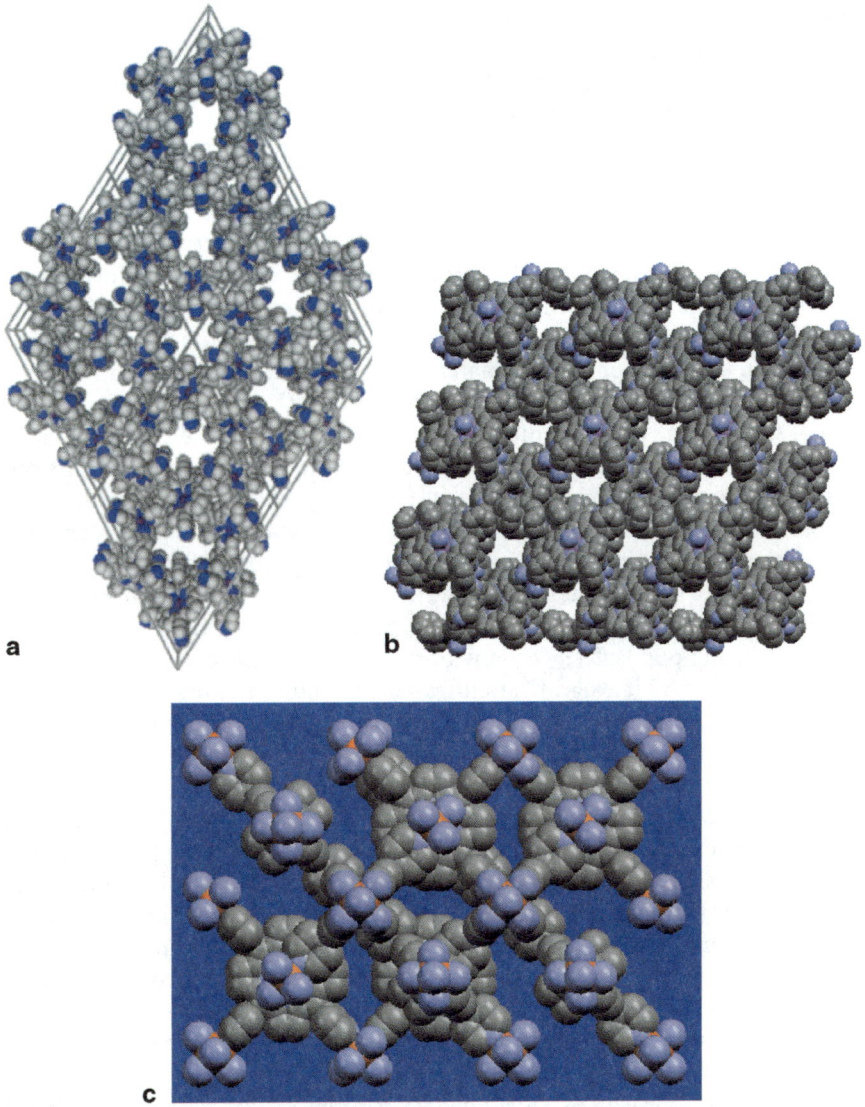

Fig. 9a–c Space-filling illustration of network assemblies based on the tetrapyridylporphyrin chromophore. **a** Homogeneous coordination polymer with three-dimensional connectivity and pore cross-section of 0.6 nm [117]. **b** Structure consisting of stacked two-dimensional polymers of this chromophore with Mn-TPP, and perforated by 0.45 nm-wide channels [122]. **c** Octahedral network grids of iron-tetrapyridylporphyrin, in which each iron ion (*red*) is surrounded by six ligating nitrogens (*light blue*). The core metals coordinated to three different porphyrin units while the exocyclic linker to six neighboring entities [125]

nitrogens) to a large variety of metal ions. Noteworthy here is the pioneering work of Robson et al. on the formation of three-dimensional frameworks of the tetrapyridyl chromophores sustained by linear cadmium(II) and tetrahedral copper(I) exocyclic linkers, which bridge by coordination between the pyridyl groups of adjacent porphyrins [123, 124]. The former yielded condensed network arrangements, while the latter afforded porous architectures with >1 nm-wide channels propagating through the crystals (which were found to collapse, however, upon removal of the included solvent). This attractive supramolecular chemistry was soon followed by numerous researchers, who employed other metal ion linkers and diversified the preparative procedures from conventional crystallizations to hydrothermal reactions between the reacting components. As a result, extended polymeric materials of different topologies and varying porosity have been prepared utilizing e.g., Cu, Fe (Fig. 9c) [125], Cd, Hg, Pb [126, 127] and Ag [128, 129] metal ions as bridging auxiliaries. In some cases the interporphyrin coordination took place through metal ions residing both within and between the chromophores [125]. In others, different coordination topologies (e.g., linear chains, two-dimensional square grids and interpenetrating three-dimensional nets) were obtained with the same bridging ion [128, 129]. It was also shown that the porphyrin cores and the interporphyrin bridging sites could be populated by different metal ions [126, 127]. From the crystal-engineering point of view, many of the above structures are quite robust, although they exhibit a significant intra-lattice void space. Yet the voids accessible to guest components are affected by the need to incorporate counter ions (in the form of either simple, or metal oxide, anions) within the crystal lattice to balance the charge introduced by the metal ion linkers, which limits their potential function as solid state receptors.

In the latter respect, the use of the tetracarboxyphenyl porphyrin in formulations of similar hybrid organic-inorganic coordination polymers is more advantageous than that of the tetrapyridyl chromophore. The acidic groups can be readily deprotonated in basic environments, thus providing the desired charge balance in the structure and eliminating the need to incorporate other ions. This also leads to enforced coordination synthons between the anionic chromophores and the cationic linkers, and enhances the propensity to form crystalline architectures with better-defined nano-porosity. Moreover, the carboxyporphyrins have an extended metal coordination capacity, and can readily form mononuclear as well as polynuclear metallo-carboxylate complexes. Since the first report on this issue [115], different external auxiliaries have been successfully applied in formulations of metal-tessellated coordination polymers based on the tetracarboxy chromophore. These include, e.g., the Na^+ [59, 107, 115], K^+ [130], Co^{2+} [131], Ca^{2+} [132], Zn^{2+} [115], Cu^{2+}, Pd^{2+} and Pt^{2+} ions and ion clusters [133], as well as metal complexes such as $[Cu(NH_3)_6]^{2+}$ [133].

The free-base and metalloporphyrins with metal centers that do not attract axial ligands self-assemble through the exocyclic linkers into flat two-dimensional networks, which stack-overlap one on top of the other along the third dimension, as in the hydrogen-bonding assemblies. The bridging ionic auxiliaries connect between the carboxylic/carboxylate functions of four (of a given layer) or more (from adjacent layers) chromophores at each binding node. The stacked, layered networks are perforated usually by ~0.4–0.5 nm-wide channels which propagate through the layers and accommodate non-interacting guest components [59, 133]. The size of the pores can be increased (e.g., to ~0.65 nm) by application of metal ion complexes rather than simple metal ions [133]. Formulation of diamondoid structures with interweaved networks and similar porosity features, by using linking sites with tetrahedral coordination directionality (e.g., Zn^{2+}), has also been reported (Fig. 10a) [115]. The notion that higher enthalpic stabilization of the network architecture can be achieved by bridges consisting of metal ion clusters has recently been confirmed by the hydrothermal synthesis of a single-framework three-dimensional inter-coordinated architecture, which consists of bent metalloporphyrin entities tessellated to each other by trimeric clusters of the cobalt ions [131]. This material is thermally robust, and contains tridirectional channels with cross sections of nearly nanometric dimensions, thus representing a useful zeolite analogue (Fig. 10b). It was found to reveal a very high affinity for water and amines, and to effectively remove water (by sorption into the porphyrin lattice) from common organic solvents.

Organic ligands can also provide a very attractive tool for the tessellation of multiporphyrin assemblies by axial bridging between the centers of adjacent metalloporphyrin units by metal-ligand coordination, and/or by lateral hydrogen-bonding between the peripheral functional groups of the chromophore. Bidentate amines, particularly those of the bipyridyl type, reveal a high affinity for transition metals as well as for hydrogen-bonding. "Shish-kebab"-type linear polymers with alternating Mn-porphyrin [Mn(III) represents a six-coordinate metal center] and bipyridyl species provide representative examples [79, 119]. Among those, chain polymers composed of cationic por-

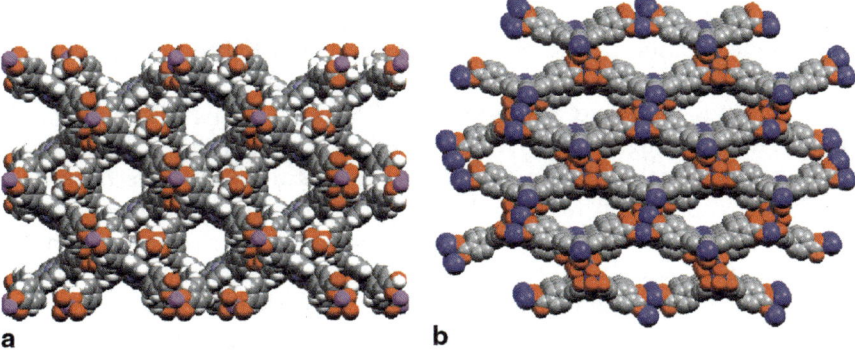

Fig. 10a, b Porphyrin sieve structures assembled by the tetra(carboxyphenyl) chromophore with the aid of exocyclic metal ion auxiliaries. **a** Diamondoid assembly tesselated by tetrahedral zinc ions with hexagonally shaped channels [115]. **b** Three-dimensional porous architecture sustained by trimeric clusters of cobalt ions [131]

phyrin and bridging radical anions were shown to function as organic magnets [134–137]. Open network assemblies in which the porphyrin chromophores are bridged equatorially by cooperative hydrogen-bonding have been reported as well [102, 103].

4.4
Hybrid Approaches

The use of two or more interaction synthons of the hydrogen-bonding and/or coordination type in a single material greatly expands the possibilities to organize porphyrins in crystalline solids. The following examples illustrate porphyrin network formation by expression of a concerted mechanism of intermolecular interactions in directions parallel and perpendicular to the porphyrin core. Thus, reactions of zinc tetracarboxyphenylporphyrin with 4,4′ bipyridyl ligands yielded extended assemblies in which axial coordination of the ligands to the zinc ions of neighboring porphyrins combined with the required equatorial interporphyrin hydrogen-bonding provided the organizing force for the network formation [102, 103]. The hydrogen-bonding linkage may take place either directly between the hydroxy functions of the porphyrin entities, or through the bipyridyl bridges, with their N-sites acting as proton acceptors from the hydroxyls. These lattices are characterized by full utilization of the self-complementary molecular recognition elements during crystallization, as well as by their porous nature.

Another variant of effective simultaneous coordination-polymerization and hydrogen-bonding was achieved in the reaction of zinc-tetra(carboxyphenyl)-porphyrin with a (2)pseudorotaxane ligand 1,2-bis(4,4′-bipyridinium)ethane-dibenzo-24-crown-8 [107]. The latter represents a linear doubly charged species containing pyridyl fragments at both ends. The crown ether ring is threaded on the molecular axle of the bidentate bipyridinium ligand, imparting bulkiness and geometric rigidity to the system. The charge balance comes from deprotonation of two of the carboxylic groups in the porphyrin. Indeed, the (2)pseudorotaxane dication turned out to be an excellent axial bridge between the metalloporphyrin cores, by coordinating from both sides to their zinc ion centers, and thus leading to the formation of linear multiporphyrin polymers which extend through the crystal. Simultaneously, the porphyrin units associate in the equatorial plane into continuous open networks, through cooperative charge-assisted strong hydrogen-bonding between the carboxylic and carboxylate groups. These hydrogen-bonding networks are characterized by an approximate square-planar symmetry, where every chromophore unit in the net is linked to four neighboring molecules at the corners of the square framework, and a wide-open architecture. The linear coordination polymers that propagate in a direction roughly perpendicular to the porphyrin plane interpenetrate through the hydrogen-bonded porphyrin network. The (2)pseudorotaxane ligands that are embedded within a given layer link to porphyrin units of neighboring layers located above and below. In other words, the linear co-

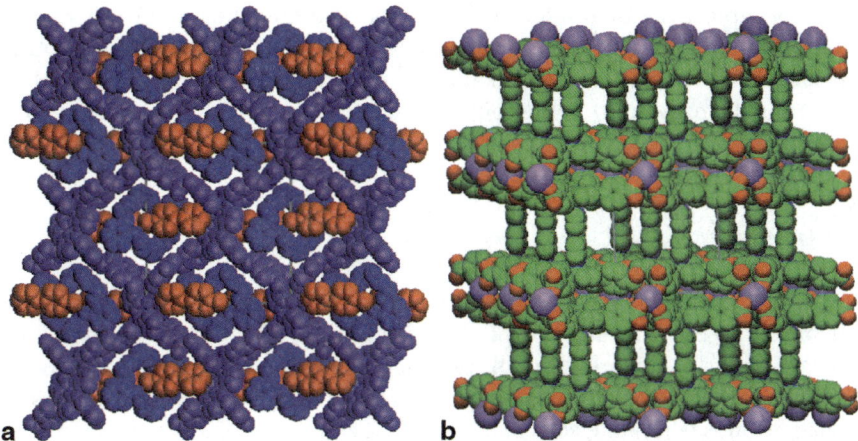

Fig. 11a, b Supramolecular crystalline assemblies of the tetracarboxyphenyl metalloporphyrin formulated by hybrid molecular recognition algorithms. **a** Non-porous architecture consisting of hydrogen-bonded framework of the Zn-porphyrin chromophores (*violet*) that are interlocked by coordination through the pseudorotaxane ligands (*red and blue*) [107]. **b** Porous zeolite analog perforated by open channels. It is composed of Zn-porphyrin bilayers, bridged by the sodium ions (*violet*), which are multiply intercoordinated in a vertical direction by bipyridyl axial ligands [130]

ordination polymers sustained by the pseudorotaxane bridging ligands interlock into the multiporphyrin molecular networks, which in turn are held together by cooperative hydrogen-bonding (Fig. 11a) [107]. While this material beautifully demonstrates an effective crystal-engineering strategy for the formulation of robust network solids by a combination of ion-pairing, coordination and hydrogen-bonding forces, it lacks porosity due to the bulkiness of the (2)pseudorotaxane ligand.

We conclude, then, with an elegant formulation of a nanoporous material tailored by cooperative metal-ligand coordination in all three dimensions, which represents another zeolite analogue. This solid was synthesized by reacting zinc-tetra(carboxyphenyl)porphyrin with 4,4′-bipyridyl (as axial bridging auxiliaries), in the presence of sodium ions (as equatorial linkers to the carboxylic/carboxylate functions). Thus-assembled inter-coordinated architecture reveals a fascinating single-framework polymeric structure (Fig. 11b) [130]. It consists of porphyrin bilayers held together by the sodium ion bridges. The latter coordinate to the carboxylate groups of several porphyrin units, an effective interaction assisted by interporphyrin hydrogen-bonding in the equatorial plane. The successively displaced bilayers are then cross-linked in the axial directions from both sides by the bipyridyl ligands, which coordinate to the metalated porphyrin cores. The lattice thus formed is perforated by nearly 1 nm-wide open galleries that extend between the bipyridyl pillars in different directions parallel to the plane of the porphyrin bilayers. The lattice framework occupies less than 40% of the crystal volume, and the channel voids that propagate through the crystal can accommodate a wide range of other species. Yet it exhibits a remarkable thermal stability up to about 150 °C, reflecting the robustness of the cooperative interaction synthons applied in this case.

4.5
Electrostatic

The contribution of electrostatic forces in combination with coordination chemistry to the formulations of coordination polymers has been discussed above. Similarly, network assembly of suitably functionalized porphyrinoids by *anionic* (rather than cationic) bridges can be envisioned [138]. Supramolecular aggregates wherein electrostatic attractions play the major role have been also known for decades. They are readily available by reacting tetra-cationic porphyrins such as tetra-N-methylpyridiniumporphyrin, with tetra-anionic porphyrins such as the tetra-4-sulfonatophenylporphyrin. However, these derivatives tend to form complex amorphous solids or powders, rather than uniform crystalline materials.

5
Concluding Remarks

There is still much to be learned in the field of synthetic supramolecular chemistry – especially in terms of tertiary and quaternary structure – as the weak and oftentimes non-directional interactions between supermolecules that are responsible for the ordering in a crystal or aggregate structure are difficult to predict routinely. Much of the initial progress and insights in this arena have used porphyrinic systems. The suitability of the above supramolecular systems for commercial applications depends not only on the activity of the material, but also on the cost of producing it compared to other technologies, demand, etc. The notable progress in the synthetic chemistry of porphyrinoids [18, 74], including a "green" procedure for porphyrins that uses neither solvent nor man-made catalyst [139], as well as their robustness, assures of a bright future for porphyrinic materials.

Acknowledgements Hunter College infrastructure is supported by the NSF, NIH (including the RCMI program, 03037) and The City University of New York. I.G. and C.M.D. would like to acknowledge the Israel-US Binational Science Foundation (1999082) for support, C.M.D. and A.F. acknowledges support from The National Science Foundation (0135509, IGERT DGE-9972892), and I.S. acknowledges support from the National Institutes of Health (SCORE program).

References

1. Lehn J-M (1990) Angew Chem Int Ed Engl 29:1304
2. Lehn J-M (1994) Pure Appl Chem 66:1961
3. Lehn J-M (2002) Proc Natl Acad Sci, USA 99:4763
4. Stang PJ, Olenyuk B (1997) Acc Chem Res 30:502
5. Goldberg I (2002) Cryst Eng Comm 4:109
6. Collinson SR, Bruce DW (1999) Metallomesogens–supramolecular organization of metal complexes in fluid phases. In: Sauvage J-P (ed) Transition metals in supramolecular chemistry. Wiley, New York, p 285
7. Holliday BJ, Mirkin CA (2001) Angew Chem Int Ed 40:2022
8. Drain CM (2002) Proc Natl Acad Sci, USA 99:5178
9. Alivisatos AP, Barbara PF, Castleman AW, Chang J, Dixon DA, Klein ML, McLendon GL, Miller JS, Ratner MA, Rossky PJ, Stupp SI, Thompson ME (1998) Adv Mater 10:1297
10. Lemieux UR, Spohr U (1994) Adv Carbohydrate Chem Biochem 50:1
11. Clardy J (1999) Proc Natl Acad Sci USA 96:1826–1827
12. Pedersen CJ (1967) J Am Chem Soc 89:2495
13. Drain CM, Batteas JD, Flynn GW, Milic T, Chi N, Yablon DG, Sommers H (2002) Proc Natl Acad Sci, USA 99:6498
14. Milic TN, Chi N, Yablon DG, Flynn GW, Batteas JD, Drain CM (2002) Angew Chem Int Ed 41:2117
15. Patel SS, Picha KM (2000) Ann Rev Biochem 69:651
16. Gong X, Milic T, Xu C, Batteas JD, Drain CM (2002) J Am Chem Soc 124:14290
17. Mauzerall DC (1998) Clin Dermatol 16:195
18. Kadish KM, Smith KM, Guilard R (2000) The porphyrin handbook. Academic Press, New York
19. Drain CM, Corden BB (1987) J Chem Educ 64:441
20. Drain CM, Sable D, Corden BB (1988) Inorg Chem 27:2396
21. Drain CM, Gong X, Ruta V, Soll CE, Chicoineau PF (1999) J Comb Chem 1:286
22. Pasetto P, Chen X, Drain CM, Franck RW (2001) Chem Commun 81–82
23. Retsek JL, Drain CM, Kirmaier C, Nurco DJ, Medforth CJ, Smith KM, Sazanovich IV, Chirvony VS, Fajer J, Holten D (2003) J Am Chem Soc 125:9787
24. Drain CM, Gentemann S, Roberts JA, Nelson NY, Medforth CJ, Jia S, Simpson MC, Smith KM, Fajer J, Shelnutt JA, Holten D (1998) J Am Chem Soc 120:3781
25. Drain CM, Kirmaier C, Medforth CJ, Nurco DJ, Smith KM, Holten D (1996) J Phys Chem 100:11984
26. Norwich JJ (1989) Byzantium: the early centuries. Knopf, New York
27. Fischer H, Stern A (1940) Die chimie des pyrroles. Akademische Verlagsges, Leipzig
28. Fischer H, Orth H (1937) Die chimie des pyrroles. Akademische Verlagsges, Leipzig
29. Ungashe SB, Groves JT (1994) Adv Inorg Biochem 9:317
30. Merlau ML, Mejia MDP, Nguyen ST, Hupp JT (2001) Angew Chem Int Ed 40:4239
31. Drain CM, Hupp JT, Suslick KS, Wasielewski MR, Chen X (2002) J Porph Phthal 6:241
32. Wienke J, Goosens A, Schaafsma TJ (1999) J Phys Chem B 103:2702
33. Cherian S, Wamser CC (2000) J Phys Chem B 104:3624
34. Drain CM, Mauzerall DC (1992) Biophys J 63:1544
35. Drain CM, Mauzerall DC (1992) Biophys J 63:1556
36. Drain CM, Mauzerall DC (1990) Bioelectrochem Bioenerg 24:263
37. Drain CM, Christensen B, Mauzerall DC (1989) Proc Natl Acad Sci, USA 86:6959
38. Hong FT (1989) J Mol Elect 5:163
39. Wagner RW, Lindsey JS, Seth J, Palaniappan V, Bocian DF (1996) J Am Chem Soc 118:3996

40. Burrell AK, Wasielewski MR (2000) J Porph Phthal 4:401
41. Aratani N, Osuka A (2001) Macromolecules 22:725
42. Burrell AK, Officer DL, Plieger PG, Reid DCW (2001) Chem Rev 101:2751
43. Kumble R, Palese S, Lin VS-Y, Therien MJ, Hochstrasser RM (1998) J Am Chem Soc 120:11489
44. Osuka A, Tanabe N, Zhang R, Maruyama K (1993) Chem Lett 9:1505
45. Smith KM, Jaquinod L, Vicente MGH (1999) Chem Commun 1771
46. Holten D, Bocian DF, Lindsey JS (2002) Acc Chem Res 35:57
47. Benites MR, Johnson TE, Weghorn S, Yu L, Rao PD, Diers JR, Yang SI, Kirmaier C, Bocian DF, Holten D, Lindsey JS (2002) J Mater Chem 12:65–80
48. Roth KM, Lindsey JS, Bocian DF, Kuhr WRG (2002) Langmuir 18:4030
49. Schweikart K-H, Malinovskii VL, Diers JR, Yasseri AA, Bocian DF, Kuhr WG, Lindsey JS (2002) J Mater Chem 12:808–828
50. Tour JM (2000) Acc Chem Res 33:791
51. Lent CS (2000) Science 288:1597
52. Ellenbogen JC, Love JC (2000) Proc IEEE 88:386
53. Lee TH (2002) Sci Am 286:52
54. Livingston JN, Bliss WG (1999) IEEE Trans Magn 35:2298
55. Lieber CM (2001) Sci Am 285:58
56. Belanger S, Hupp JT (1999) Angew Chem Int Ed 38:2222
57. Zhang J, Williams ME, Keefe MH, Morris GA, Nguyen ST, Hupp JT (2002) Electrochem Solid-State Lett 5:E25
58. Diskin-Posner Y, Goldberg I (1999) Chem Commun 1961
59. Diskin-Posner Y, Goldberg I (2001) New J Chem 25:899
60. Ogawa K, Zhang T, Yoshihara K, Kobuke Y (2002) J Am Chem Soc 124:22
61. Screen TEO, Thorne JRG, Denning RG, Bucknall DG, Anderson HL (2002) J Am Chem Soc 124:9712
62. Epstein AJ (2000) MRS Bull 33
63. Prins L, Reinhoudt D, Timmerman P (2001) Angew Chem Int Ed 40:2382
64. Kobko N, Paraskevas L, Rio Ed, Dannenberg JJ (2001) J Am Chem Soc 123:4348
65. Steiner T (2002) Angew Chem Int Ed 41:48
66. Archer EA, Gong H, Krische MJ (2001) Tetrahedron 57:1139
67. Krauss N, Schubert W-D, Klukas O, Fromme P, Witt HT, Saenger W (1996) Nat Struct Biol 3:965
68. McDermott G, Prince SM, Freer AA, Hawthornthwaite-Lawless AM, Papiz MZ, Cogdell RJ, Isaacs NW (1995) Nature 374:517
69. Drain CM, Fischer R, Nolen E, Lehn J-M (1993) Chem Commun 243
70. Drain CM, Russel KC, Lehn J-M (1996) Chem Commun 337
71. Wasielewski MR (1993) Modeling primary electron transfer events in photosynthesis using supramolecularstructures. In: Deisenhofer J, Norris J (eds) The photosynthetic reaction center, vol. II. Academic, New York, p 465
72. Balaban TS, Leitich J, Holzwarth AR, Schaffner K (2000) J Phys Chem B 104:1362
73. Drain CM, Shi X, Milic T, Nifiatis F (2001) Chem Commun 287 (Amendment published 17th July 2001)
74. Shi X, Barkigia KM, Fajer J, Drain CM (2001) J Org Chem 66:6513
75. Wilkinson G (1987) Comprehensive coordination chemistry: the synthesis, reactions, properties, and applications of coordination compounds. Pergamon, New York
76. Fujita M (1998) Chem Soc Rev 27:417
77. Drain CM, Lehn J-M (1994) Chem Commun 2313
78. Stang PJ (1998) Chem Eur J 4:19
79. Kumar RK, Balasubramanian S, Goldberg I (1998) Chem Commun 1435

80. Drain CM, Nifiatis F, Vasenko A, Batteas JD (1998) Angew Chem Int Ed 37:2344
81. Micali N, Mallamace F, Romeo A, Purrello R, Scolaro LM (2000) J Phys Chem B 104:5897
82. Micali N, Romeo A, Lauceri R, Purrello R, Mallamace F, Scolaro LM (2000) J Phys Chem 104:9416
83. Sagawa T, Fukugawa S, Yamada T, Ihara H (2002) Langmuir 18:7223
84. Xu W, Guo H, Akins DL (2001) J Phys Chem B 105:1543
85. Hwang KC, Mauzerall D (1993) Nature 361:138
86. Sun K, Mauzerall DC (1996) Biophys J 71:295
87. Sun K, Mauzerall DC (1996) Proc Natl Acad Sci, USA 93:10758
88. Sun K, Mauzerall DC (1996) Biophys J 71:309
89. Akins DL, Zhu H-R, Guo C (1996) J Phys Chem 100:5420
90. Byrn MP, Curtis CJ, Hsiou Y, Khan SI, Sawin PA, Terzis A, Strouse CE (1996) Porphyrin-based lattice clathrates. In: MacNicol DD, Toda F, Bishop R (eds) Comprehensive supramolecular chemistry, vol. 6. Elsevier, New York, p 715
91. Byrn MP, Curtis CJ, Khan SI, Sawin PA, Tsurumi R, Strouse CE (1990) J Am Chem Soc 112:1865
92. Byrn MP, Curtis CJ, Goldberg I, Hsiou Y, Khan SI, Sawin PA, Tendick SK, Strouse CE (1991) J Am Chem Soc 113:6549
93. Byrn MP, Curtis CJ, Hsiou Y, Khan SI, Sawin PA, Tendick SK, Terzis A, Strouse CE (1993) J Am Chem Soc 115:9480
94. Kumar RK, Balasubramanian S, Goldberg I (1998) Inorg Chem 37:541
95. Hunter CA, Sanders JKM (1990) J Am Chem Soc 112:5525
96. Hunter CA, Meah MN, Sanders JKM (1990) J Am Chem Soc 112:5773
97. Etter MC (1991) J Phys Chem 95:4601
98. Goldberg I (2000) Chem Eur J 6:3863
99. Brunet P, Simard M, Wuest JD (1999) J Am Chem Soc 119:2737
100. Zaworotko MJ (1997) Nature 386:220
101. Choi HJ, Lee TS, Suh MP (1999) Angew Chem Int Ed 35:1405
102. Diskin-Posner Y, Patra GK, Goldberg I (2002) Chem Commun 1420
103. Diskin-Posner Y, Patra GK, Goldberg I (2002) Cryst Eng Comm 4:296
104. Goldberg I, Krupitsky H, Stein Z, Hsiou Y, Strouse CE (1995) Supramol Chem 4:203
105. Kobayashi K, Koyanagi M, Endo K, Masuda H, Aoyama Y (1998) Chem Eur J 4:417
106. Bhyrappa P, Wilson SR, Suslick KS (1997) J Am Chem Soc 119:8492
107. Diskin-Posner Y, Patra GK, Goldberg I (2001) Eur J Inorg Chem 10:2515
108. Diskin-Posner Y, Kumar RK, Goldberg I (1999) New J Chem 23:885
109. Dastidar P, Stein Z, Goldberg I, Strouse CE (1996) Supramol Chem 7:257
110. Schareina T, Kempe R (2000) Z Anorg Allg Chem 626:1279
111. Dahal S, Goldberg I (2000) J Phys Org Chem 13:382
112. Fleischer EB, Shachter AM (1991) Inorg Chem 30:3763
113. Shachter AM, Fleischer EB, Haltiwanger RC (1988) Chem Commun 960
114. Fleischer EB (1962) Inorg Chem 1:493
115. Diskin-Posner Y, Dahal S, Goldberg I (2000) Chem Commun 585
116. Gunter MJ, McLaughlin GM, Berry KJ, Murray KS, Irving M, Declercq PE (1984) Inorg Chem 23:283
117. Krupitsky H, Stein Z, Goldberg I, Strouse CE (1994) J Incl Phenom Macrocyclic Chem 18:177
118. Lin K-J (1999) Angew Chem Int Ed 38:2730
119. Diskin-Posner Y, Patra GK, Goldberg I (2001) Dalton Trans 19:2775
120. Pan L, Kelly S, Huang X, Li J (2002) Chem Commun 2334
121. Barkigia KM, Battioni P, Riou V, Mansuy D, Fajer J (2002) Chem Commun 956
122. Kumar RK, Goldberg I (1998) Angew Chem Int Ed 37:3027

123. Abrahams BF, Hoskins BF, Robson R (1991) J Am Chem Soc 113:3606
124. Abrahams BF, Hoskins BF, Michail DM, Robson R (1994) Nature (London) 369:727
125. Hagrman D, Hagrman PJ, Zubieta J (1999) Angew Chem Int Ed 38:3165
126. Sharma CVK, Broker GA, Huddleston JG, Baldwin JW, Metzger RM, Rogers RD (1999) J Am Chem Soc 121:1137
127. Pan L, Noll BC, Wang X (1999) Chem Commun 157
128. Carlucci L, Ciani G, Proserpio DM, Porta F (2003) Angew Chem Int Ed 42:317
129. Kondo M, Kimura Y, Wada K, Mizutani T, Ito Y, Kitagawa S (2000) Chem Lett: 818
130. Diskin-Posner Y, Dahal S, Goldberg I (2000) Angew Chem Int Ed 39:1288
131. Kosal ME, Chou J-H, Wilson SR, Suslick KS (2002) Nat Mater 1:118
132. Kosal ME, Chou J-H, Suslick KS (2002) J Porph Phthal 6:377
133. Shmilovits M, Vinodu M, Goldberg I (2003) Cryst Growth Des 3:855
134. Brandon EJ, Rogers RD, Burkhart BM, Miller JS (1998) Chem Eur J 4:1938
135. Brandon EJ, Rittenberg DK, Arif AM, Miller JS (1998) Inorg Chem 37:3376
136. Miller JS, Calabrese JC, McLean RS, Epstein AJ (1992) Adv Mater 4:498
137. Miller JS (2001) Polyhedron 20:1723
138. Vinodu M, Goldberg I (2003) Cryst Eng Comm 5:204
139. Drain CM, Gong X (1997) Chem Commun 2117

Discrete Organic Nanotubes Based on a Combination of Covalent and Non-Covalent Approaches

Marco A. Balbo Block[1] · Christian Kaiser[1] · Anzar Khan[1] · Stefan Hecht[1,2] (✉)

[1] Institut für Chemie/Organische Chemie, Freie Universität Berlin, Takustr. 3, 14195 Berlin, Germany
shecht@chemie.fu-berlin.de
[2] Max-Planck-Institut für Kohlenforschung, Kaiser-Wilhelm-Platz 1, 45470 Mülheim an der Ruhr, Germany
hecht@mpi-muelheim.mpg.de

1	Introduction	90
2	Tubular Design Principles	92
3	Tubes from Hollow Helical Backbones	94
3.1	Inherently Stiff [n]Phenylene Helices	95
3.2	Helical Backbones Based on Intramolecular H-Bonding	96
3.2.1	Alternating D,L-α-Peptides	96
3.2.2	Aromatic and Heteroaromatic Amides	97
3.3	Helical Backbones Based on the *trans*-2,2'-Bipyridine Helicity Codon	101
3.3.1	Alternating Pyridine–Pyrimidine Oligomers	102
3.3.2	Alternating Pyridine-Pyridazine Oligomers	104
3.3.3	Alternating Naphthyridine–Pyrimidine Oligomers	105
3.4	Helical Backbones Based on Solvophobic Forces	105
3.5	Covalent Stabilization by Intramolecular Helical Crosslinking	110
4	Tubes from Self-Assembled Helices and Sheets	113
4.1	Multi-stranded Helices	113
4.2	Stacking Helices	114
4.3	Bundles	114
4.4	Barrels	115
5	Tubes from Stacked Macrocycles	117
5.1	Tubular Structures in the Solid State	117
5.1.1	Cyclic Peptides	117
5.1.2	Cyclic Oligosaccharides	120
5.1.3	Phenylene Macrocycles	121
5.1.4	Phenylene Ethynylene Macrocycles	121
5.2	Tubular Structures at Interfaces	125
5.3	Tubular Structures in Mesophases	125
5.4	Tubular Structures in Solution	126
5.4.1	Phenylene Ethynylene Macrocycles	126
5.4.2	Phenylene Diethynylene Macrocycles	126
5.4.3	Coil-Ring-Coil Block Copolymers	128
5.4.4	Covalent Stabilization by Crosslinking	129
5.4.5	Cyclodextrins	129
5.4.6	Cored Dendronized Polymers	130
5.4.7	Cyclic Alternating D,L-α-Peptides	130

6	Tubes from Stacked Rosettes	132
6.1	G-Quartets and Related Motifs	133
6.2	Melamine–Cyanuric Acid Motifs	135
7	Tubes from Self-Assembled Amphiphiles	138
7.1	Wedge-Shaped Amphiphiles	138
7.2	Linear Amphiphiles and Block Copolymers	140
8	Conclusions and Outlook	143
	References	145

Abstract Tubular organic nanostructures offering precise control over inner and outer surface functionality represent attractive building blocks for the bottom-up approach in nanotechnology, as well as for the development of future materials and biological applications. In this review, the current state of the art in creating organic nanotubes is summarized, focusing on the chemistry behind tubular structure formation. The discussion of general design principles, which are altogether inspired by Nature and combine covalent and non-covalent synthesis, is followed by detailed treatments of the individual approaches including helices and their assemblies, stacked rings and rosettes, as well as cylindrical micelles and rolled sheets. The authors provide a critical evaluation of past accomplishments, promising current developments, and future challenges in this fascinating and highly interdisciplinary research topic.

Keywords Materials science · Nanotubes · Self-assembly · Helices · Macrocycles

Abbreviations
Aib α-Aminoisobutyric acid
Cys(Acm) Acetamidomethyl cysteine
Hag Homoallylglycin
Htrp Homotryptophane
N-MeAla N-methyl alanine
RCM Ring closing (olefin) metathesis
Triglyme Tri(ethyleneglycol)
XR X-ray specular reflectivity

1
Introduction

Continuing miniaturization of electronic devices [1] represents the main driving force at the dawn of the nanotechnological era [2]. As established top-down methods, most notably photolithography, approach sub-100 nm resolution, emerging alternative bottom-up approaches based on the assembly of molecular building blocks promise to revolutionize nanofabrication [3]. In order to realize this ambitious goal, two major scientific challenges have to be mastered: (1) a molecular construction kit involving a diverse set of functional units for generating, transporting, and converting signals is needed, in which the individual modules possess a certain shape and can be organized in a defined arrangement, preferably by self-assembly; and (2) the generated functional nanostructures have to be integrated into actual electronic devices, and therefore the connection to the macroscopic outside world is critical. In addition, important aspects concern the physical and economical feasibility of the proposed scheme involving questions related to, for instance, effective local cooling and cost-efficient parallel fabrication of such devices. From a chemist's perspective it is gratifying to note that the first main aspect truly belongs to the realm of synthesis. The unprecedented control over atomic placement controlling molecular shape and properties that can be achieved by the mature art of organic synthesis, in combination with the ever increasing understanding of supramolecular design principles [4], enables the synthetic chemist to play a – perhaps the – key role in the emerging bottom-up approach [5].

Tubular structures (Fig. 1) are among the most versatile functional modules in such nanoconstruction kits [6]. Their inherent geometric features, notably defined inner and outer surfaces as well as termini, give rise to a controlled spatial segregation. This property is associated with a cylindrical dimensionality, which encodes both the directionality necessary for transport processes and the addressability needed for generation of more complex architectures by self-assembly [7] or rational manipulation [8]. Nanotubes can potentially perform a variety of tasks associated with transport of charges and neutral/ionic/chiral species, as well as site-isolation of conducting and/or emitting cores, while at the same time providing the mechanical strength needed for efficient scaffolding. With regard to nanoelectronics, one of the obvious goals is the generation of addressable, insulated nanowires. It should be noted that transport processes could also be coupled to chemical transformations, i.e., catalysis, affording a flow-through nanoreactor for high-end analytic applications.

The synthesis of such functional tubular nanoobjects remains challenging and can be divided into two types of approaches. While the first class is based on the utilization of artificial growth, template, or shape transformation

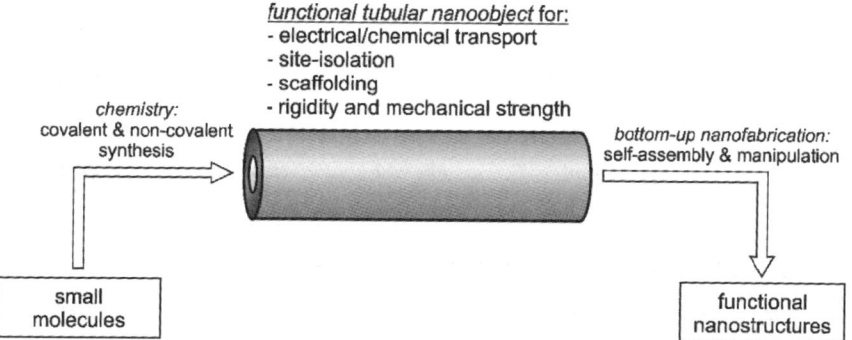

Fig. 1 Versatility of functional tubular nanoobjects

processes giving rise to the efficient production of carbon and inorganic nanotubes, the second class is founded on a bio-inspired, purely molecular basis. After a brief discussion of general tubular design principles, this review focuses on the latter class of molecule-based synthetic approaches to organic nanotubes [9]. This article primarily covers aspects of covalent and non-covalent synthesis as well as the resulting structural characteristics. For the most part, methods for generating *individual* tubes are presented here, since the nanoobjects have to be addressed separately in order to construct artificial nanostructures. It should be emphasized that due to the fundamental character of this type of research, a detailed discussion of (potential) applications cannot be given. Furthermore, this treatment of the topic can by no means be comprehensive; instead, informed by their personal experiences, the authors have tried to emphasize concepts and present original and significant contributions to this exciting field of research.

2
Tubular Design Principles

The information stored in molecular building blocks, such as shape and linkage geometry as well as local solvophilicity/solvophobicity and recognition sites, can be used to generate open-ended hollow tubular objects via a variety of conceptually different approaches (Fig. 2). All approaches use a combination of both covalent and non-covalent syntheses. While defined chemical bonds

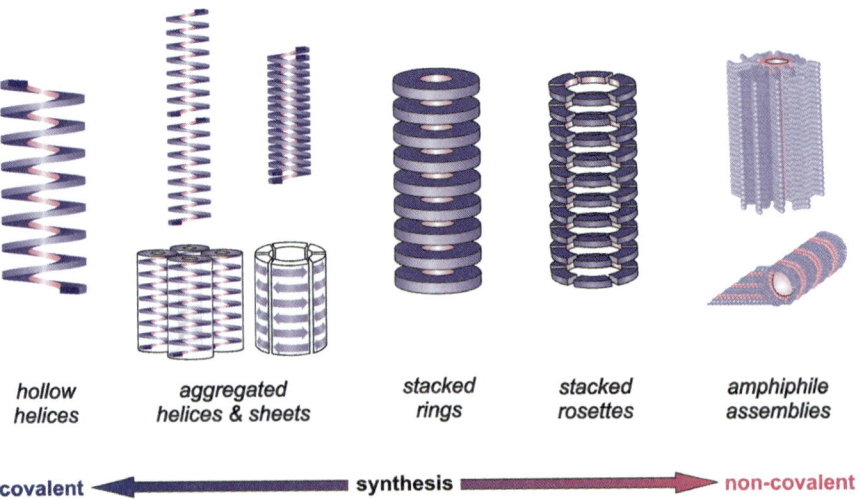

Fig. 2 General strategies for the design of organic nanotubes based on hollow helical scaffolds and their self-assembly, stacks of macrocycles and rosettes, as well as cylindrical micelles and rolled bilayer sheets

provide both increased stability and control over molecular conformation, supramolecular interactions allow for ease of synthesis and repair of defects. These advantages are accompanied by significant challenges concerning, for instance, efficient and large-scale covalent synthesis on the one hand and predictability of structure formation in self-organizing systems on the other.

The approach with the uppermost covalent character relies on generating helices containing an internal void channel. Hereby, non-covalent interactions are only used to control secondary helical structure formation in the case of folding backbones. Multiple helical entities can be organized in three different modes, namely, they can intertwine to form multi-stranded helices, stack along their axis in continuous fashion to afford tubes, or align parallel to each other to yield bundles. A related motif arises from the aggregation of β-sheets to form barrels. Reduction of the helical to a cyclic motif and introduction of face-to-face non-covalent interactions leads to tubes based on stacked rings. By establishing additional orthogonal non-covalent interactions, the cyclic moiety itself can be formed by self-assembly from sector-shaped units, and the resulting rosettes arrange in a stacked fashion. Finally, the ultimate self-assembly approach is based on either cylindrical micelles or rolled sheets arising from certain amphiphiles and curved bilayers, respectively. Regioselective covalent crosslinking [10] of the tubular architectures proves to be a valuable means to permanently stabilize the frequently rather fragile objects. In principle, the approaches outlined above allow for precise control over dimensions and local chemical functionality of the resulting tubular structures. This is of the utmost importance for incorporating such nanoobjects into more complex nanostructures.

Before discussing these molecular approaches to obtain organic nanotubes in this order, alternative processes for generating carbon [11, 12] and inorganic [13] nanotubes should briefly be mentioned. The many possible preparation strategies are mainly based on one-dimensional growth processes, template-directed syntheses, and shape transformations [13]. Polymer nanotubes have also been prepared using various template-based approaches [14, 15]. In an analogy to the creation of chiral tubes from folded bilayer sheets (*vide infra*), it is interesting to note that the different electronic properties of single-walled carbon nanotubes can be related to the directions in which a graphene sheet is folded. It should be emphasized that currently, carbon (and inorganic) nanotubes represent perhaps the most promising building blocks for the future bottom-up approach [16] because of the tubes' exceptional properties as well as their commercial availability. Current drawbacks consist mainly of separation/purification issues and the solely statistical, i.e., not regioselective, functionalization of the tubes' peripheries [17, 18].

3
Tubes from Hollow Helical Backbones

The helix is among the most abundant structural motifs in Nature due to its inherent chirality and advantageous geometric features (*vide supra*). Oligonucleotide helices in the form of double-stranded DNA are responsible for storing and copying the genetic information, and therefore carry out one of the most essential tasks in the organization of life. Peptide helices perform diverse biological functions including: (1) molecular recognition at the helix surface owing to the well-defined arrangement of amino acid side chains, for instance mediating the Bak peptide – Bcl-x_L-protein interaction that regulates cell apoptosis related to the development of cancer [19]; (2) selective chemical transport through the helical interior due to transmembrane-channel-forming properties, illustrated by the gramicidin A channel [20]; (3) mechanical strength due to formation of supramolecular superhelical fibers, for instance in the triple helical, bundled collagen cable [21]; and (4) scaffolding due to its inherent rigidity and well-defined binding sites, for instance used for the arrangement of the different bacteriochlorophyll pigments in the light-harvesting complex II in bacteria [22]. The astonishing variety of naturally occurring helical motifs serves as inspiration for the design of artificial helices.

Helical backbones are either inherently stiff due to molecular arrangements such as linkage geometry and steric requirements, or they represent thermodynamically favored helical conformations (foldamers). The first class mostly consists of helical rods having no internal void, reflected in well-known helical polymers [23] such as poly(isocyanide)s or poly(phenylacetylene)s. In the case of foldamers [24], the conformational preference is dictated by non-covalent interactions between neighboring and/or non-neighboring repeat units as well as entropic factors. The reversible helix-coil transition allows population of the global minimum structure and therefore provides a means to repair local defects. The existing fine balance between secondary structural elements is one of the key requirements for the hierarchical structure evolution in Nature. However, from a materials standpoint, this reversibility might also constitute a disadvantage with regard to structural integrity and hence device stability/lifetime under a variety of environmental conditions.

Synthesis of such helical scaffolds is accompanied by their critical structural characterization. X-ray crystallography has proven to be among the most useful tools for elucidating helical structures in the solid state, while in solution a variety of techniques have been employed. In addition to various NMR spectroscopic methods that provide important information on neighboring interactions, CD spectroscopy has been instrumental in monitoring (excess) helicity. In certain cases, UV/vis absorption as well as fluorescence techniques have been used as complementary analytical tools.

In this section, *hollow* helical backbones are reviewed according to their design features responsible for helix formation. For comprehensive treatments of helical polymers and foldamers, respectively, the reader is referred to the excellent reviews by Nakano and Okamoto [23] as well as Moore and coworkers [24].

3.1
Inherently Stiff [n]Phenylene Helices

The classical helicene structure [25, 26] containing no inner cavity can be extended by introduction of cyclobutadiene fragments into the *meta*-annelated benzene ring system to afford angular, and finally helical, [n]phenylenes ($n \geq 6$). The internal diameter of the helix measures 4.0 Å [27]. It should be noted that this backbone is inherently stiff and helical owing to strong covalent interactions introduced by the mode of ring annelation. Vollhardt and coworkers recently realized the synthesis of several helical [n]phenylenes derivatives with $n=7-9$ [28, 29]. The elegant syntheses are based on either twofold or threefold intramolecular cobalt-catalyzed [2+2+2] cyclotrimerization reactions of various phenylacetylene derivatives (Fig. 3).

These moderately air-stable, orange-colored [n]phenylenes can be crystallized, for instance [8]phenylene adopts right- and left-handed helical conformations in the solid (Fig. 4). For [7]phenylene 1, variable temperature NMR experiments indicate a surprisingly low activation energy of 12.6 kcal/mol for helix reversal in solution rendering an enantiomeric separation infeasible.

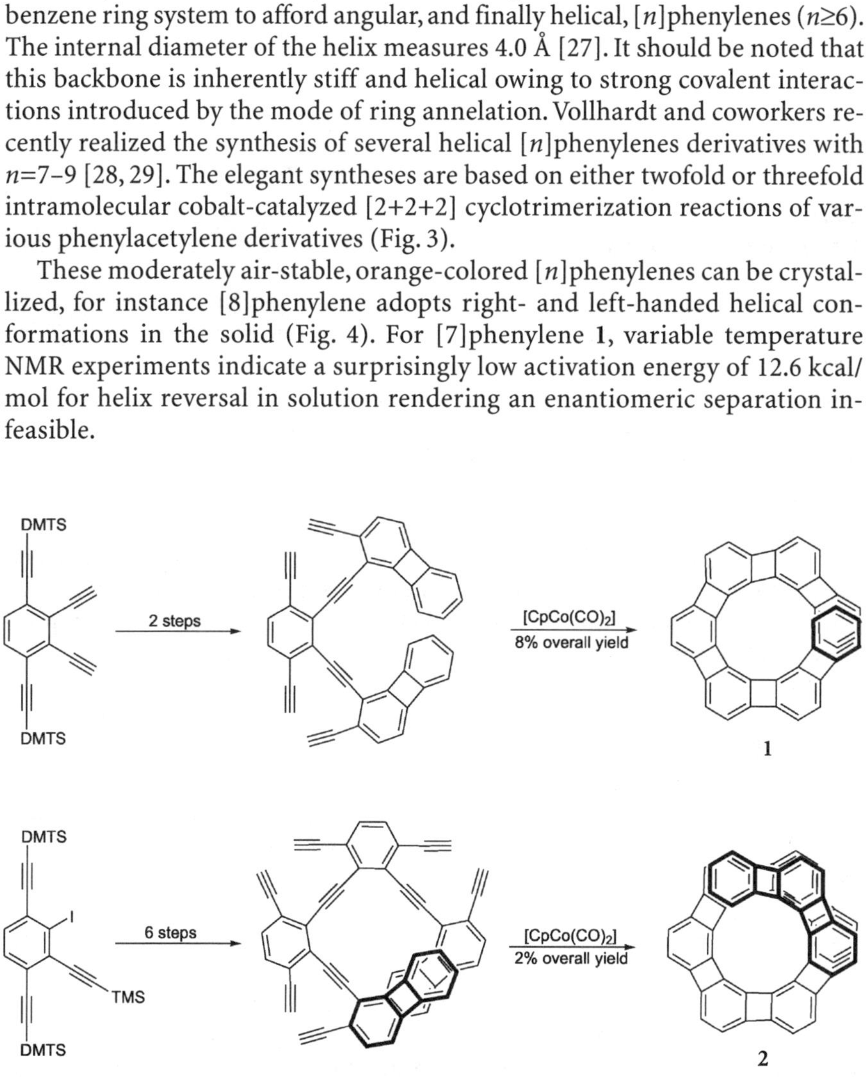

Fig. 3 Vollhardt's synthesis of [7]phenylene 1 and [9]phenylene 2 by multiple intramolecular [2+2+2]cyclotrimerization reactions. *DMTS* Dimethylthexylsilyl, *TMS* trimethylsilyl

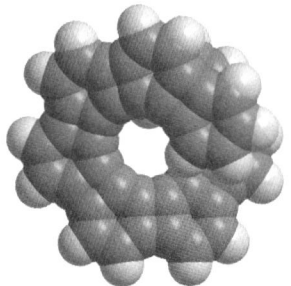

Fig. 4 Crystal structure of [8]phenylene (CCDC-183203)

3.2
Helical Backbones Based on Intramolecular H-Bonding

3.2.1
Alternating D,L-α-Peptides

Secondary structural preferences in peptides are governed by the inherent rigidity of the amide bonds, which significantly reduces the number of accessible conformations, as well as directing and stabilizing intramolecular H-bonding interactions. The antiparallel amide segments are connected by the asymmetric carbon atom carrying the amino acid side chain, and therefore two important dihedral angles Φ and Ψ are defined (Fig. 5). The most abundant secondary structural motifs found in peptide sequences are the β-sheet and the α-helix.

In an idealized α-helical conformation ($\Phi=-57°$, $\Psi=-47°$), the side chain residues radiate outward and 3.6 repeat units constitute one turn. This is owing to an intramolecular H-bond between the carbonyl oxygen atom in position i and the amide proton in position $i+4$ giving rise to a 13-membered ring, and the α-helix is hence referred to as a 3.6_{13}-helix. The inner walls of an α-helix are in van der Waals contact so that the structure contains no inner void. Slightly different dihedral angles ($\Phi=-60°$, $\Psi=-60°$) cause the formation of an alternative 3_{10}-helix having a smaller outer diameter.

In an idealized β-strand composed of all L-amino acids ($\Phi_L=-120°$, $\Psi_L=+120°$), the arrangement of adjacent residues causes the direction of the peptide backbone to alternate in a sinusoidal fashion. Therefore, a peptide composed of alternating D- and L-α-amino acids can adopt a stable β-helical conformation stabilized by β-sheet-type H-bonding [30]. In an idealized β-helix ($\Phi_L=-133°$, $\Psi_L=+117°$; $\Phi_D=+118°$, $\Psi_D=-130°$) the amino acid side chains are pointing outward, leaving a central pore running along the helix axis. The diameter of the inner channel is strongly influenced by the helix periodicity and the number of residues per turn. Computational studies show that single stranded β-helices with 4.8, 6.2, and 8.2 residues per turn should display pore diameters of 2.3, 3.3, and 4.7 Å.

Discrete Organic Nanotubes

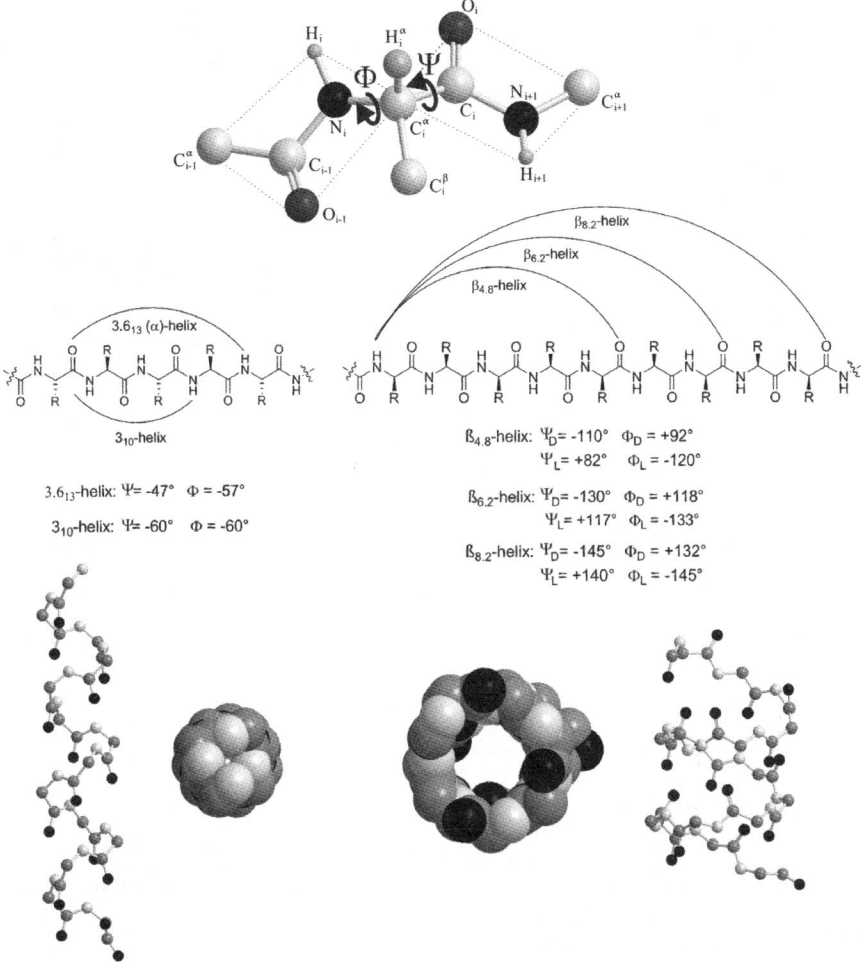

Fig. 5 Conformational preferences of an α-peptide backbone giving rise to either an α- or β-helix depending on the chirality of the amino acid building blocks. Models show top and side views of an idealized α-helix (*left*) and the β-helix of gramicidin A (*right*) (pdb code: 1MAG; O-atoms are *black*, C-atoms are *gray*, N-atoms are *light*; H-atoms are omitted for clarity)

In addition, a variety of alternative helical conformations exist in these and other peptides composed of non-natural amino acids.

3.2.2
Aromatic and Heteroaromatic Amides

In contrast to peptide helices, H-bonding interactions between adjacent repeat units can be used to organize the backbone into a (continuously) kinked con-

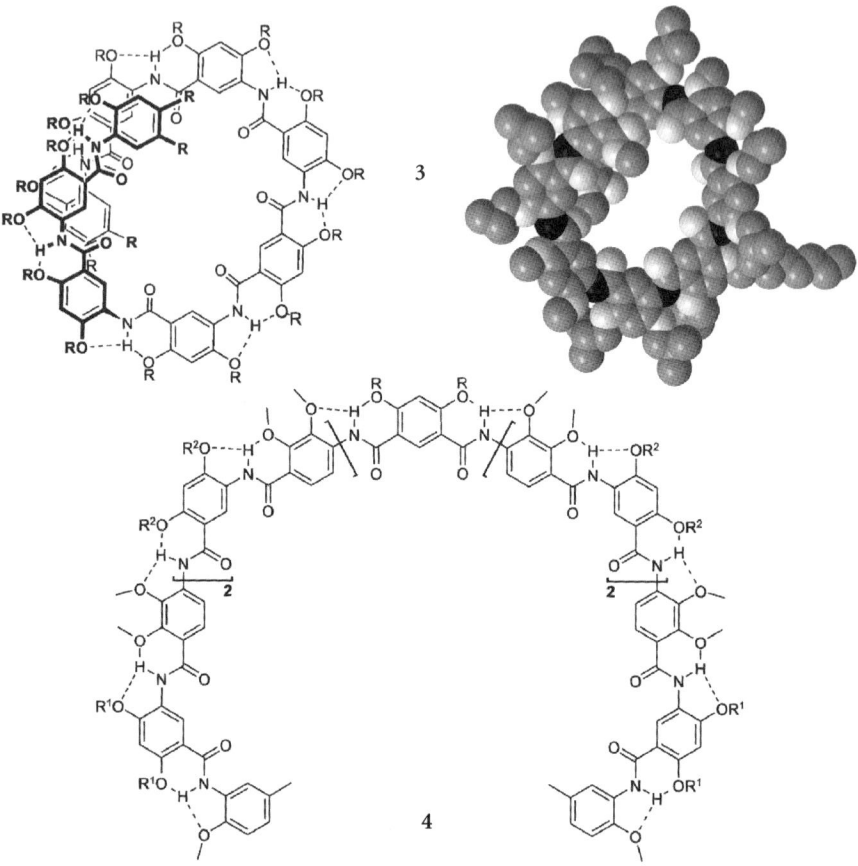

Fig. 6 Gong's helical diarylamides of varying diameter depending on their mode of connectivity. Crystal structure of *meta*-linked diarylamide **3** (CCDC-188141; O-atoms are *light*, C-atoms are *gray*, N-atoms are *black*; H-atoms are omitted for clarity)

formation that at a certain backbone length affords a helical structure. In the case of aromatic amides [31], further stabilization arises from π,π-stacking interactions between overlapping (non-neighboring) repeat units of adjacent turns.

The first reported systems by Hamilton and coworkers are based on a sequence of alternating anthranilamides and 2,6-pyridinedicarboxamides [32–34]. Several oligomers were shown to adopt stable helical conformations in the crystal and in solution. However, these foldamers do not contain a significant internal cavity due to their internal H-bonding motif.

In order to obtain hollow crescent and helical structures (Fig. 6), Gong and coworkers investigated *meta*-linked diarylamides **3** sharing three-center H-bonds at the external rim [35]. One turn consists of six repeat units and the inner hydrophilic cavity has a diameter of 8.2 Å [27], as confirmed by X-ray crystallographic analysis (Fig. 6) [36]. In solution, 2D NMR NOESY experiments

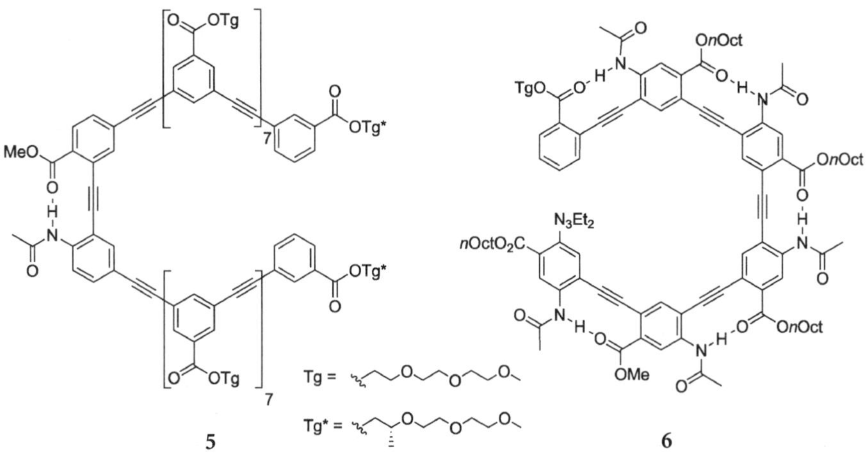

Fig. 7 Stabilization of helical oligo(*meta*-phenyleneethynylene)s by either single or multiple peripheral H-bonding interactions

show strong interactions between the terminal methyl groups and both aromatic and amide protons of the overlapping turn in the case of the nonamer. The internal diameter can be modulated by insertion of *para*-connected repeat units [36]. The resulting alternating *meta-/para*-diarylamide oligomers **4** possess an increased pore diameter of ~25–30 Å. Inserting additional *para*-connectors further increases the inner cavity yielding an astonishing inner diameter of ~50 Å in the case of an alternating *meta-/para-/para*-diarylamide backbone. This strategy to tune the helix diameter by linkage geometry [37] resembles Lehn's approach to various *trans*-2,2′-bipyridine-based helical backbones (*vide infra*).

The helical conformation of *meta*-phenylene ethynylene oligomers can be stabilized using an external H-bonding motif between adjacent repeat units. This has been accomplished by incorporating either single [38] or multiple [39] H-bonding interactions (Fig. 7). When one H-bonding unit was present, the helix stability of backbone **5** carrying polar triglyme side chains was increased by ~1 kcal/mol in acetonitrile, as indicated by solvent denaturation experiments. Introduction of these stabilizing interactions between every repeat unit of the same backbone having non-polar side chains (**6**) leads to helical structure formation in chloroform, as indicated by the observed NOE signals of end-to-end contacts. The cavity of the folded structures in both cases resembles the one of the amphiphilic *meta*-phenylene ethynylene foldamer family (*vide infra*) and amounts to ~7 Å [27].

Huc, Lehn, and coworkers have carried out detailed studies on alternating 2′,6′-diaminopyridine 2,6-pyridinedicarboxamide oligomers **7** (Fig. 8) [40–42]. In this case, the helical structure has an inner pore of 3.0 Å [27] and is caused by internal H-bonding interactions between the pyridine moieties and the amide groups connecting them. The heptamer can dimerize to form double-

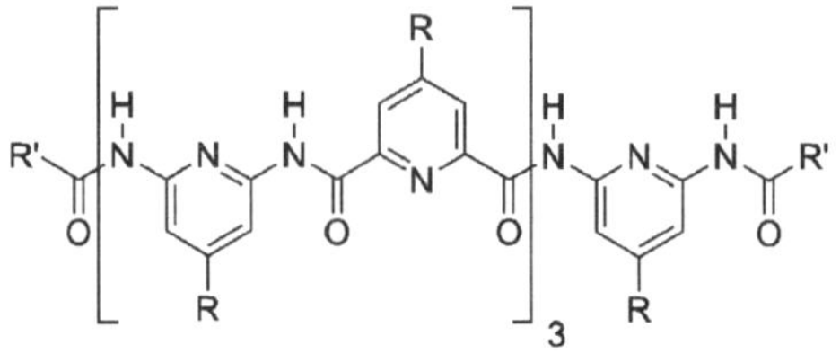

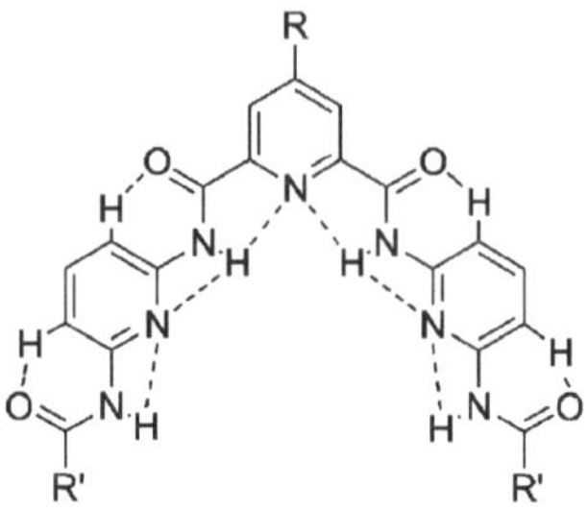

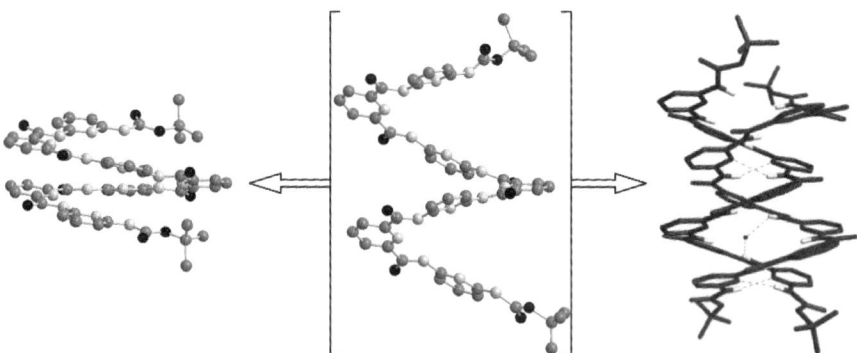

Fig. 8 Huc's and Lehn's alternating 2′,6′-diaminopyridine 2,6-pyridinedicarboxamide oligomers 7 and favored kinked conformation due to H-bonding

stranded helices, which are primarily stabilized by additional intermolecular aromatic stacking interactions, shown by the up-field shift of the aromatic protons in NMR. While a single helix has been crystallized from a rather polar solvent mixture, the corresponding double helix has been obtained by crystallization from a less polar solvent mixture and shows both aromatic stacking as well as H-bonding interactions between the strands (Fig. 9). The system has also been investigated by stochastic dynamic simulations that suggest an enhanced duplex stability with increasing oligomer length. The conformational preference of the backbone is dependent on the protonation of the pyridine units and can therefore be controlled by varying the pH [43, 44]. Recently, an extended quinoline-derived foldamer backbone has been reported by Huc et al. [45]; however owing to the internal H-bonding motif, only a small inner diameter was obtained.

In the case of alternating 2′,6′-diaminopyridine isophthalamides, helical structure formation can also be templated by a suitable guest such as cyanuric acid [46]. Molecular recognition is based on triple H-bonding between the

Fig. 9 Crystal structures of 7 from acetonitrile/DMSO showing a single helix (CCDC-142810) and from heptane/nitrobenzene showing the helix stretching to the double-stranded form and the actual duplex (CCDC-142811; O-atoms are *black*, C-atoms are *gray*, N-atoms are *light*; H-atoms are omitted for clarity). Reproduced in part from [40]

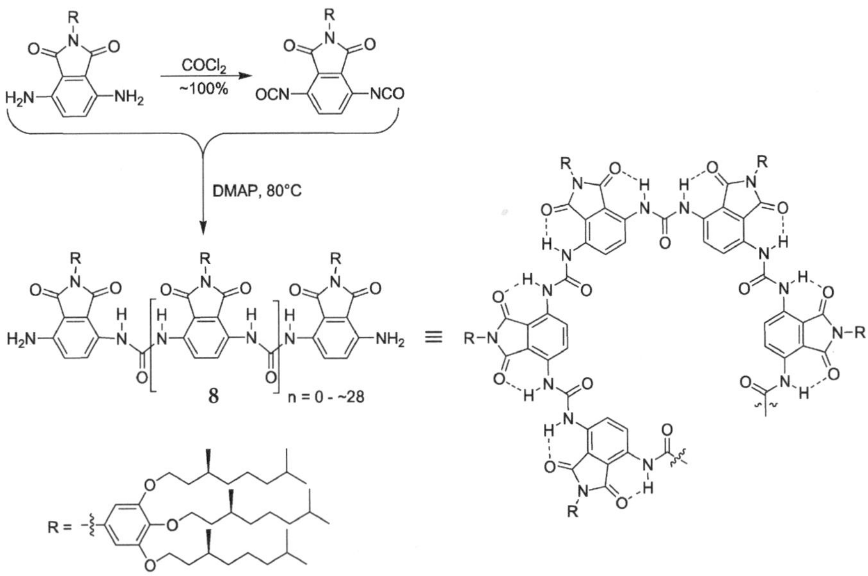

Fig. 10 Meijer's helically folding ureidophthalimides 8. *DMAP* Dimethylaminopyridine

guest and the inner rim of the helix as suggested by NMR spectroscopy. Interestingly, the binding process is cooperative, leading to an enhanced affinity for the second guest.

Very recently, ureidophthalimide oligomers and polymers 8 carrying chiral non-polar side groups (Fig. 10) have been described by Meijer et al. [47]. The backbone adopts a helical conformation due to H-bonding interactions between neighboring *anti*-oriented urea and 3,6-disubstituted phthalimide units in combination with π,π-stacking between adjacent turns. The helix consists of six to eight repeat units per turn giving rise to an internal cavity of ~8–9 Å in diameter [27]. The chemical shift of the urea protons at 9 ppm indicates intramolecular H-bonding. In CD spectroscopy, oligomers/polymers having more than six repeat units show a Cotton effect in THF. In a competitive H-bonding acceptor solvent such as chloroform no Cotton effect was observed, again illustrating the importance of H-bonding for helical structure formation.

3.3
Helical Backbones Based on the trans-2,2′-Bipyridine Helicity Codon

The preference of 2,2′-bipyridine to adopt a *transoid* conformation has been utilized extensively by Lehn's group to create polyheterocyclic strands with a helical secondary structure. According to computational results, the *cisoid* conformation of 2,2′-bipyridine is destabilized by 5.7 kcal/mol due to repulsion between the *ortho*-protons and lone pairs, while the *transoid* form allows for favorable electrostatic interactions (Fig. 11). Varying connectivity and placement

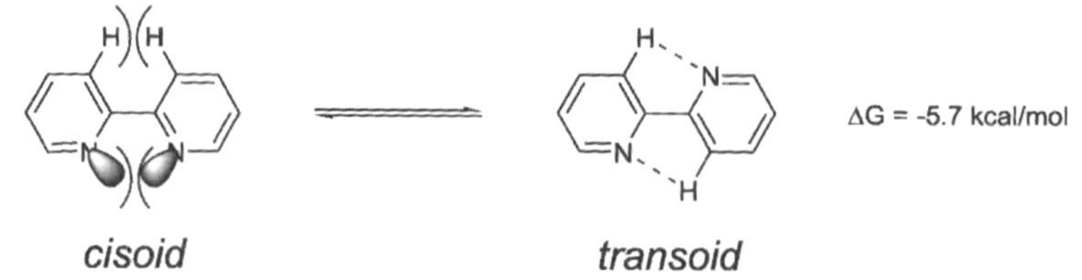

Fig. 11 Equilibrium between *cisoid* and *transoid* conformations of 2,2′bipyridine

of the heteroatoms, several foldamer families have been obtained by exploring this design feature, also referred to as 'helicity codon'.

3.3.1
Alternating Pyridine–Pyrimidine Oligomers

Oligo(pyridine-*alt*-pyrimidine)s constitute the first helical backbone of this type and were described in 1995 [48]. The combination of the above-mentioned preference of 2,2′-bipyridine to adopt a *transoid* conformation, the alternating sequence of pyridine and pyrimidine units with a kinked *meta* connectivity, and intramolecular stacking of overlapping aromatic moieties cause formation of the helical structure. Several oligomers **9** (Fig. 12, left) have been synthesized and investigated. Several X-ray crystal structures were resolved and showed the expected helical conformation [49–52]. In the case of the nonadecamer, the data reveal three complete turns and an interior cavity of 2.5 Å in diameter [27]. Owing to the dynamic helix reversal, chiral crystals contain only one enantiomer.

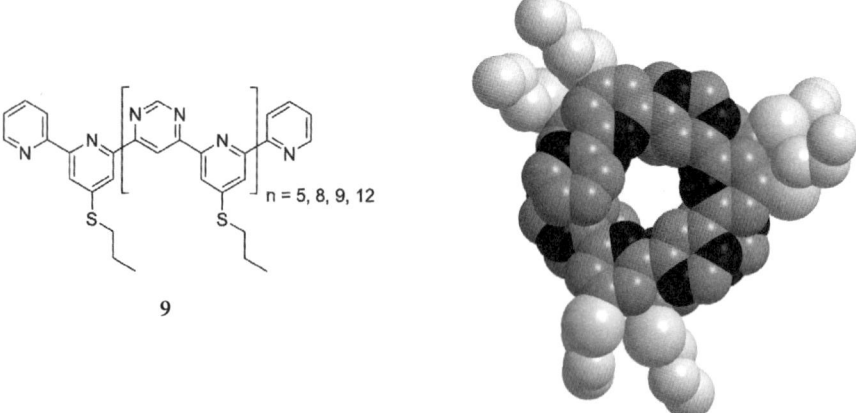

Fig. 12 Lehn's oligo(pyridine-*alt*-pyrimidine)s **9** and crystal structure of 13mer (*n*=5) (CCDC-100440; N-atoms are *black*, C-atoms are *gray*, side chains are *light*; H-atoms are omitted for clarity)

To investigate the conformational behavior in solution, several techniques have been employed [50, 52]. Extensive NMR studies monitoring up-field chemical shifts of overlapping aromatic protons, measuring NOE signals of interior protons, and utilizing ROESY and COSY methods are all consistent with a helical conformation in solution. UV/vis absorption spectroscopy shows a hypochromic effect indicating stacked chromophores; however, no bathochromic shift was observed. The stacking interaction between chromophores in a helical arrangement is also reflected in the characteristic broad, excimer-like fluorescence emission occurring in oligomers having more than one turn.

The interconversion of the enantiomeric helices has been reported to be fast on the NMR timescale at room temperature in solution. Coalescence temperatures have been used to determine the free energies of activation for helix reversal. Interestingly, they seem to be rather independent of chain length ($12.3 \leq \Delta G^{\ddagger} \leq 13.6$ kcal/mol), suggesting a stepwise mechanism for helix inversion [50, 52].

Upon addition of metal ions, the 2,2'-bipyridine-containing helices can undergo a significant structural reorganization, i.e., expansion, forming grid-type architectures. This transition can be reversed by addition of a stronger chelating ligand, which can be either masked or activated by decreasing or increasing the pH. Therefore, a system capable of reversible molecular motion potentially suitable for designing nanosized machines can be realized [53, 54].

The pyridine heterocycle within the helicity codon can be mimicked by a hydrazone unit that, in analogy, stabilizes the *transoid* conformation yet offers the advantage of facilitating synthesis. Incorporation of hydrazone units has been

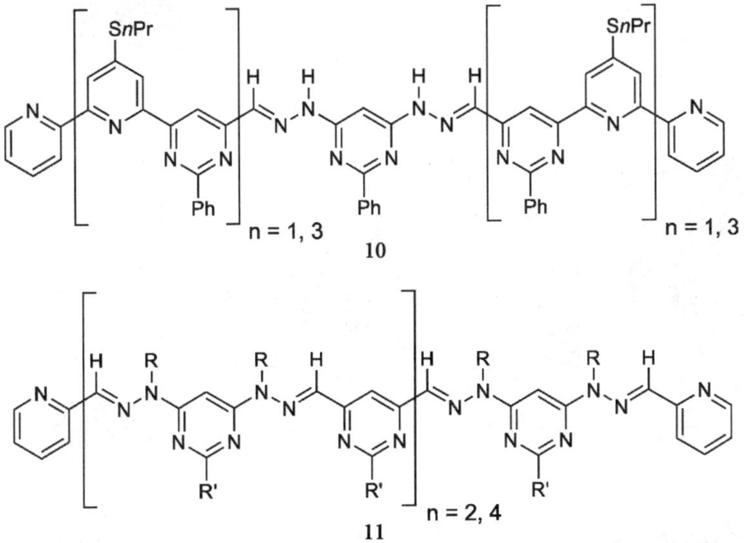

Fig. 13 Incorporation of hydrazone units to mimic the pyridine heterocycle in the center (**10**) or throughout the sequence (**11**)

accomplished either in the center of the backbone [55] or throughout the entire sequence (Fig. 13) [56]. Again, NMR, UV/vis absorption as well as X-ray crystallographic studies demonstrate the helical structure in solution and in the solid state. In the case of oligo(hydrazone-*alt*-pyrimidine)s **11**, phenyl substituents (R' in Fig. 13) provide additional aromatic stacking interactions and therefore stabilize the helical structure.

3.3.2
Alternating Pyridine-Pyridazine Oligomers

In order to increase the diameter of the helix, *para*-connected building blocks were introduced and the position of the heteroatoms carefully chosen. The resulting oligo(pyridine-*alt*-pyridazine)s **12** (Fig. 14, top) adopt a helical conformation with 12 heterocyclic units per turn, having a central cavity measuring ~8–9 Å in diameter [27, 57]. In solution, up-field-shifted NMR signals indicate

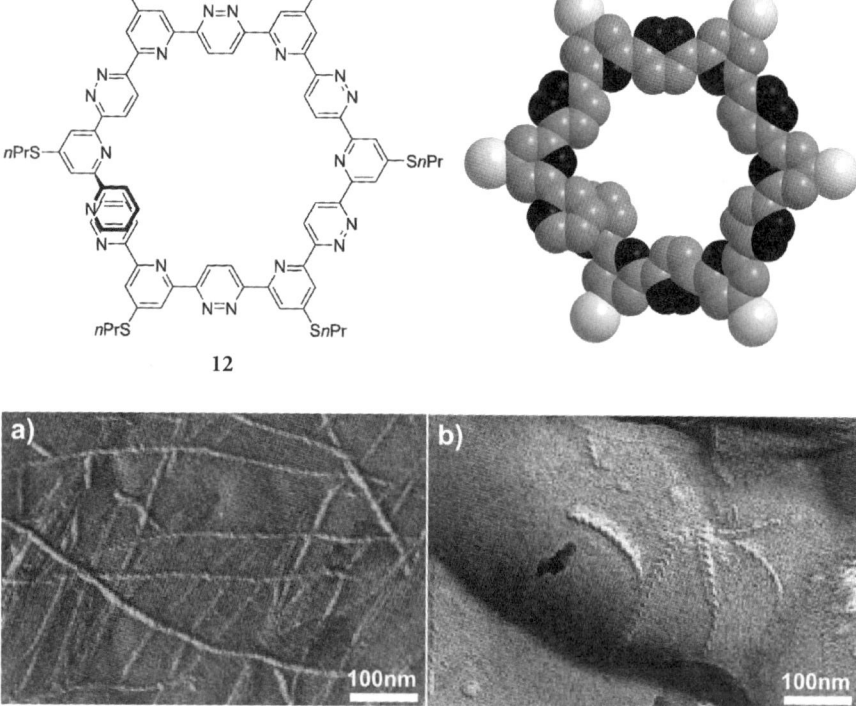

Fig. 14 Extending the helix diameter using the *para*-connected pyridazine heterocycle to afford oligo(pyridine-*alt*-pyridazine)s **12**. Model (N-atoms are *black*, C-atoms are *gray*, S-atoms are *light*, H-atoms and side chains are omitted for clarity) and freeze-fracture electron micrographs of **12** in CH_2Cl_2 (*a*) and pyridine (*b*) showing fiber network formation with helical textures. Reproduced in part with permission from [57]

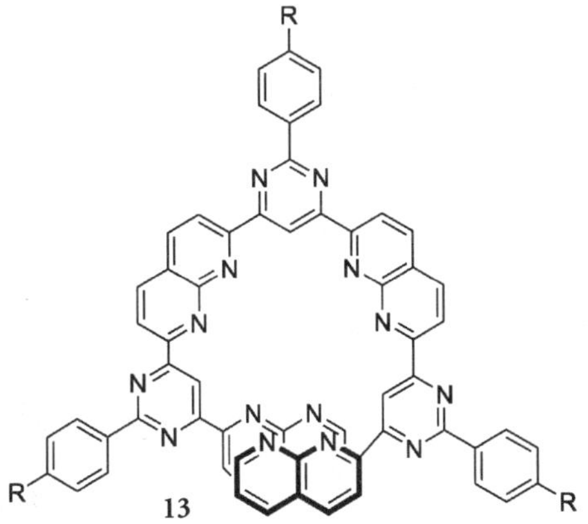

Fig. 15 Extended oligo(1,8-naphthyridine-*alt*-pyrimidine)s **13** carrying internal chelating ligands

formation of the helical conformation. Interestingly, the aromatic signals are further shifted with increasing concentration pointing to aggregation of the helices. Vapor-pressure osmometry confirms formation of higher aggregates, namely duplexes, in a variety of solvents. Freeze-fracture electron microscopy shows an extensive network of linear and intertwined chiral fibers of high aspect ratio, i.e., micrometer length and ~80 Å diameter (Fig. 14, bottom). The fibers are composed of coiled-coil bundles of two or three individual supramolecular helices. The self-aggregation to supramolecular stacks and superhelices is based on intra- and intermolecular π-π-stacking interactions between overlapping and neighboring aromatic helix turns.

3.3.3
Alternating Naphthyridine–Pyrimidine Oligomers

An alternative way to extend the helical structure is to replace the pyridine with naphthyridine units. The corresponding oligo(1,8-naphthyridine-*alt*-pyrimidine)s **13** (Fig. 15) have been synthesized and shown to adopt a helical conformation both in solution and the solid state [58]. The diameter (~4–5 Å [27]) and polarity of the inner cavity allow for interaction with cationic species. Indeed using NMR and ESI MS, it could be shown that cations rapidly induce aggregation of the helical folds in a stacked fashion, forming long interdigitated fibers as evident from TEM.

3.4
Helical Backbones Based on Solvophobic Forces

In the cases discussed so far, non-covalent interactions inherent in the backbone have been driving helical structure formation. Inspired by the hydro-

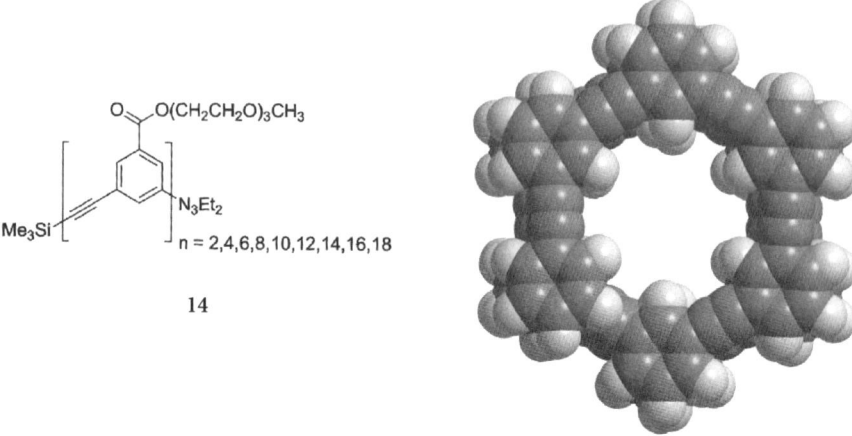

Fig. 16 Moore's amphiphilic oligo(*meta*-phenyleneethynylene)s **14** and model of the octadecamer (side chains are omitted for clarity)

phobic effect used in Nature to stabilize folded biomacromolecules, amphiphilicity can also be implemented in the backbone to stabilize its helical conformation in solution. Thereby, external solvent leads to segregation of the amphiphilic structure, exposing solvophilic and hiding solvophobic regions of the molecule. In recent years, this design principle has been impressively demonstrated for the case of the amphiphilic oligo(*meta*-phenyleneethynylene) foldamer family **14** (Fig. 16) pioneered by Moore and his group [59].

The helical secondary structure in polar organic solvents such as acetonitrile is caused by the *meta* connectivity of the backbone and π,π-stacking interactions between aromatic moieties in adjacent turns, as well as segregation of the polar triglyme side chains and the non-polar phenylacetylene scaffold. In solution, an ESR study investigating line broadening as a function of the number of connecting repeat units between two spin labels suggests that each turn is comprised of six repeat units [60], as expected on the basis of the hexagonal symmetry of related macrocycles (see 5.1.4 and 5.4.1). This conformation gives rise to a tubular cavity measuring ~7 Å in diameter [27] (Fig. 16, right). For the structural organization in the solid state see section 4.2.

Given the *meta* connectivity of the backbone, several parameters have been varied and their influence on helix stability investigated. The strength of the π,π-stacking interactions can be adjusted by changing the electron density in the aromatic ring system [61, 62]. Only in the case of the electron-deficient ester substituents favoring aromatic stacking helical structure formation could be observed [63]. Additional non-covalent interactions such as H-bonding (see 3.1.2) or interior metal ion binding (*vide infra*) can be implemented in the backbone. Also, the amphiphilicity of the system can be reversed, as illustrated by incorporation of non-polar alkyl side chains that lead to folding in hydro-

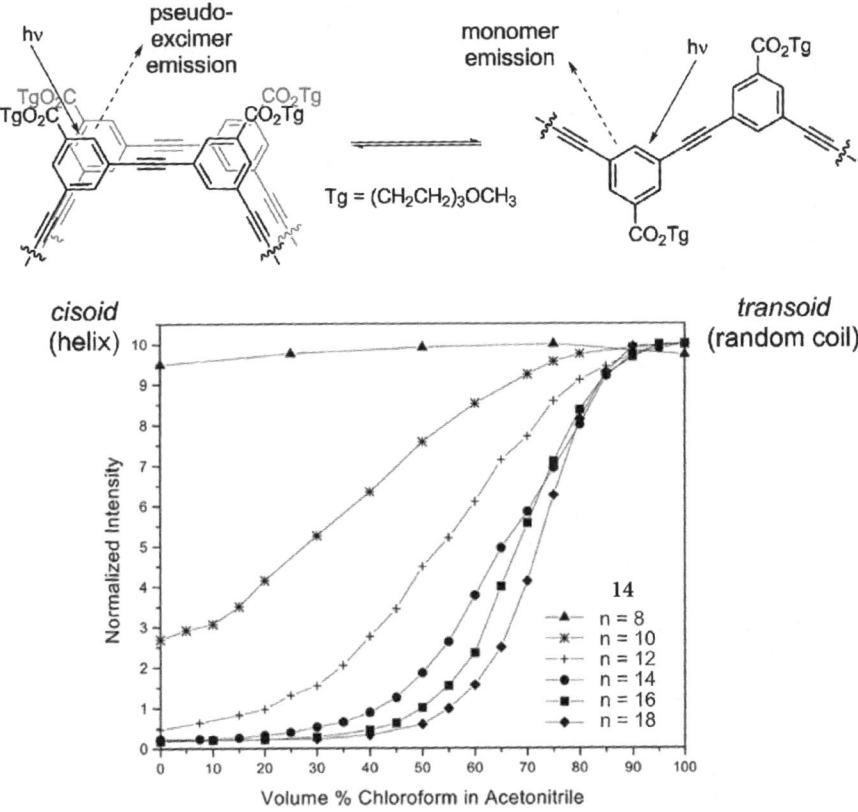

Fig. 17 *Cisoid* and *transoid* conformations reflecting the helix-coil transition (*top*). Solvent denaturation experiments demonstrate a chain length dependent, cooperative folding transition in oligomers **14** (*bottom*). Reproduced in part with permission from reference [66]

carbon solvents [64]. Solvents in general play a critical role and their ability to fold as well as unfold backbone **14** has been studied in detail [65].

Convenient optical monitoring of the conformational transition by means of UV/vis absorption as well as fluorescence spectroscopies has been established for this backbone type (Fig. 17, top) [59, 66]. While *cisoid* and *transoid* conformations display different absorption spectra, isolated and stacked cross-conjugated monomer units show direct or excimer-like emission. The helix-coil transition is temperature-dependent, but is more sensitive to changes in solvent composition. Solvent denaturation experiments show a clear chain-length-dependence of the folding process (Fig. 17, bottom). Starting at the dodecamer, the helix-coil transition both sharpens and shifts to larger amounts of denaturant for higher oligomers. Quantitative analysis shows a linear relationship between chain length and helix stabilization energy, as expected by the helix-coil theory [67]. An incremental stabilization energy $\Delta G \sim 0.7$ kcal/mol per repeat unit has been calculated from the data [66]. The observed high degree of cooperativity was quantified and is comparable to that of α-helix formation in protein folding.

The bias of the twist sense of the helical conformation has been achieved by introducing chiral units, either in the main or side chains (Fig. 18). One of the two resulting diastereomeric helices is energetically favored and causes excess helicity, conveniently monitored by CD spectroscopy. While enantiopure binaphthol segments [68] as well as tartrate-based linkers [69] have been incorporated in the main chain, chiral polar [70] and non-polar side chains [64] have

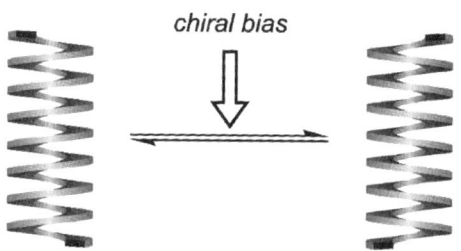

Fig. 18 Biasing the helical twist sense by covalent incorporation of chirality in the backbone or the side chains. *TMS* Trimethylsilyl

been attached to the backbone as well. Remarkably, a single extra methyl group in the side chain of **17** is able to transfer its chirality to the backbone in a highly cooperative fashion. In "sergeants and soldiers" experiments it was found that the intensity of the CD signal is not linearly dependent on the amount of chiral side chains [71]. Although only a small amount of chiral side chains is needed to induce significant Cotton effects, the values displayed by the corresponding homochiral oligomers could not be achieved.

Host-guest interactions involving guests of complementary size and shape that fit into the inner cavity of the helically folded host have been extensively explored (Fig. 19). For example, the foldamer receptor has been decorated with nitrile functionalities that bind two silver ions at the interior to provide an additional driving force for helical structure formation [72]. Since the coordinating group in **19** was introduced at every other repeat unit, three nitriles per turn can form a favored trigonal planar complex. In another example, induction of chirality has been achieved in polar solvents by binding either +α-pinene or −α-pinene in a 1:1 ratio to the internal hydrophobic surface of dodecamer **14** [73]. The resulting enantiomeric complexes have characteristic and opposite CD spectra. Also, the binding of rod-like guest molecules such as cis-(2S,5S)-2,5-dimethyl-N,N′-diphenylpiperazine **21** within the tubular cavity was found to be chain length-dependent, indicating that maximizing the area of contact determines affinity [74]. Interestingly, both affinity and chain length specificity for binding the dumbbell-shaped rod **22** were higher than for **21**, illustrating the dynamic nature of the foldamer receptor [75].

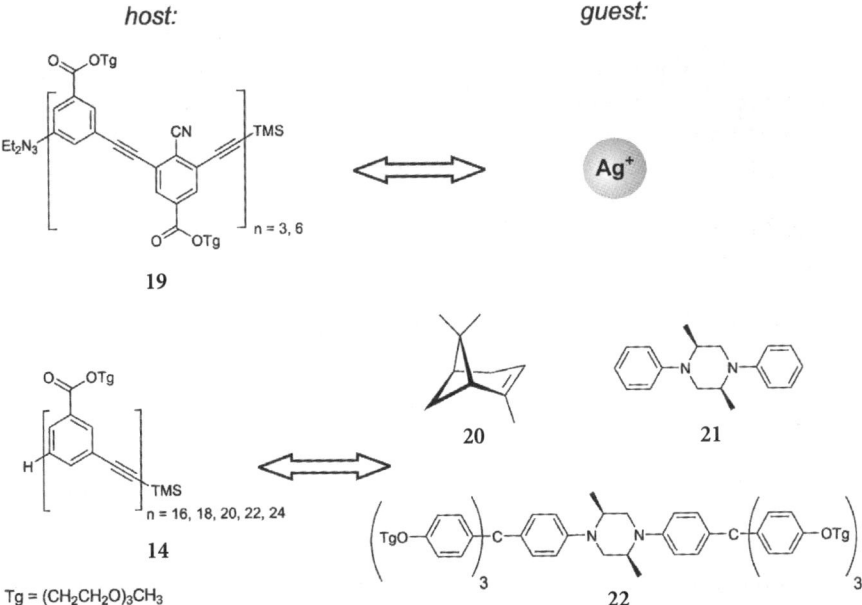

Fig. 19 Molecular recognition within the interior of dynamic foldamer receptors

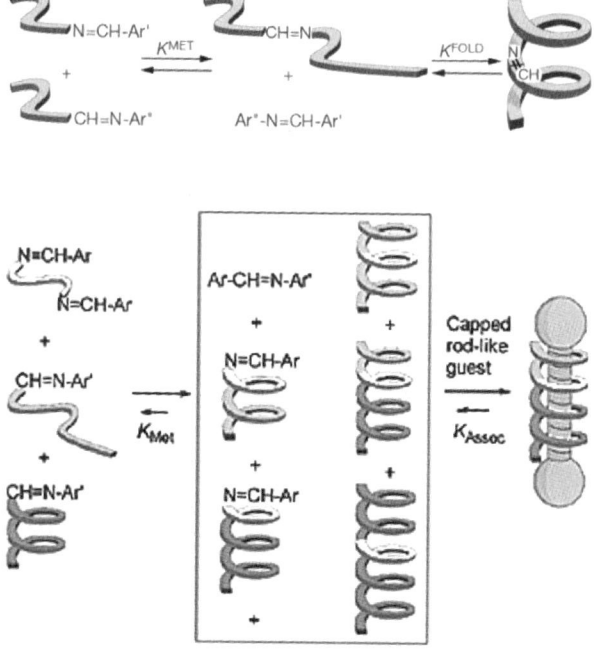

Fig. 20 Dynamic covalent chemistry by imine metathesis for the synthesis of oligomers via a purely folding-driven ligation reaction (*top*) and a templated, size-selective synthesis (*bottom*). Reproduced with permission from references [77] and [79]

Combining this length-specific binding with dynamic covalent synthesis [76] by taking advantage of reversible imine metathesis, template-controlled, size-selective oligomer synthesis of foldamer receptors has been demonstrated (Fig. 20) [77]. Utilizing further elongated templates, this supramolecular approach should in principle allow for the synthesis of well-defined oligomers of tremendous length. Under equilibrium conditions and in the absence of a template, monomer addition can be driven solely by the free energy gain owing to helix formation [78], as was demonstrated for the case of folding-driven synthesis of oligomers [79, 80] and polymers by imine metathesis [81, 82].

3.5
Covalent Stabilization by Intramolecular Helical Crosslinking

In order to utilize hollow helical structures in devices that have to operate under changing and sometimes demanding environmental conditions, it might be advantageous to covalently lock the reversibly folding structure by intramolecular crosslinking. Regioselectively crosslinked nanostructures [10] of various topologies are of particular interest in several fields, for instance for the generation of hollow capsules for controlled release applications [83]. However, the

Discrete Organic Nanotubes

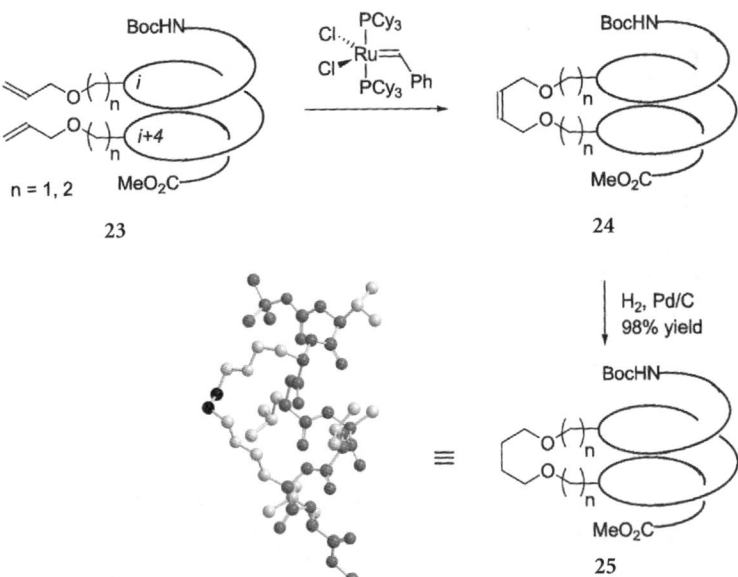

Fig. 21 Grubbs' introduction of a single covalent crosslink by RCM to stabilize an α-helical peptide conformation. Crystal structure of **25** (CCDC-101810; ethylene linkages are *black*, helical backbone are *gray*, side chains are *light*, H-atoms are omitted for clarity)

covalent capture of hollow helical conformations has not been explored until very recently.

In one case, Blackwell and Grubbs utilized RCM to introduce a single covalent linkage between two amino acid residues in a short peptide, and thereby stabilize its α-helical conformation (Fig. 21) [84, 85]. With regard to the context of this review, it is important to note that the α-helix does not contain an inner void (see 3.2.1). In the heptapeptide sequence Boc-Val-X-Leu-Aib-Val-X-Leu-OMe **23**, the allylated residues X representing serine or homoserine in positions i and $i+4$ are located on the same side of the helix, enabling efficient generation of the covalent crosslink. The obtained mixture of olefinic isomers **24** (~5:1 E/Z) was subsequently converted to the respective saturated analogue **25**. CD spectra of the linear (**23**) and the cyclic (**25**) peptides in trifluoroethanol do not show noticeable differences, indicating a preorganized helical peptide backbone and negligible structural reorganization during crosslinking. In the case of the shorter, and hence less flexible tether, the X-ray structure revealed a slight conformational change from the initial α-helix to a 3_{10}-helix. It should be mentioned that Woolley and coworkers have utilized a single azobenzene-containing linker to turn an α-helical peptide conformation either on or off [86, 87].

In order to obtain stable organic nanotubes, Hecht and Khan recently reported the intramolecular crosslinking of helically folding, hollow, amphiphilic poly(*meta*-phenylene ethynylene)s (Fig. 22) [88]. The lengthy and defect-free

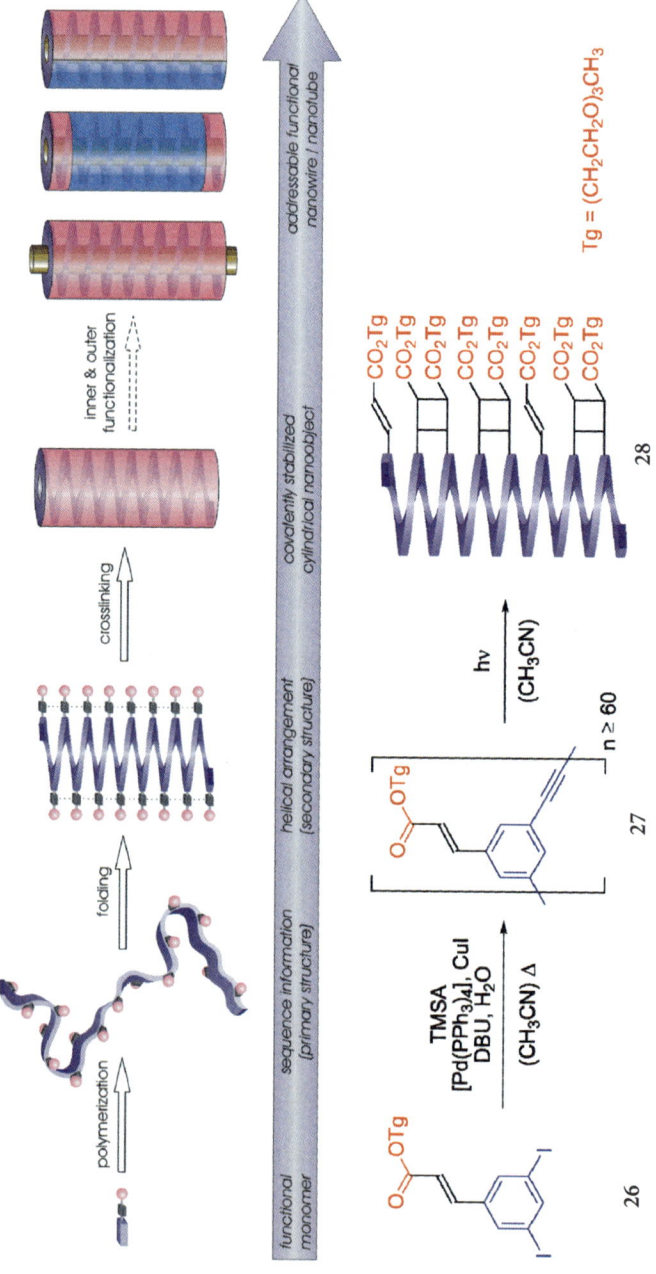

Fig. 22 Intramolecular crosslinking of helically folded polymers as a route to stable functional organic nanotubes illustrated for poly(*meta*-phenyleneethynylene)s containing photocrosslinking cinnamate groups. *TMSA* Trimethylsilylacetylene, *DBU* 1,8-diazabicyclo[5.4.0]undec-7-ene

polymer **27** was prepared from diiodomonomer **26** via a newly developed A_2+BB' polycondensation protocol [89]. Taking advantage of the short distances between repeat units in adjacent turns, crosslinking groups were introduced as cinnamates to allow for efficient [2+2] photodimerization. Using the diagnostic tools for the helix-coil transition, well established for the parent oligomer series (see 3.4), stabilization of the helical conformation by means of covalent capture could be demonstrated. Solvent denaturation experiments using UV/vis absorption as well as fluorescence spectroscopies show a typical reversible cooperative folding process for the polymer **27**. Subsequent irradiation of the folded structure of **27** under high dilution conditions leads to incorporation of approximately 20–30% of cyclobutane moieties and the resulting crosslinked helical polymer **28** cannot be unfolded. NMR data indicate the formation of β-truxinates, i.e., *syn* head-head photodimers, and therefore suggest some degree of topochemical control, yet certain structural changes during the crosslinking process cannot be avoided and are evident from the UV/vis titration data and fluorescence spectra. By developing narrow disperse polymerization and introducing proper postfunctionalization, this approach in principle allows for the generation of cylindrical nanoobjects with defined dimensions as well as controlled inner and outer functionality, such as segregated tubes and insulated wires.

4
Tubes from Self-Assembled Helices and Sheets

The state of the art in hollow helical design documented in the last section provides a respectable degree of control over relatively small, unimolecular tubes. However, in order to obtain larger structures, additional non-covalent interstrand interactions have to be implemented, for instance by intertwining, stacking, or aligning individual helices. In Nature, helical multiplexes are quite common; however, only a few artificial analogues exist. In this section, some examples of helical aggregation modes again focusing on hollow tubular structures are briefly discussed. An excellent review on (supramolecular) helical programming has been given by Nolte and coworkers [90].

4.1
Multi-stranded Helices

The most compact mode of helical aggregation arises from intertwining two or more helices to afford double or multi-stranded helices. Double-stranded DNA constitutes the prototype of a helical duplex. Both strands are held together by interstrand H-bonding interactions between the complementary base pairs that largely determine the stability of the duplex. Although it may seem obvious, the location of the bases at the helix interior precludes the use of DNA as a tube with an exploitable inner cavity. Another naturally occurring example involves double-stranded gramicidin D complexes that, although displaying a rather high pitch, form channels of considerable diameter (4.8 Å) [91]. Artificial systems such as Lehn's helicates, i.e., artificial duplexes based on coordination of two ligand strands around a central string of metal ions, are not hollow and the reader is referred to the numerous reviews on this topic [92–94]. One example of double helix formation has already been presented in the case of 7 (see 3.2.2), but again the inner void is rather small. To our knowledge, intertwined multiplexes generally do not contain a considerable unoccupied inner cavity and hence will not be discussed in this section.

4.2
Stacking Helices

In order to increase the aspect ratio of covalently synthesized helices, where the length that can realistically be achieved is limited by the efficiency of chemical bond formation, helices can be stacked on one another in a continuous fashion to generate hollow tubes. The gramicidin A dimer represents a naturally occurring example of such a self-assembled stacked helix motif that spans the cell membrane and acts as an ion channel [20]. Both helices are held together by end-to-end H-bonding contacts that direct aggregation. Stacking of helices has been noted in some artificial systems including Lehn's alternating pyridine-pyrimidine (see 3.3.1) and pyridine-pyridazine oligomers (see 3.3.2) as well as Moore's *meta*-phenylene ethynylene foldamers (see 3.4). In the latter case, it was shown that the mode of aggregation in the solid state is dominated by packing effects [95–97]. While the parent oligomer series **14** aggregates in a lamellar structure to minimize the void volume that would be created by stacking helices, the *endo*-methylsubstituted series **29** forms a hexagonal columnar structure since the helix interior is occupied (Fig. 23).

Helical aggregation of oligomers **17** also occurs in aqueous acetonitrile solution and plays an important role in the context of hierarchical chirality transfer [71]. The side chains induce excess helicity and (chain length dependent) aggregation into chiral columns results in the reversal of overall chirality.

4.3
Bundles

Tubular structures with, in general, relatively small inner channels are realized by ordered α-helical bundles. Peptitergents, vertically segregated, amphiphilic α-helical peptides, can arrange in a columnar fashion to generate helix bundles having either hydrophilic or hydrophobic inner cavities. In Nature, these bundles are frequently constructed in an intramolecular fashion from individual helical segments of a single large protein that are connected by loops [98]. Perhaps the most investigated example is the potassium channel (Fig. 24, top left) with a void diameter of ~6 Å at the selectivity filter region, i.e., the narrowest segment [99]. Rational design of the amino acid sequence enables fine-tuning

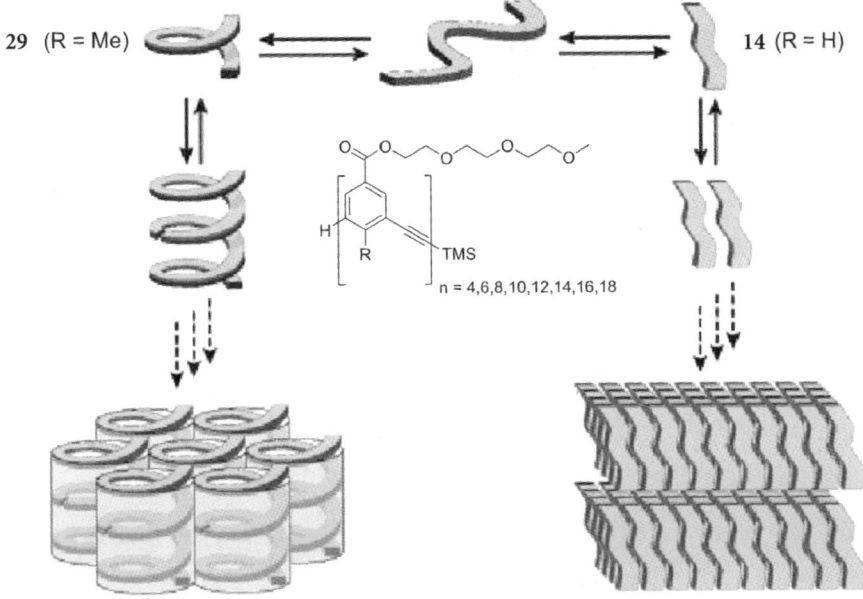

Fig. 23 Balance between lamellar and columnar solid-state aggregation in oligo(*meta*-phenyleneethynylene) series **14** and **29**. Reproduced in part with permission from reference[97]

of the axial segregation ratio and local electrostatics and therefore allows for the design of tubular objects with varying numbers of α-helices and defined inner and outer surfaces. For example, four-helix bundles with lengths of 3.5 nm display inner void diameters of only 1.5 Å [100]. In contrast, arrays of a larger number of helices can exhibit inner diameters beyond 1 nm [101].

4.4
Barrels

In addition to forming α-helix bundles, peptide-based systems assemble into related tubular motifs such as β-barrels [102]. β-barrels are constructed from staves consisting of β-sheets. In biological systems, the assembled staves are arranged in a slightly twisted manner and usually, as the name indicates, contain a barrelhead. Most barrels are formed by a single large protein having several interconnected β-strand segments [103, 104]. For example, α-hemolysin (Fig. 24, top right) consists of a heptameric barrel with a large inner diameter of 2.5 nm [105]. Artificial rigid-rod β-barrels **30** based on stiff oligo(*para*-phenylene)s carrying short peptide side chains have been described by Matile and coworkers (Fig. 24, bottom) [106]. Covalent preorganization of the stave's shape by using a rigid rod-like backbone combined with the formation of an interdigitating antiparallel H-bonding array between the peptide fragments

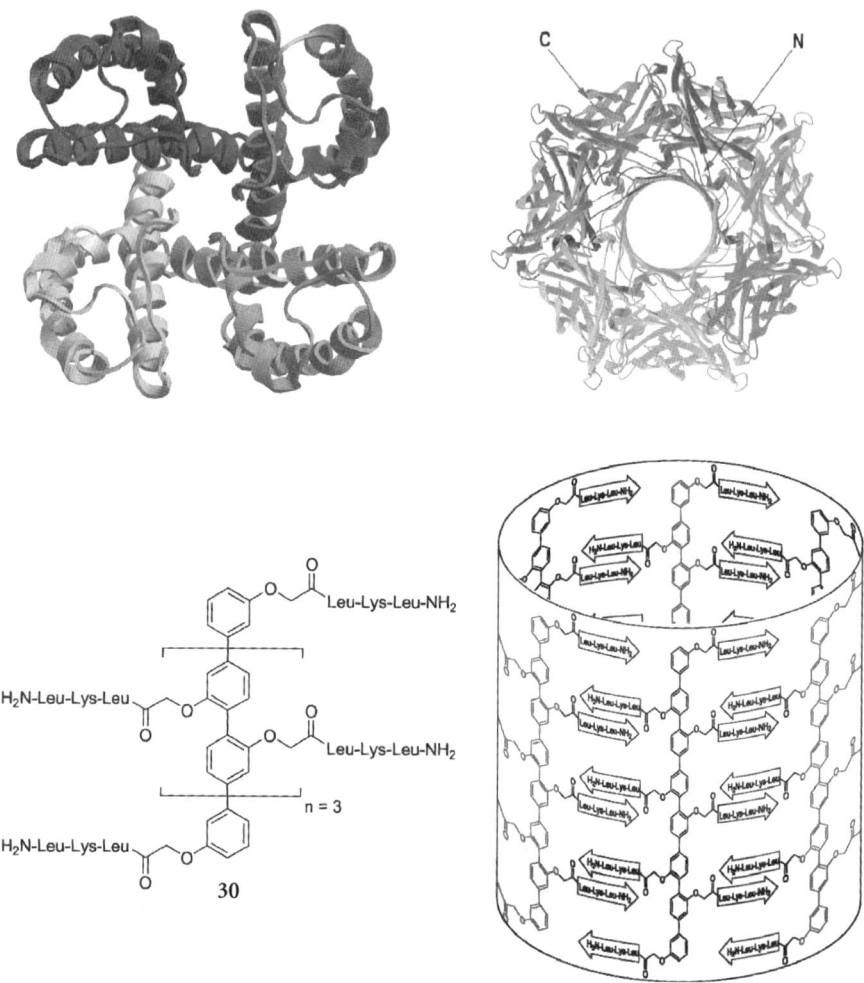

Fig. 24 Crystal structures of naturally occurring bundles and barrels: the potassium channel (pdb code: 1BL8, *top left*) and α-hemolysin (*top right*). Matile's rigid-rod β-barrels **30** composed of oligo(*para*-phenylene) rods with β-strand forming short peptide side chains (bottom). Reproduced in part with permission from reference [105]

proves to be an efficient strategy for creating hollow nanostructures with defined interior functionality. Most likely, hexameric barrels measuring 3.4 nm in length and ~1 nm in internal diameter are formed [107]. The barrel dimensions can be tuned by the number of staves as well as by the length of the peptide side chains; for instance, 2–6 staves encode inner diameters of 2–25 Å [106]. The transmembrane activity of rigid-rod β-barrels has recently been exploited for designing a high-throughput fluorometric detection system of enzyme activity [108].

5
Tubes from Stacked Macrocycles

One way to create hollow tubular structures is the self-assembly or stacking of rings. In this design concept, covalent bonds encode the inner and outer diameters of the tube while non-covalent interactions control the length of the object. Intermolecular interactions responsible for continuing association in a stacked fashion include H-bonding, aromatic π,π-stacking, and solvophobic forces. It is interesting to note that although this ring-stacking motif has been utilized extensively to design artificial tubular systems, related cases are not common in Nature. In this section, the tubular structures are discussed depending on environmental conditions, i.e., phases, and backbone types. Furthermore, covalent stabilization of the assembled objects will be described. Due to the topological resemblance, tubes based on cored, dendronized polymers have been included in this section. Related metal complexes of some cyclic ligands and their discotic self-assembly have been omitted since the metal occupies the inner void of the system. For further information on (hetero)aromatic macrocycles, the reader is referred to recent comprehensive reviews by the groups of Höger [109], Moore [110], and Schlüter [111].

5.1
Tubular Structures in the Solid State

5.1.1
Cyclic Peptides

As early as 1974, DeSantis et al. recognized that peptides having an even number of alternating D- and L-amino acids displaying β-type dihedral angels could in theory form closed rings, which can generate hollow cylinders by stacking through intermolecular H-bonding interactions [30]. Almost two decades later Ghadiri and coworkers demonstrated this concept as they designed, synthesized, and studied alternating cyclic peptides **31** (Fig. 25) [112, 113].

Upon acidification of alkaline solutions of *cyclo*-[-(L-Gln-D-Ala-L-Glu-D-Ala)$_2$-], rod-shaped crystals having a high aspect ratio with average dimensions of 10–30 μm×100–500 nm were formed. Each crystal consists of an organized bundle of hundreds of tightly packed nanotubes. A variety of experimental methods including TEM, cryo-TEM, and electron diffraction as well as molecular modeling show that the tubular stack has an inner diameter of 7 Å and a 4.7 Å axial periodicity. This intersubunit distance corresponds to an ideal antiparallel β-sheet conformation, which is further supported by the observed N–H-stretching frequency at 3,277 cm^{-1}, indicative of strong H-bonding interactions. In this β-sheet-type arrangement, all amino acid side chains are located at the periphery, affording a polar inner cavity. It should be noted that these peptide crystals display good mechanical and chemical stability.

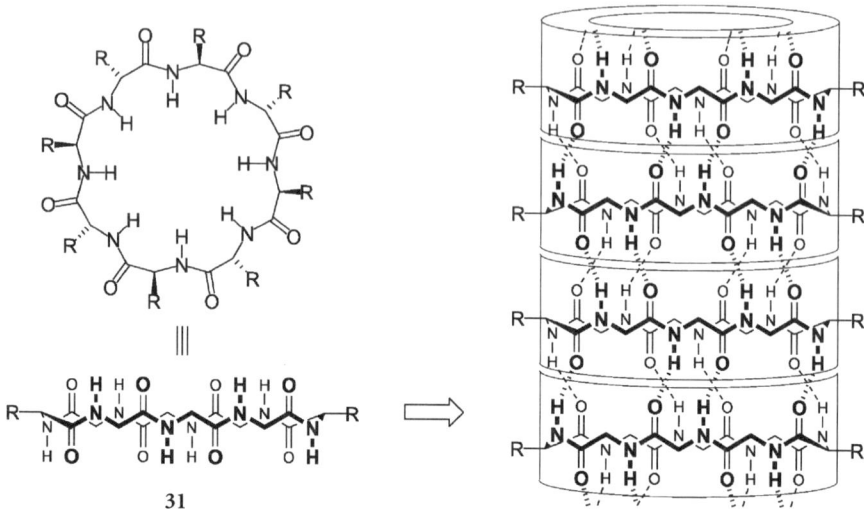

Fig. 25 Ghadiri's self-assembling nanotubes based on cyclic α-peptides **31** consisting of alternating D- and L-amino acids

The versatility of Ghadiri's approach arises from the fact that the internal diameter (at least within a certain range) and the properties of the outer surface can easily be modified by adjusting the ring size and introducing various amino acid residues. For example, cyclopeptides containing either 10 or 12 repeat units give rise to tubular assemblies with internal diameters of 10 and 13 Å, respectively [114, 115]. When the periphery is rendered hydrophobic, such tubular aggregates can be inserted into lipid bilayers to create transmembrane channels [116]. For example, channels consisting of cyclo-[-(L-Trp-D-Leu)$_n$-L-Glu-D-Leu-] with $n=3$ possessing an inner diameter of 7 Å show transport activity for potassium and sodium ions close to that of the natural gramicidin A. Increasing the inner channel diameter to 10 Å by using an extended cyclopeptide with $n=4$ leads to glucose transport activity [114]. This remarkable selectivity is based solely on pore size and points to potential applications in controlled drug delivery. Finally, the inner pore could recently be rendered hydrophobic by incorporating triazole-containing ε-amino acids [117].

Seebach et al. have found that the cyclic tetramers of 3-aminobutanoic acid **32–34** (Fig. 26) adopt tubular structures in a fashion similar to those of α-peptides described earlier [118]. X-ray powder diffraction data suggest that all three β^3-peptide isomers exhibit tubular crystal packing with non-linear H-bonding. Presumably, the additional methylene group provides the rings with more flexibility and allows them to adopt a conformation suitable for stacking.

Homochiral cyclic β^3-peptides such as cyclo[-(β^3-Htrp)$_4$-] are able to form transmembrane ion channels with pore sizes of 2.6–2.7 Å [119, 120]. The nanotubes contain a parallel β-sheet-like structure in contrast to their alternating α-peptide counterparts described above. The parallel arrangement of amide

Discrete Organic Nanotubes

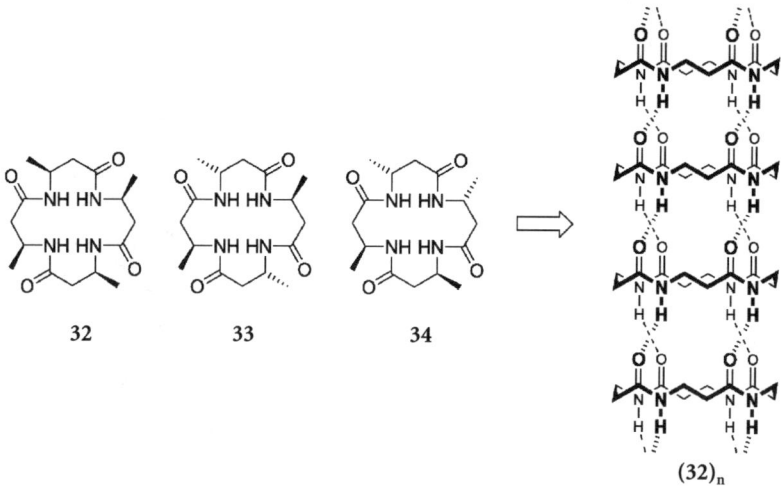

Fig. 26 Seebach's cyclic β-peptide tetramers 32–34 and mode of self-association for 32 (side chains are omitted for clarity)

units causes formation of a net dipole in the resulting nanotube, which might affect channel conductance by voltage gating and display current rectification behavior.

The use of spirobicyclic peptides affords tubes that are covalently linked. This has been demonstrated by Ranganathan et al. who synthesized cystinospirane cycle 35, which displays a stacked tubular structure due to progressing H-bonding interactions in the crystal (Fig. 27) [121].

Serinophanes, which belong to the class of cyclodepsipeptides and contain alternating repeats of serine and aromatic units in the cyclic framework, have

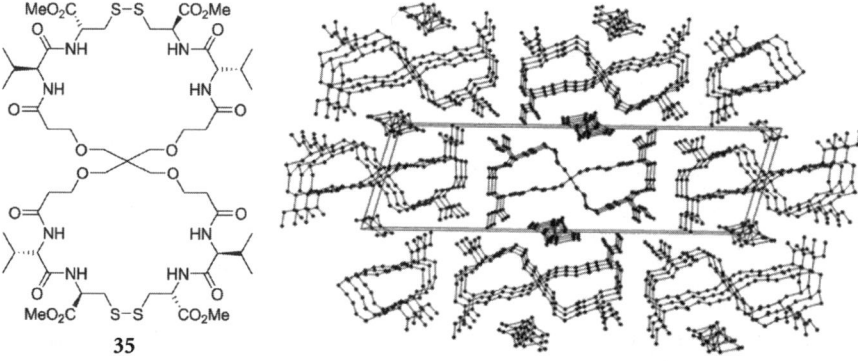

Fig. 27 Ranganathan's fused tubes from spirobicyclic peptide 35. Reproduced in part with permission from reference[121]

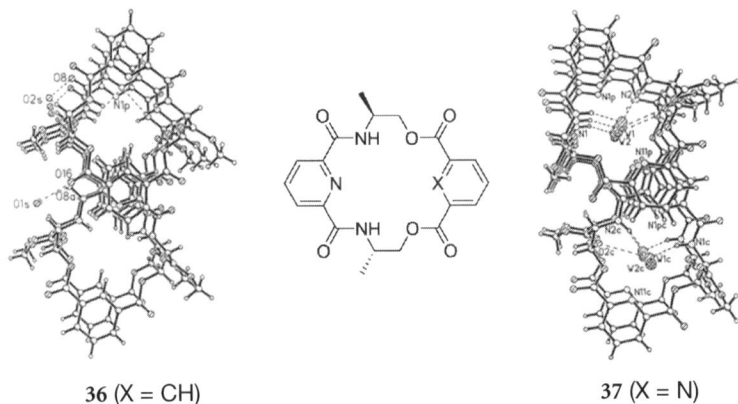

Fig. 28 Serinophanes 36 and 37 forming different types of channels in the solid state. Reproduced in part with permission from reference [122]

been shown to generate tubular assemblies (Fig. 28) [122]. In this case, H-bonding is used for conformational stabilization in an intramolecular fashion only, while intermolecular π,π-stacking is responsible for the preferred tubular packing arrangement. Both macrocycles **36** and **37** form stacks with interdigitating aromatic units. In contrast to **37**, retained water molecules in **36** cannot be accommodated inside the tubes due to steric crowding of internal benzene protons, and are located in between sheets instead.

Tubes with an inner diameter of 3 Å have been obtained from an 18-membered lactam, i.e., *cyclo*-(NHCH$_2$CH=CHCH$_2$CO)$_3$ [123]. The dipolar anisotropy arising from the same orientation of the amide groups amounts to a strong dipole along the tubular axis, which is amplified in the bulk to give a highly anisotropic crystalline material. Macrocyclic bisureas and bisamides, so called hybrid cyclopeptides, have been explored by Ranganathan [124]. Their tubular stacking in the solid state reveals the expected intermolecular H-bonding interactions between individual rings.

5.1.2
Cyclic Oligosaccharides

Cyclic oligosaccharides **38**, having alternating (1→4)-linked α-D- and α-L-rhamnopyranose residues, and **39**, having alternating (1→4)-linked α-L-rhamnopyranose and mannopyranose residues, have been described by Stoddart and coworkers (Fig. 29) [125–127]. X-ray crystallography shows that the cycles stack in a head-to-tail fashion to form tubular channels with inner diameters of ~10 Å. Interestingly, intermolecular H-bonds do not stabilize the individual stacks along their tubular axis but rather connect adjacent columns.

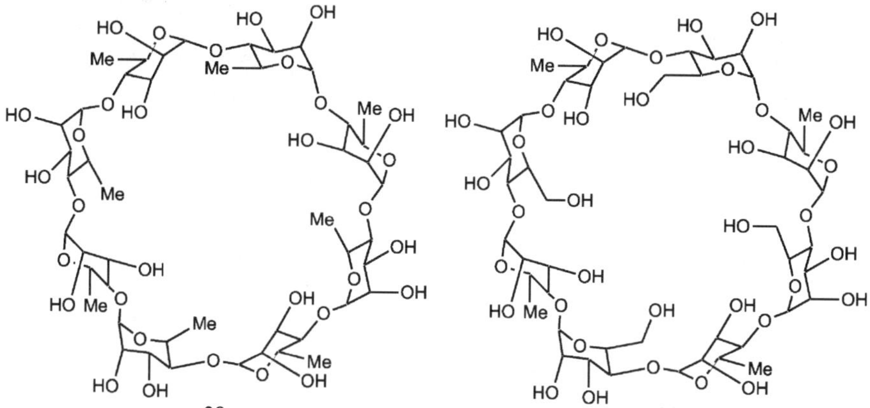

Fig. 29 Stoddart's stacked cyclic oligosaccharides **38** and **39**

5.1.3
Phenylene Macrocycles

A variety of phenylene-based macrocycles have been synthesized and crystallographically characterized by Schlüter's group [111, 128]. While most macrocycles crystallize in an "interpenetrating" manner, thereby avoiding channel formation, one example of a tubular stacking mode has been described [129]. The giant macrocycle **40** has been shown to adopt a cyclohexane-type chair conformation (Fig. 30). In the crystalline lattice, these chairs stack in such way that the inner cavity is continued, to give rise to a tubular structure having a huge internal diameter of ~27 Å. These large inner cavities are most likely filled with solvent ($CHCl_3$).

5.1.4
Phenylene Ethynylene Macrocycles

Moore and coworkers have been utilizing *meta*-connected phenylacetylenes, not only for creating a foldamer family (see 3.4) but also a variety of macrocycles (and dendritic architectures) [110, 130, 131]. For example, they reported the crystal structure of macrocycle **41** (Fig. 31), carrying peripheral phenolic groups [132]. In the solid state, hexagonally close-packed, two-dimensional H-bonded networks form layers that stack while approximately maintaining the alignment between the internal voids to afford extended channels with diameters of ~9 Å (Fig. 32). In analogy to graphite, the sheets are separated by 3.33 Å and are arranged in a repeating ..ABCABC.. sequence of cubic close packing.

Macrocycles functionalized with bulky peripheral *t*-butyl groups show different organizations in the crystalline lattice depending on their internal sub-

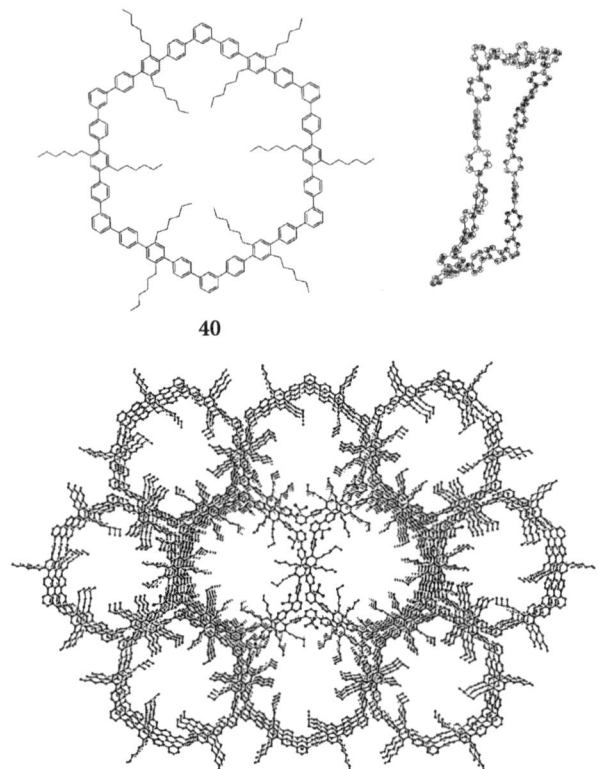

Fig. 30 Schlüter's large phenylene macrocycle **40** and its crystal structure showing a chair-like conformation and columnar stacking to generate large inner channels. Reproduced in part with permission from reference [129]

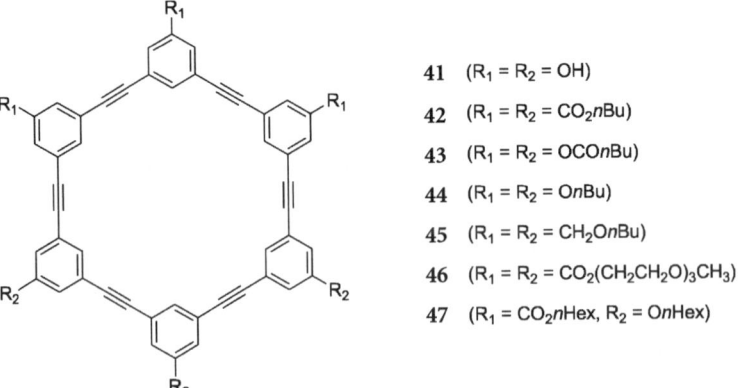

Fig. 31 Several members of Moore's shape-persistent phenylene ethynylene macrocycles

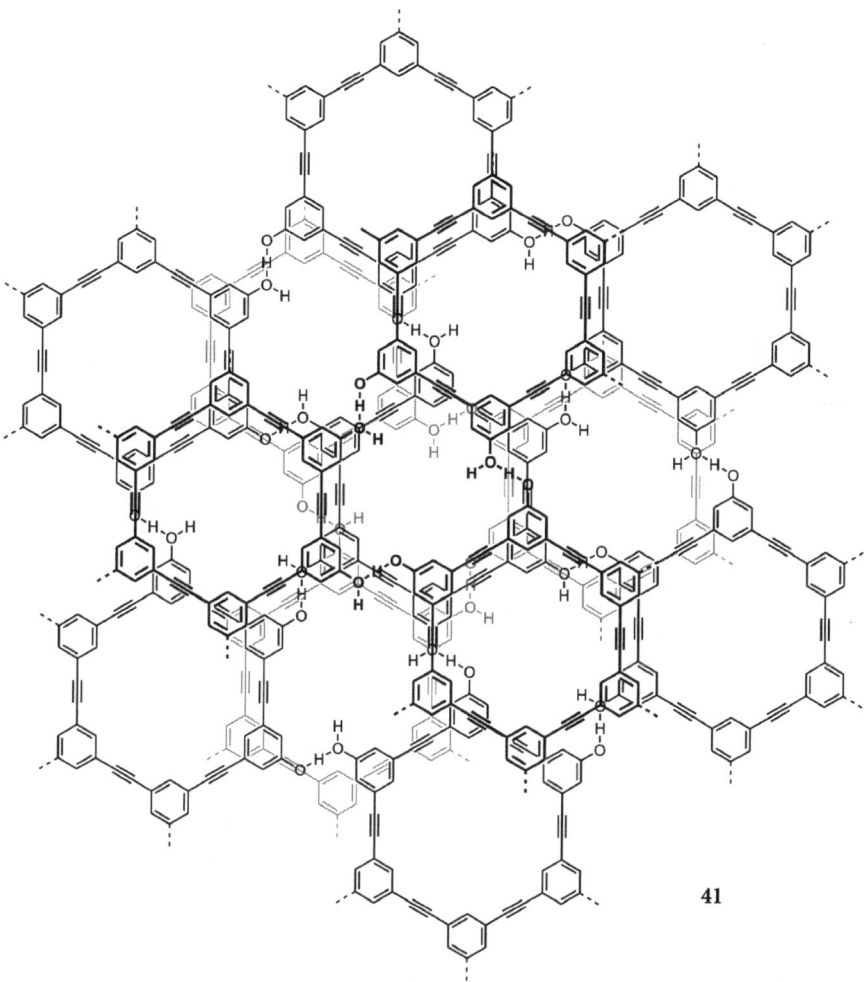

Fig. 32 Arrangement of macrocycle **41** in the crystalline lattice yielding a mesoporous material

stitution pattern (Fig. 33). Bunz and coworkers reported macrocycle **48**, which does not form a truly tubular arrangement in the solid state due to a large tilt of the aromatic rings with respect to the column axis [133]. In contrast, the corresponding derivative **49** with internal methoxy-groups, described by Oda's group, exhibits the characteristic tubular packing arrangement, thereby generating open channels with pore sizes of 4.5 Å [134]. In this case, a 3.7 Å intersheet separation was found.

Solvent-triggered conformational change was observed for macrocycle **50** (Fig. 34) [135]. Crystallization from polar solvents such as pyridine resulted in a conformation where two phenolic and two hexyloxy groups occupy the inte-

48 (R = H) 49 (R = OMe)

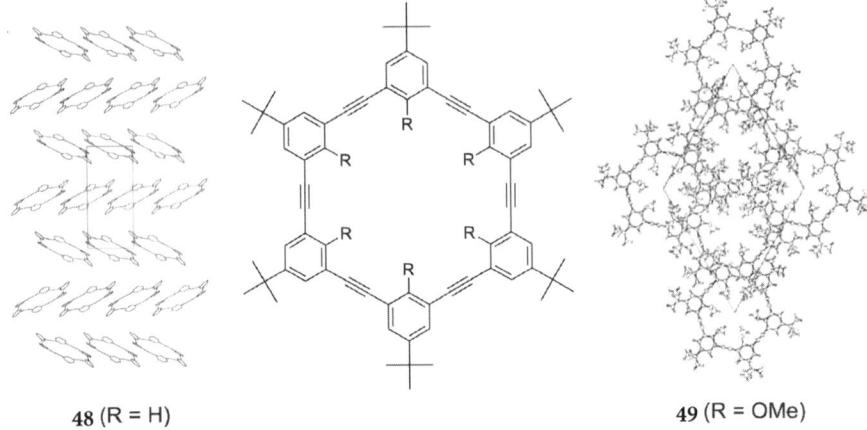

Fig. 33 Bunz' and Oda's hexameric phenylene ethynylene macrocycles **48** and **49** show different packing behavior dependent on their internal substitution pattern. Reproduced in part with permission from references [133] and [134]

50

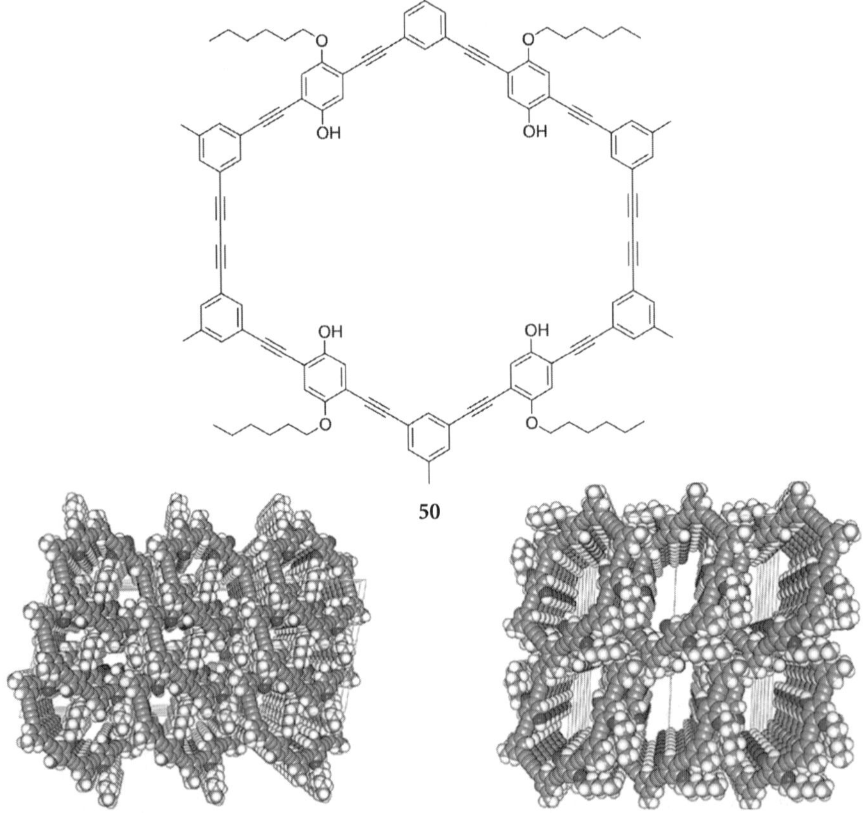

Fig. 34 Höger's macrocycle **50** exhibits different aggregation modes in the solid state depending on the crystallizing conditions, by exposing and hiding different parts of its flexible amphiphilic substituents. Reproduced in part with permission from reference [135]

rior cavity leaving channels with a pore size of ~4–5 Å. However, when the same macrocycle was crystallized from THF, a less polar solvent, all hexyloxy side chains were exposed to the outside and therefore a channel structure with much larger oval inner pores of 8×12 Å was obtained.

5.2
Tubular Structures at Interfaces

The rigid macrocycles are able to form highly ordered monolayers at the air/water interface. In these compact monolayers, macrocycle **47** adopts an edge-on orientation with respect to the subphase while individual cycles stack to afford a tubular channel running parallel to the subphase normal, as shown by GIXD and XR [136]. The order in the layers can be either enhanced by addition of KCl or disrupted by adding CsCl. Assuming cation-binding to the cycle's periphery or side chains, it was speculated that the smaller potassium ions primarily provide additional stabilization to the aggregate, while the larger cesium ions predominantly hinder stacking due to their significant spatial requirements.

In related work, the self-assembly of *cyclo*-[-(L-Phe-D-N-MeAla)$_4$-] at the air/water interface has been investigated [137]. Due to N-methylation blocking H-bonding interactions on one face of the cyclopeptide, only dimeric aggregates can be formed. With the help of GIXD, it was shown that three layers of peptide dimers adopt a face-on orientation with respect to the subphase. By changing the amino acid residues to yield *cyclo*-[-(L-Trp-D-Leu-)$_3$-L-Ser-D-Leu-], crystalline monolayers of nanotubes lying parallel to the water surface in an edge-on orientation were obtained. Interestingly, replacing the hydrophilic serine with another tryptophan residue affords a cyclopeptide that exhibits a very low tendency to form such ordered monolayers. This finding clearly demonstrates the dramatic effect of side chains substitution on peptide self-assembly at interfaces.

5.3
Tubular Structures in Mesophases

Owing to the rigid frame work and side chains mobility, shape-persistent macrocycles can be envisaged as mesogens of a tubular mesomorphic state, in which the chemically defined interior of the formed channels provides the opportunity to create amendable porous materials. Zhang and Moore observed and characterized the tubular liquid crystalline (LC) mesophase of macrocycle **47** by optical microscopy, DSC, and powder X-ray diffraction [138]. Later on, the mesophase was doped with silver ions, which were intercalated within the tubes without destroying the LC order as shown by high-resolution X-ray diffraction [139]. This study illustrates the potential use of tubular LC phases from macrocycles for preparing optoelectronic materials.

Besides forming a mesophase, the macrocycle can also be organized within a LC phase. This has recently been demonstrated by Dory and coworkers, who utilized self-assembly of a lactam (see 5.1.1) in a nematic phase to generate aggregates of nanotubes having micrometer diameters and millimeter lengths [140].

5.4
Tubular Structures in Solution

5.4.1
Phenylene Ethynylene Macrocycles

As early as 1992, Zhang and Moore reported on the ability of macrocycle **42** to self-aggregate in solution at higher concentrations [141]. NMR titration experiments monitoring the up-field shift of aromatic protons with increasing concentration indicate cooperative π,π-stacking interactions between the aggregating macrocycles. Since then, a number of macrocycles have been studied and it was found that the self-association behavior varies with the structure of the backbone as well as the side chains [110]. In agreement with Hunter and Sanders [61, 62, 142], electron-poor aromatic systems such as benzoates in macrocycles **42** favor aggregation, while in the case of electron-rich moieties such as phenolates (**43**), phenyl ethers (**44**), and benzyl ethers (**45**), no self-association was found [141]. The association constants were deduced from NMR and VPO studies, which confirmed that no aggregates larger than dimeric were formed in chloroform. Later on, the same group introduced an additional solvophobic driving force by substitution with polar triglyme side chains (**46**) and as a result, higher association constants leading to formation of larger (than dimeric) aggregates were obtained [63]. Again, aggregation could be correlated to electronic requirements for π,π-stacking in analogy to the solvophobically driven helical structure formation in the related foldamer family (see 3.4). Surprisingly, the same macrocycle (**46**) also aggregates in benzene. The association constant can be further improved by internal methyl substitution [143], owing to reduction of void volume.

5.4.2
Phenylene Diethynylene Macrocycles

Similar to the *meta*-phenylene ethynylene macrocycles, Tobe et al. investigated the self-aggregation of a series of macrocycles having *meta*-phenylene diethynylene backbones (of different sizes) bearing polar or non-polar side chains [144]. Incorporation of the diacetylene linkage renders the aromatic rings more electron-deficient and therefore results in the higher self-association constants of **51** and **52** when compared to the corresponding *meta*-phenylene ethynylene macrocycles. Electrostatic repulsion among pyridine or cyano

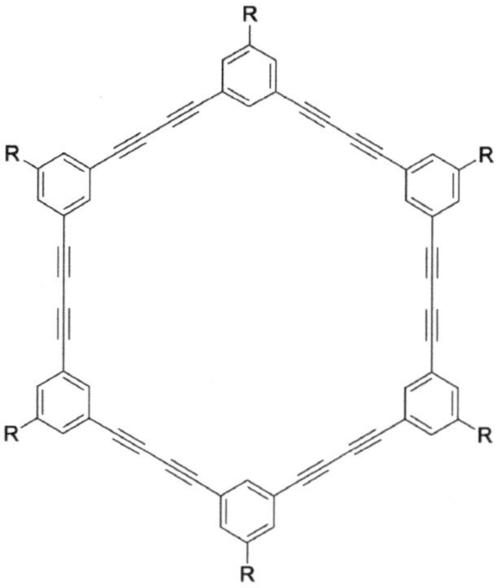

51 (R = CO$_2$nOct)
52 (R = CO$_2$(CH$_2$CH$_2$O)$_3$CH$_3$)

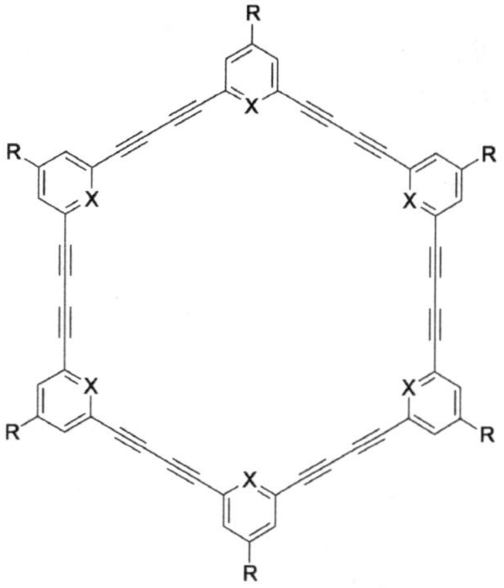

53 (X = N, R = CO$_2$nOct)
54 (X = C-CN, R = CO$_2$nOct)

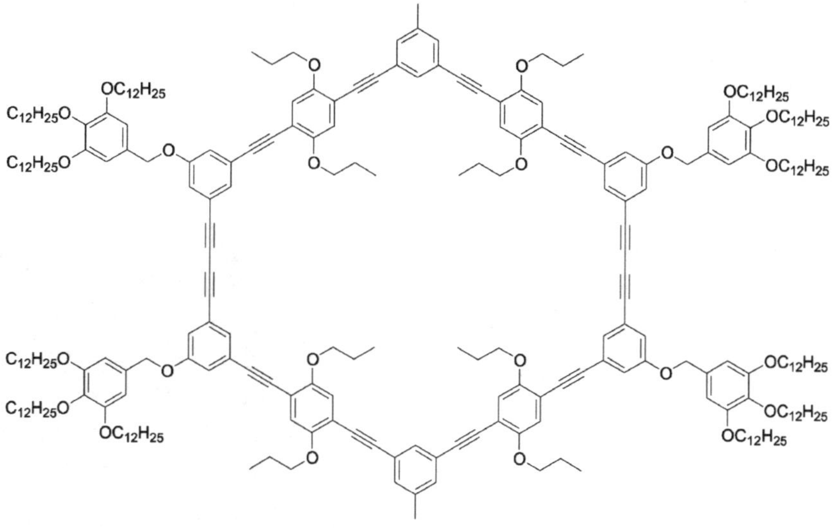

55

Fig. 35 Tobe's phenylene diethynylene macrocycles **53** and **54** that do not self-associate but form columnar heteroaggregates with macrocycle **51**. Höger's macrocycle **55**, having an electron-rich aromatic backbone surrounded by a non-polar corona, self-aggregates in non-polar solvents

groups prohibits self-aggregation in the case of macrocycles **53** and **54**, respectively (Fig. 35) [145, 146]. However, these compounds can form heteroaggregates with macrocycle **51**, presumably driven by favorable dipolar interactions between the phenylene and electron-deficient benzonitrile/pyridine moieties.

Introduction of large hydrophobic substituents leads to reversal of amphiphilicity in macrocycle **55**, (Fig. 35) and hence self-association occurred at higher concentrations in non-polar solvent mixtures [147]. Interestingly, the rather electron-rich phenyl ether backbone units do not seem to interfere with aggregation.

5.4.3
Coil-Ring-Coil Block Copolymers

A remarkable study by Höger and coworkers describes the self-assembly of coil-ring-coil block copolymers **56**, in which the macrocycle is linked in between two narrow disperse polystyrene blocks of various lengths (Fig. 36) [148]. Upon cooling in cyclohexane, the stacking of the cyclic moieties driven by the amphiphilic structure results in formation of hollow cylinders with an average length of ~500 nm. A combination of DLS, X-ray scattering, TEM, and AFM experiments confirms the existence of such long cylindrical aggregates. The high persistence length of this supramolecular polymer of ~100 nm indicates the surprisingly high rigidity of the structure.

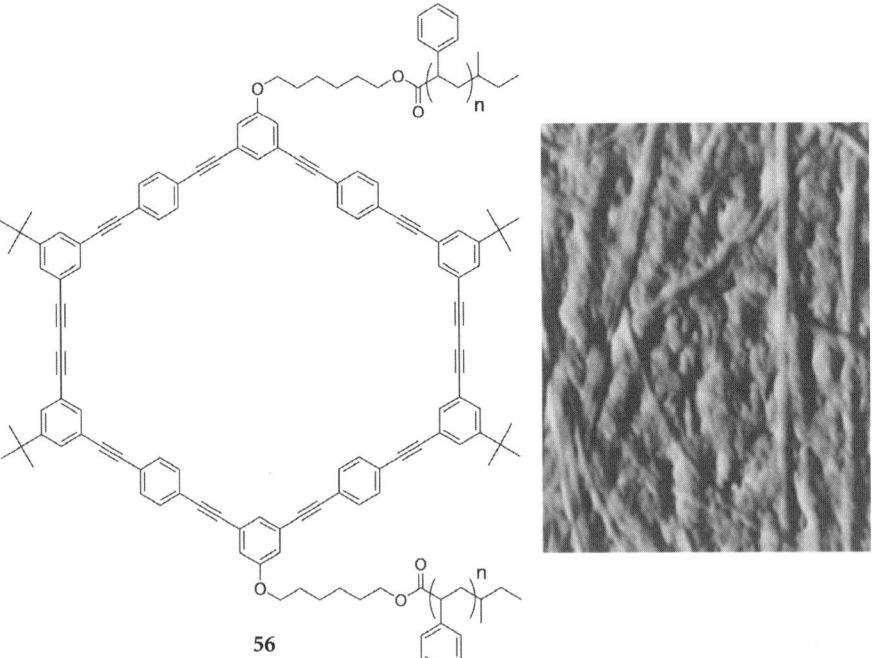

Fig. 36 Höger's coil-ring-coil block copolymers **56** forming long aggregates of stacked rings surrounded by a polystyrene sheath. AFM image of 0.15 wt% **56** in cyclohexane on mica. Reproduced in part with permission from reference [148]

5.4.4
Covalent Stabilization by Crosslinking

Although self-assembly utilizing various non-covalent interactions offers significant advantages such as minimization of synthetic efforts, high efficiency, assembly control via information-rich backbones among others, the resulting supramolecular structures often exhibit low kinetic stability due to rapid assembly-disassembly processes. Thus, covalently locked structures can be advantageous in view of certain future materials applications. In this section, the few related examples, where the target structure is formed by self-assembly of the subunits and then covalently captured, are discussed.

5.4.5
Cyclodextrins

The well-known ability of cyclodextrins to form inclusion complexes with a wide range of guest molecules in aqueous solution due to their cylindrical

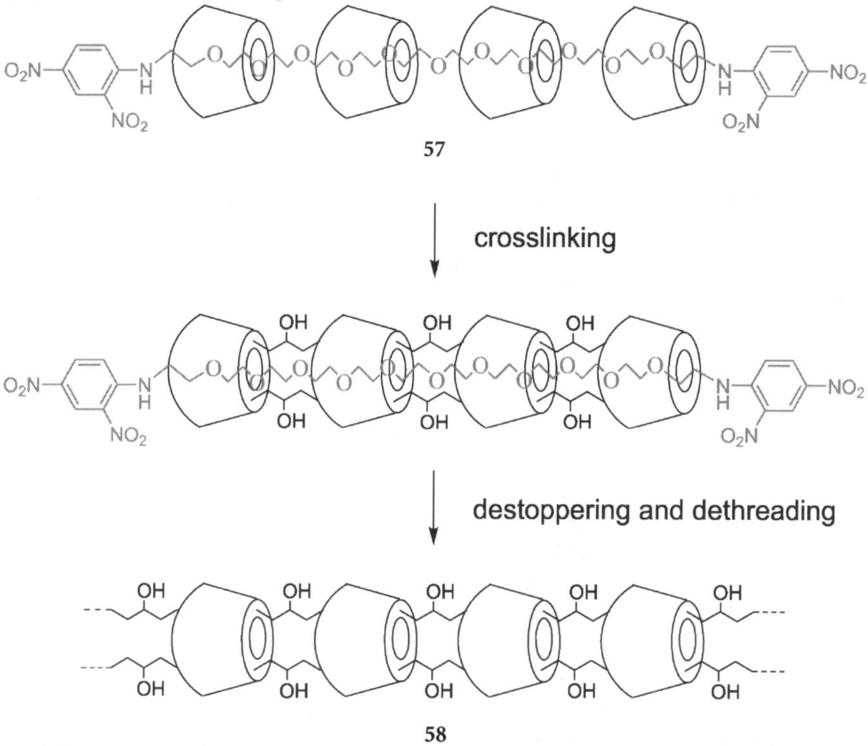

Fig. 37 Harada's approach to cyclodextrin-based nanotubes **58** via formation, subsequent covalent crosslinking, and final template removal from polyrotaxane **57**

0.7 nm-deep cavities having diameters of 0.45, 0.7, and 0.85 nm for the α-, β-, and γ-forms, respectively, has been exploited by many researchers for the construction of polyrotaxanes. In 1992, Harada et al. reported the threading of approximately 20 cyclodextrin molecules onto a diamino-terminated polyethylene glycol to afford a pseudopolyrotaxane, which was capped using Sanger's reagent to afford polyrotaxane **57** of lengths up to 20 nm (Fig. 37) [149]. Subsequent crosslinking involving the reaction of adjacent cyclodextrin hydroxyl groups with epichlorohydrin followed by removal of the polymeric template via de-stoppering and de-threading, yielded the desired macromolecular tubes **58** [150]. The tubes are soluble in water and polar organic solvents. In an analogous approach, Liu et al. prepared tubular dimers from β-cyclodextrin derivatives linked by selenium and platinum complexes [151].

5.4.6
Cored Dendronized Polymers

In order to extend their unimolecular imprinting concept [152] from cored spherical dendrimers [153] to a cylindrical topology, Zimmerman and coworkers realized a three-step sequence of assembly, crosslinking, and core removal [154]. This approach is related to Harada's synthesis of cyclodextrin nanotubes (see 5.5.1); however, the crosslinking step occurs in an intramolecular rather than an intermolecular fashion. Tin porphyrin **59**, carrying four polybenzylether dendrons attached via ester moieties and containing multiple olefin terminal groups, represents the key building block of the synthesis (Fig. 38). Addition of succinic acid serving as bridging ligand yields relatively unstable coordination oligomers with, on average, five face-to-face-oriented tin porphyrin repeat units. Subsequent RCM of the peripheral olefin groups leads to an extensively crosslinked periphery with low degrees of interoligomer crosslinking at high dilution. Finally, the internal porphyrin core was removed using transesterification of the internal ester linkages.

5.4.7
Cyclic Alternating D,L-α-Peptides

Taking advantage of the well-defined H-bonding pattern in stacking cyclic peptides (see 5.1.1), Ghadiri and coworkers have successfully captured the preorganized dimeric peptide assemblies by intermolecular covalent bond formation (Fig. 39). In both cases, association was restricted to one face of the cyclic peptide by means of N-methylation (*vide supra*). Specifically, olefin-containing *cyclo*[-(L-Phe-D-N-MeAla-L-Hag-D-N-MeAla)$_2$-] **60** undergoes regioselective intermolecular RCM to afford covalently stabilized β-sheet peptide dimer **62** [155]. In another example, the dimeric complex of *cyclo*[-(L-Phe-D-N-MeAla-L-Cys(Acm)-D-N-MeAla)$_2$-] **61** was locked by oxidative disulfide formation [156]. So far, this promising approach has been limited to covalent stabilization of dimeric aggregates only.

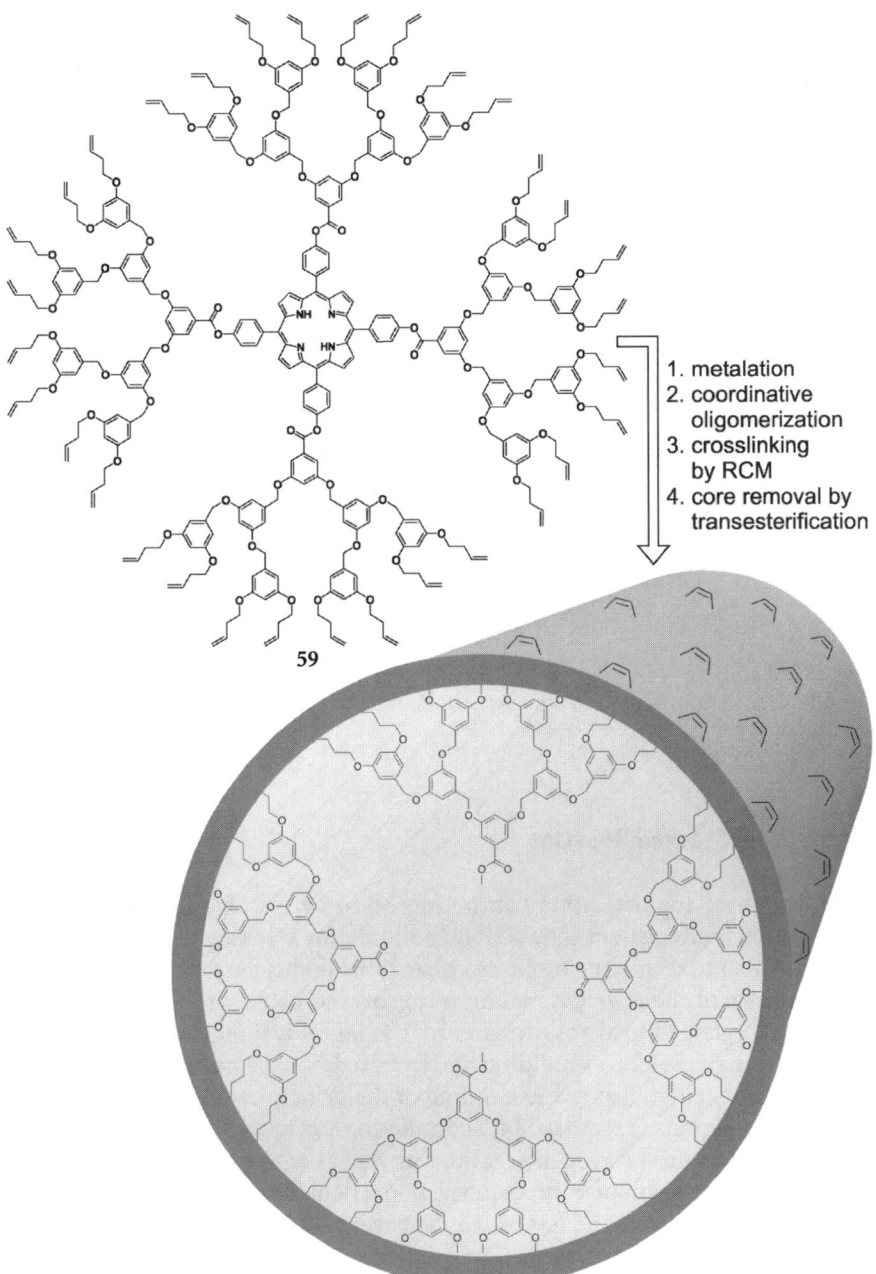

Fig. 38 Zimmerman's approach to organic nanotubes via cored dendronized polymers based on **59**

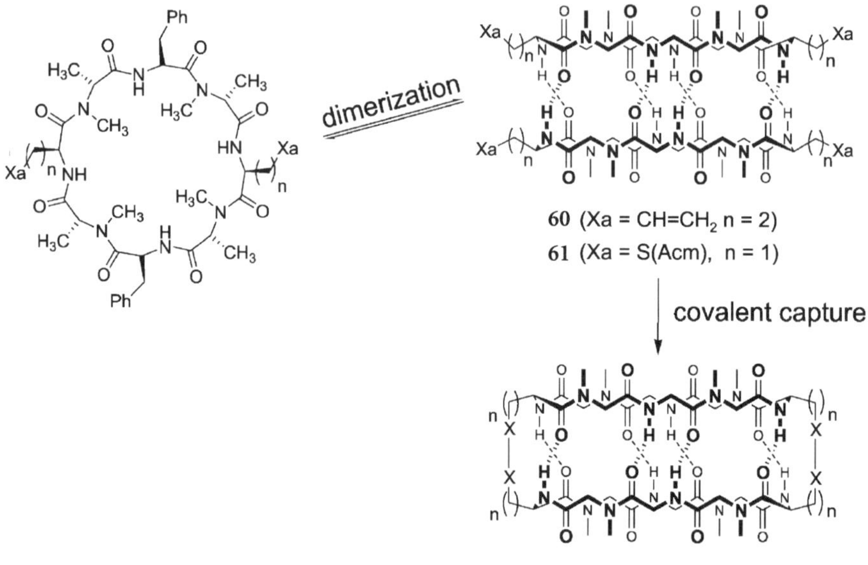

60 (Xa = CH=CH$_2$ n = 2)
61 (Xa = S(Acm), n = 1)

62 (X = CH)
63 (X = S)

Fig. 39 Ghadiri's covalent capture of defined cyclopeptide dimers **60** and **61** using RCM or disulfide formation. Reproduced in part with permission form reference [156]

6 Tubes from Stacked Rosettes

The degree of sophistication can be increased by orthogonal self-assembly, which uses non-covalent interactions not only for stacking along the tubular axis, but also to create the cyclic cross-sections themselves. These self-assembled mimics of the covalent macrocycles covered in the previous section are frequently referred to as rosettes. In order to take advantage of the 3D organization of small molecular building blocks into defined complex supramolecular structures, a detailed understanding of the principles of self-assembly and its key interactions is required [4, 7]. In the context of this section, H-bonding and aromatic π-π-stacking interactions as well as solvophobic and electrostatic effects are of particular importance. In particular, the H-bond is a common motif found in Nature and has been extensively explored for the generation of artificial rosette-like structures. The overview given in this section covers two important rosette families, focusing on assemblies in solution and liquid crystals. Related structures that are only known in the solid state [157] as well as coordination assemblies incorporating metal ions in the tube walls have been omitted [158].

6.1
G-Quartets and Related Motifs

One of the most studied examples of artificial rosettes involves guanine **64**, which is known to form tetrameric self-assembled disk-like structures, so-called G-quartets, around metal ions in water as well as in the LC solid states. Multiple H-bonds between the guanine units and additional metal coordination to the templating cations in the central cavity (0.45 nm diameter) render the assembly remarkably stable (Fig. 40) [159]. Lipophilic analogs, such as bis-decanoyl-deoxyguanosine **65** investigated by Gottarelli and coworkers, also assemble in organic solvents, taking advantage of the solvophobic effect (*vide supra*). The addition of alkaline ions (1/8 eq.) promotes stacking of these G-quartets into discrete octameric sandwich-like complexes with the metal coordinated to eight carbonyl-oxygens [160, 161]. Upon further addition of cations, the G-quartets continue to stack to form columnar polymeric aggregates. These supramolecular ion channels possess a stoichiometry of guanosine:cation=4:1 [162]. The introduction of bulky substituents in the 8-position, as in compound **66**, even allows assembly into stable quartets without the need

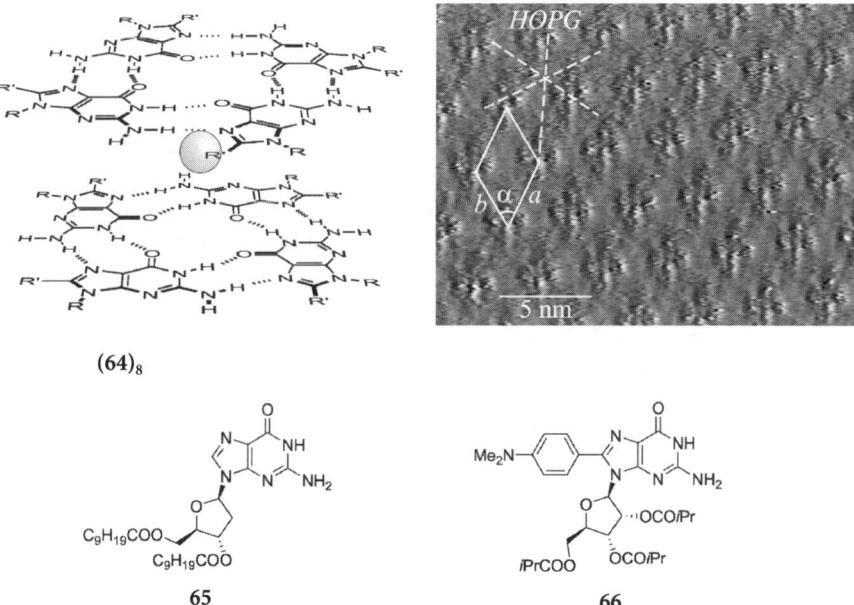

Fig. 40 Gottarelli's and Sessler's G-quartets based on guanine **64**. The first step towards columnar stacks is represented by two G-quartets intercalated by a potassium ion. The STM image shows a hexagonal array of 8-oxoguanosine G-quartets at the solid-liquid interface. Amphiphilic guanosine derivative **65** displaying solvophobicity-assisted assembly and bulky guanosine derivative **66** that assembles without cation mediation. Reproduced in part with permission from [164]. *HOPG* Highly ordered pyrolitic graphite

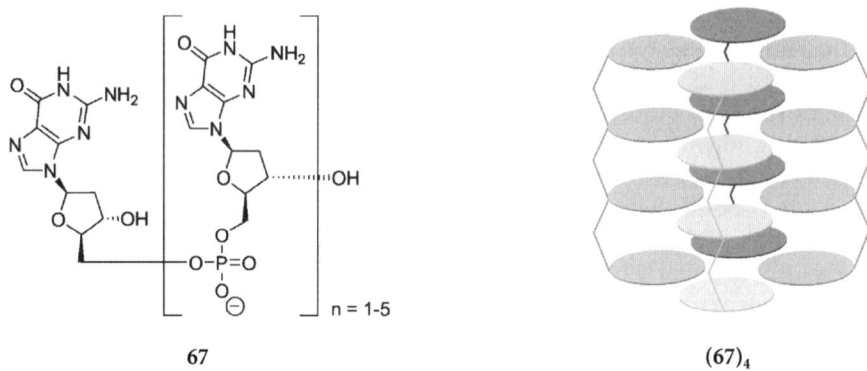

Fig. 41 Oligoguanosine **67** and cartoon of barrel-forming oligoguanosine **67** tetramer

of templating cations, whereas other assembling motifs become sterically unfavorable, as shown by Sessler [163]. Oxyguanosine derivatives that assemble in the absence of cations into tubular helical arrays in similar fashion as stacked G-quartets have also been reported recently [164].

Oligomers of deoxyguanosine phosphate also self-organize into columnar stacks forming LC phases. Deoxyguanosine oligomers **67** (dimer to hexamer) were shown to assemble into discrete tetrameric barrels, which can stack to afford columnar nanoobjects (Fig. 41) [165].

The guanine self-organization properties were exploited to assemble supramolecular tubes of calixarenes [166]. Here, the repeating unit is a calix[4]arene with four guanosine moieties attached to it in a 1,3-alternating fashion. The addition of cations triggered the assembly of G-quartets arranging their parent calix[4]arenes into ordered columns in a highly cooperative way.

Isoguanosines form even stronger G-quartets than guanine-based systems and they also display cation-mediated sandwich-type octamer formation [167]. However, in this case the resulting assemblies are not planar and thus columns cannot be obtained. Interestingly, cesium cations are capable of templating expanded pentameric discs that dimerize to sandwich-type complexes [168]. In the presence of photoreactive thymidine residues in synthetic DNA strands, such discotic pentamers could successfully be crosslinked by [2+2] photodimerization [169].

Retaining the double H-bonding motif of guanine and expanding its five-membered heterocycle to the homologue six-membered ring yields the pterine heterobicyclic system. In analogy to the guanine-based systems, folate **68** (Fig. 42) assembles into tetrameric planar discs that pile to form columns in solution as well as in the LC and solid phases [170]. In solution, the column length can be tuned by changing concentration and temperature as well as by adding salts, thereby yielding a maximum average length of 8.5 nm corresponding to about 25 piled discs. The column radii are around 1.45 nm and their inner voids

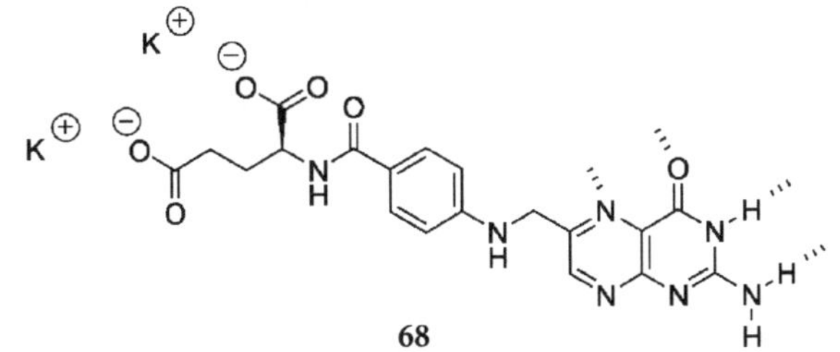

Fig. 42 Gottarelli's folate **68** as a guanine analogue with same array of H-bond-donors and acceptors and therefore displaying G-quartet formation and stacking

can host potassium and sodium ions as well as water molecules [171]. In related approaches, Zimmerman et al. extended the heteroannulene feature to pyridoquinoline [172] and 1,8,9-triazaanthracene [173] derivatives that assemble in solution via multiple H-bonding interactions to form either stable trimers or hexamers; however, no stacking of these rosettes has been reported.

6.2
Melamine–Cyanuric Acid Motifs

The cyclic melamine–cyanuric acid H-bonding motif was introduced by Whitesides and coworkers and intensively applied to the construction of supramolecular rods of stacked rosette-like assemblies with inner diameters of 5.9 Å [174]. Alternative assembly modes giving rise to ill-defined linear aggregates were excluded by controlling the steric crowding at the periphery of the subunits. Using complementary building blocks of tongs-shaped bisisocyanurates **69** and bismelamines **70**, self-assembled cylinders with a diameter of 5 nm are spontaneously formed in organic solvents [175]. Further aggregation yields the corresponding cylinder bundles with lengths up to 1.5 μm (3,000 stacked rosettes). The design takes advantage of solvophobically driven, H-bonding-directed self-assembly between complementary tongs affording supramolecular rosettes that are covalently linked in a face-to-face fashion by the tongs' spacers (Fig. 43). An important feature involves the use of different spacer lengths in both complementary tongs in order to rule out a cage-type assembly. The same design concept was used by Timmerman, Reinhoudt et al. for generating similar nanostructures of orthogonal calix[4]arene dimelamine and calix[4]arene dicyanurate building blocks [176].

By replacing the cyanuric acid with an elongated and linear building block such as naphthalenedicarboximide, a rosette with enlarged dimensions can be obtained as shown by the work of Kunitake (Fig. 44, top) et al. [177]. The assembly of amphiphilic dialkylmelamine **71** with the complementary diimide **72** leads to the formation of columns that can further aggregate to yield fibers of micrometer length. Whether the individual columns having inner and outer diameters of 4 and 8 nm, respectively, consist of stacked rosettes or helical tapes

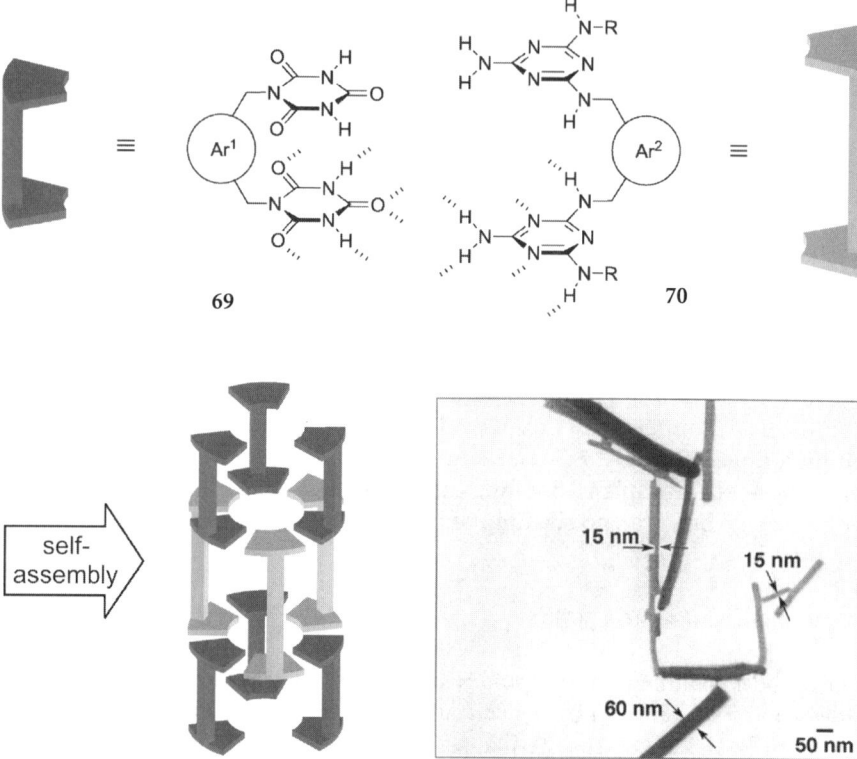

Fig. 43 Whitesides' and Reinhoudt's approaches to stacked rosettes based on the melamine-cyanuric acid cyclic H-bonding array. Tong-shaped building blocks **69** and **70** co-assemble into cage-like substructures forming tubular stacks (Ar^1, Ar^2=benzene, furane, naphthalene, calix[4]arene). *Bottom right* is a TEM image of bundled aggregates formed from tubular assemblies. Reproduced in part with permission from [175]

remains unclear. In non-polar solvents, structure formation is driven by solvophobic as well as aromatic π,π-stacking interactions and directed by the H-bonding array.

Fenniri and coworkers designed a self-complementary heteroaromatic system **73** containing guanine and cytosine H-bonding arrays [178]. The relative orientation of the orthogonal binding sites leads to formation of hexagonal rosettes with an inner diameter of 1 nm (Fig. 44, bottom). By attaching chiral side chains at the periphery, supramolecular chirality was induced in the columns consisting of stacked rosettes [179]. Tuning and switching of chirality could be realized by binding chiral analytes to crown ethers covalently attached to the rosette's periphery.

Very recently, Meijer and coworkers reported on self-assembled nanotubes in solution consisting of chiral stacks of π-conjugated rosettes [180]. Wedge-shaped non-polar oligo(*para*-phenylenevinylene)s **74** and **75** having a di-

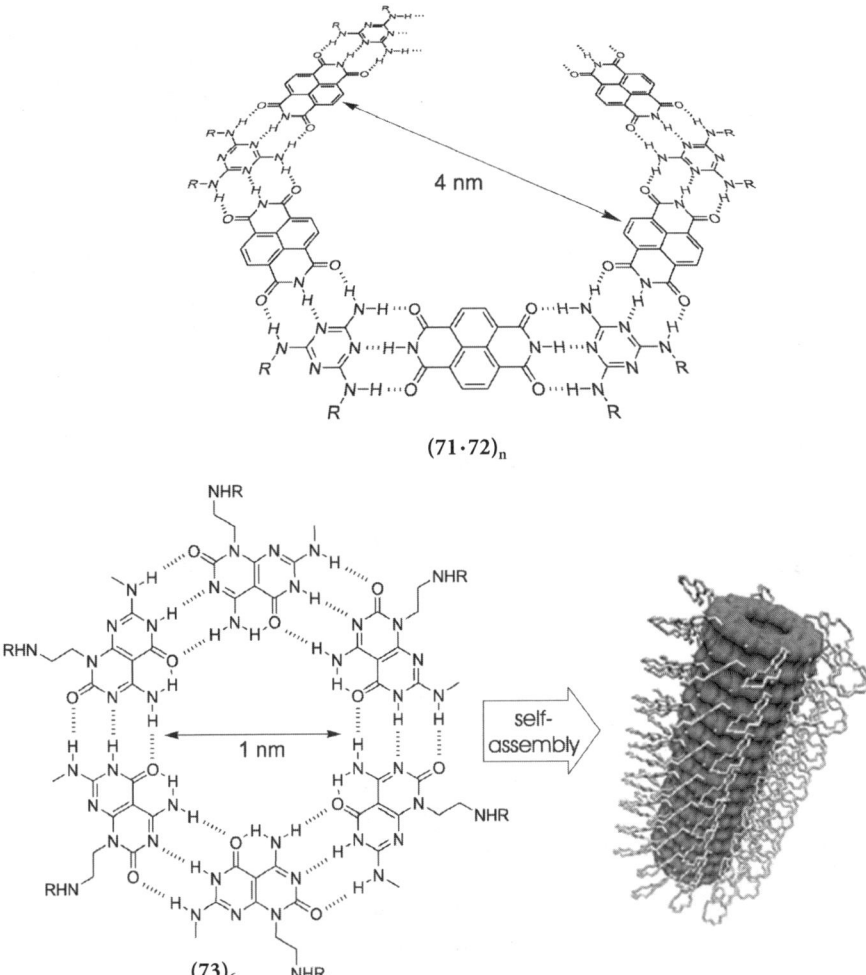

Fig. 44 Kunitake's binary melamine-diimide system assembling to hollow cylinders via cyclic or helical assembling modes (*top*). Fenniri's rosette-forming pyrimido-pyrimidinedione **73**, and 3D model of nanotube consisting of piled rosettes (*bottom*). Reproduced in part with permission from [179]

aminotriazine head group are able to form chiral cyclic hexamers with an inner void of 7 Å as visualized by STM (Fig. 45). Subsequent aggregation of the rosette subunits in non-polar solvents yields columns with lengths exceeding 185 nm in solution and 10 µm when cast on a solid substrate.

Finally, the classical H-bonding carboxylic acid dimer motif has been exploited by Hamilton and coworkers, who demonstrated the cyclic assembly of various isophthalic acid derivatives. Depending on the bulky substituent in the *meta*-position, individual or stacked rosettes were obtained in solution and in the crystal, respectively [181].

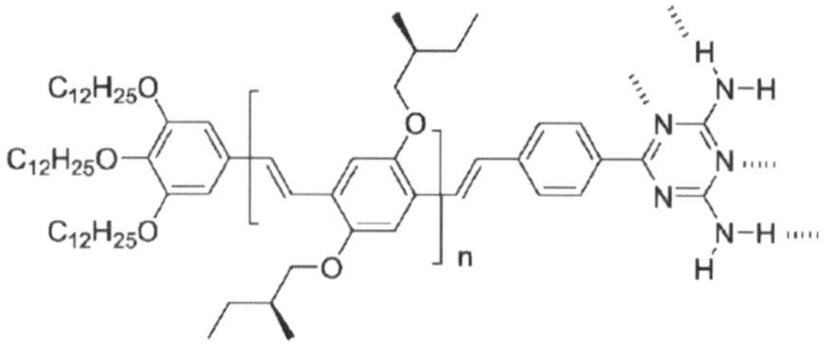

74 (n = 1)
75 (n = 2)

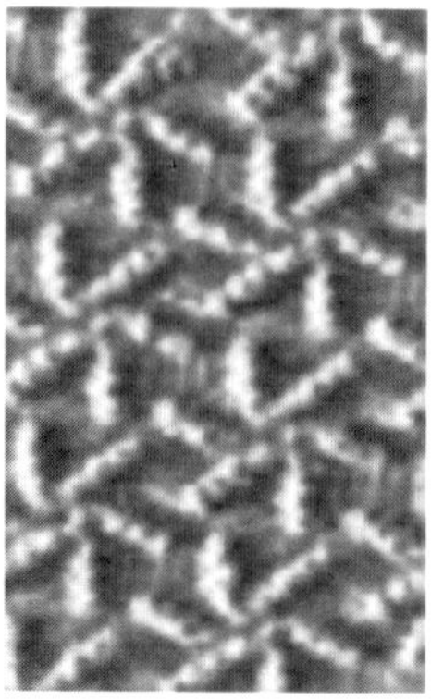

Fig. 45 Meijer's oligo(*para*-phenylenevinylene)s capable of assembling into rosettes and subsequently into tubes. STM image of 75 shows a monolayer of chiral hexameric rosettes. Reproduced in part with permission from [180]

7
Tubes from Self-Assembled Amphiphiles

Ultimately, directed non-covalent interactions and molecular rigidity/geometric constraints can be removed from the self-assembling building blocks. The resulting systems organize into higher aggregates based on their amphiphilicity and the proportions, i.e., aspect ratio, of their individual components [182]. In Nature, most cylindrical structures are not entirely founded on segregation of immiscible or solvophilic/solvophobic components but involve additional interactions. The tobacco mosaic virus (TMV) represents perhaps the most impressive and best illustration of this design concept [183]. In this case, the self-assembly of taper-shaped protein building blocks templated by viral RNA affords a cylindrical structure of controlled length, in which the subunits are held together by various non-covalent bonds. Artificial systems have been thoroughly investigated in an attempt to relate the different observed aggregation modes in solution, in the LC state, and in bulk to the amphiphile's structure. Components can vary from simple detergents [184] to complex (bio)macromolecular amphiphiles and block copolymers [185, 186]. In this section, three tubular design concepts are discussed in relations to the shape of the amphiphile. Emphasis is mostly put on the solution and LC phases.

7.1
Wedge-Shaped Amphiphiles

Most common biological amphiphiles are lipids composed of long hydrocarbon tails and polar and/or ionic head groups. The aspect ratio of these building blocks, as well as their miscibility properties, defines the assembly structure [182]. The more pronounced the volume difference of head to tail becomes, the more curvature is introduced, and the more cyclic structures are realized.

Discrete Organic Nanotubes

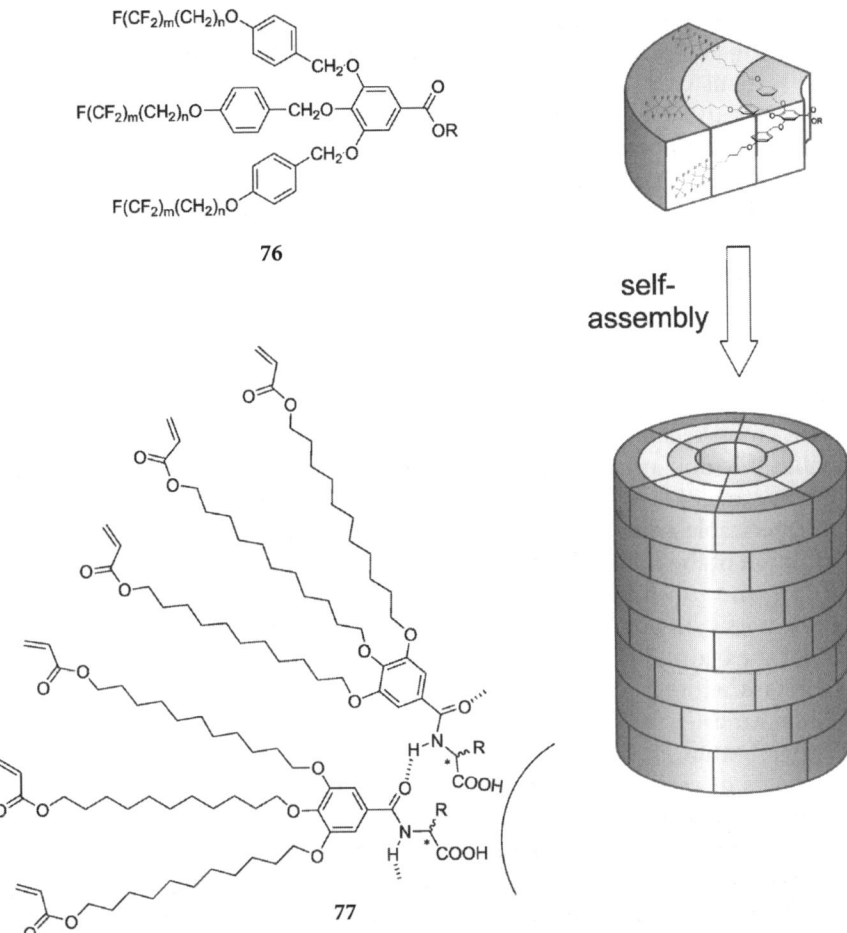

Fig. 46 Percec's wedge-shaped amphiphilic gallate **76** (*top*), cartoon illustrating assembly of amphiphile into layered tubes; Gin's analogue gallate **77** with H-bond-stabilized inner core and crosslinkable periphery (*bottom*)

Wedge-, taper-, or sector-shaped amphiphiles can readily form cylindrical shapes in solution or suspension and inverted hexagonal phases with tubular channels in the LC state. Polar head groups form the inner periphery of tubes with the central channel filled with polar solvent molecules, while the disordered non-polar tails form the outer periphery and isolate the tubes from each other. Inspired by the structural organization found in TMV, Percec and his group designed flat taper-shaped dendritic amphiphiles **76**, based on gallic acid, which form five- to six-membered discotic assemblies, which further stack into columns in solution and their corresponding columnar LC phase (Fig. 46, top) [187]. The carboxylic acid function, which can also be esterified or ionized, serves as the hydrophilic head at the interior of the assembly, whereas long linear chains, i. e., hydrocarbons or fluorocarbons, attached to a small aromatic branching unit generate bulkiness and hydrophobicity at the exterior. While the first systems only assembled via ion-mediated complexation processes [188, 189], wedge-shaped amphiphiles with semifluorinated tails also assemble solely via the fluorophobic effect [190]. In this case, the segregation of flexible perfluorinated and perhydrogenated segments forms a deformable sheath of onion-like layers within the cylindrical assembly with a rigid aromatic interior surrounding the hollow core. Upon casting on a surface, the columns further arrange into an inverted hexagonal LC phase with infinite column lengths and total column diameters reaching 6 nm [191]. Several approaches were shown to stabilize these structures. When the inner carboxylic acid functions are esterified with ion receptors such as crown ethers or ethylene glycols, lithium or sodium ions can template the self-assembly [192]. In an intriguing report, the same group recently demonstrated the packing of organic conducting units within the tubular cavity to create insulated molecular wires via this self-assembly strategy [193]. In another case, the interior has been functionalized with polymerizable groups and subsequent polymerization has led to covalent stabilization of the tubular assembly at the cost of inner void volume [194].

Gin and coworkers designed analogous taper-shaped gallic acid salt monomers with polymerizable olefin, diolefin, or acrylate units at the periphery of the hydrophobic tails allowing for crosslinking the inverted hexagonal LC phase without affecting the column's inner void [195]. Introduction of H-bond motifs, as in compound 77, further stabilized the columnar phases (Fig. 46, bottom) [196]. By the introduction of functional units such as emitting polymers or catalytically active sites, ordered nanocomposite materials have been obtained [195].

A similar crosslinkable amphiphile forming an inverted hexagonal LC phase was recently designed by Kim and coworkers [197]. The wedge-shape is realized by a polar G-2 amide dendron with a focal carboxylic acid group. Peripheral, linear hydrocarbon chains with incorporated diyne moieties control the segregation behavior of these molecules and allow for subsequent topochemical crosslinking.

7.2
Linear Amphiphiles and Block Copolymers

A radical simplification of the design of tubular structures is achieved by the use of linear amphiphiles. The ease of their synthesis and frequently commercial availability is generally counterbalanced by the limited predictability of the self-assembling process due to the high flexibility and ill-defined shape of the building blocks. Two principal aggregation modes that lead to tubular architectures involve rolled lamellar sheets and cylindrical micelles. Formation of helical fibers, coiled ribbons, and tubules is a common motif in the assembly of lamellar sheets among lipids, with changes in concentration, solvent, temperature, and addition of salts being common stimuli to induce the formation of desired structural features. The various formed architectures can be covalently stabilized by means of a crosslinking polymerization within the assembly [198, 199]. In view of their simple and large-scale preparation, such self-assembled nanotubes represent interesting candidates for potential applications in the life and materials sciences [200, 201].

Common amphiphile structures include: bolaamphiphiles [202], N-acylated amino acids [203], nucleic acid derivatives [204], double-chain ammonium salts 78 [205], unsaturated double-chain phosphatidylethanolamines 80 [206] and related diacetylenic phosphatidylcholines [207], (fluorinated) anionic glucophospholipids [208], and aldonamides or gluconamides [209, 210] (Fig. 47). Polymerizable diacetylenic aldonamides 82 assemble to structures of helical, tightly coiled bilayer sheets forming hollow and open-ended tubes (Fig. 47, bottom). Depending on the exact position of the diacetylenic function and the assembling conditions, tubes having lengths of several micrometers, as well as average inner and outer diameters of 8–10 nm and 50–70 nm, respectively, have been prepared [211]. In one case, outer diameters of up to 300 nm have been

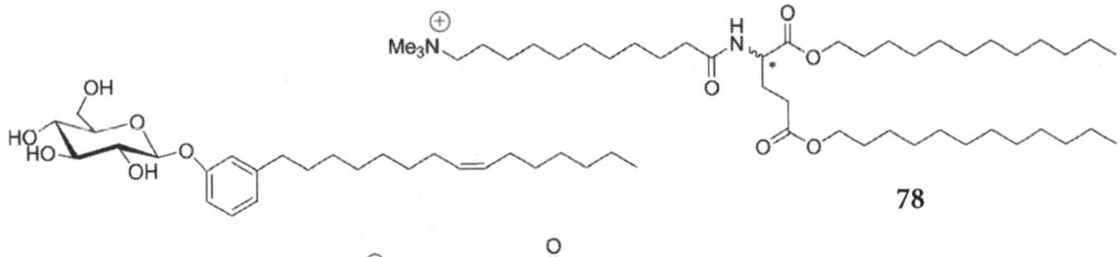

Fig. 47 Selected linear amphiphiles assembling into tubes (*top*): double-chain ammonium amphiphile (**78**), aromatic glycolipid (**79**), unsaturated double-chain phosphatidylethanolamine (**80**), H-bond enhanced bolaamphiphile (**81**); diacetylenic aldonamide (**82**) and TEM of polymerized assembly of the aldonamide as helical coiled bilayer and tightly coiled to a tubule (*bottom*). Reproduced in part with permission from [212]

realized [212]. These structures were covalently stabilized by (incomplete) topotactic polymerization of the ordered diacetylenic units forming polydiacetylene networks throughout the tube [210, 212]. Higher degrees of crosslinking and thus higher tubular stabilities were achieved with phosphatidylcholines by incorporating additional polymerizable vinyl or methacrylate groups at the chain ends [213]. All these stabilization approaches are restricted to solutions or suspensions, since crosslinking in the LC phase takes place in a highly intertubular fashion, yielding insoluble porous bulk material [206].

Fuhrhop and coworkers demonstrated the enhanced stability of tubular structures through the introduction of additional H-bonding-motifs into bolaamphiphiles [214]. Unsymmetric amphiphiles of hydrocarbon chains with amino and amino acid functionalities at the termini and amide functionalities within the chain, for instance in **81**, were shown to form suspensions in water consisting of open-ended tubular vesicles of monolayered membranes with an inner diameter of 50 nm and micrometer lengths. Unsaturated glycolipids of the type **79** were shown to assemble via a H-bonding array between sugar moieties, π,π-interactions between phenyl groups, and hydrophobic forces between hydrocarbon tails [215]. Upon cooling of hot aqueous solutions containing **79**, aggregation to tubular open-ended structures occurred giving inner diameters of 10–15 nm, lengths up to 100 µm, and a membrane thickness of 8–15 nm, corresponding to two to four lipid-interdigitated bilayers. The formation of multilamellar tube-like structures using either diacetylenic phospholipids [216] or complexes of filamentous actin with cationic lipids in the form of ammonium salts [217] has also been reported.

Huge tubular structures have been achieved by the introduction of di- and triblock copolymers serving as amphiphiles [185, 218]. The introduction of single-backbone rod–coil copolymers allowed for microphase separation, and subsequent controlled crosslinking of the resulting ordered two or three dimensional oligomer assemblies, such as "2D polymers" [219], anisotropic mushrooms [220], nanofibers [221], and ribbons [222] was reported by Stupp and coworkers. Related two-dimensional polymers based on crosslinked sheets of self-assembled ionic amphiphiles were described by Kunitake et al. [223].

Diblock copolymer amphiphiles such as polystyrene-*block*-poly(ethyleneoxide) [224] or poly(ethyleneoxide)-*block*-poly(ethylethylene) [225] were observed to form tubular bilayer aggregates in solution under appropriate conditions. Compared to small molecule assemblies, these systems display increased thermal stabilities but possess closed termini resembling cylindrical micelles. Additional stabilization by internal crosslinking of similar poly(butadiene)-*block*-poly(ethyleneoxide) copolymers forming worm-like micelles in solution was reported by Bates and his group [226]. However, the structure generated was not hollow.

A sophisticated approach towards hollow, open-ended tubes was reported by Stewart and Liu (Fig. 48) [227]. Polymers **83** composed of polyisoprene, poly(2-cinnamoylethyl methacrylate) (PCMA), and poly(*tert*-butyl acrylate) (P*t*BA)

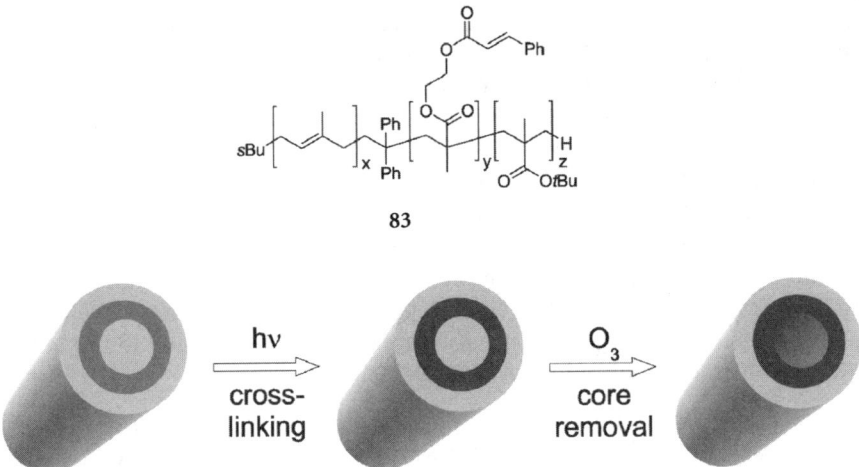

Fig. 48 Stewart's and Liu's approach to polymeric nanotubes based on block copolymer assembly followed by photocrosslinking of the poly(2-cinnamoylethyl methacrylate) (PCMA) shell and removal of the polyisoprene core by ozonolysis

blocks were demonstrated to form cylindrical micelles in solution. The ratio of the blocks determines the size and shape of the assemblies. Ratios of approximately 1:1:6 (130:130:800 repeating units) lead to cylindrical micelles having dimensions of 79 nm in diameter and a few micrometers in length with a wide length distribution. Polyisoprene formed the core surrounded by a PCMA shell and a P*t*BA corona. Photocrosslinking of the methacrylate shell was performed up to a conversion of 50% under retention of the cylindrical structure while the P*t*BA prevented intertubular crosslinking. Subsequent removal of the polyisoprene core by ozonolysis, referred to by the authors as "nanosculpturing", afforded a hollow polyacrylate tube that could be filled with guests demonstrating the opened termini.

Very recently, the assembly of amphiphilic hyperbranched multi-arm copolymers into stable and soluble tubes was reported [228]. Apparently, only a certain fraction of the polydisperse and ill-defined amphiphiles aggregates to multi-walled tubes of centimeter lengths and millimeter diameters visible to the naked eye.

8
Conclusions and Outlook

The tremendous progress that has been made in the research area outlined above has been sparked and guided by many recent advances in both covalent and non-covalent syntheses. On the one hand, the preparation of many new oligomer series necessary to deduce structure–property relationships has been facilitated by rapid-growth approaches and the use of polymeric supports and reagents/catalysts. On the other hand, supramolecular science has evolved and grasped a much deeper understanding of the fundamental principles of self-assembly. The advent of dynamic covalent chemistry taking advantage of both areas and combining them with combinatorial methodology exemplifies perhaps the most powerful recent development. Last but not least, the availability of sophisticated analytical tools, most notably scanning probe techniques, has had a significant impact on the chemical sciences in general.

However, past accomplishments cannot eclipse the many related problems that await their solutions. Besides the general quest to further develop the above-mentioned disciplines, specific future challenges related to the development of functional organic nanotubes include:

1. Creating hollow tubes by developing strategies to overcome attractive van der Waals forces, and create tubes with an exploitable interior that can be tuned by means of postfunctionalization.
2. Increasing the aspect ratio by developing helical backbones that can be prepared in a living polymerization process and generating multi-stranded polymeric helices with an inner void.
3. Controlling the growth in a self-assembly process by incorporating subtle structural distortions or utilizing templates in order to achieve length control.
4. Stabilizing individual nanotubes by developing suitable crosslinking reactions that involve negligible geometrical changes but provide significantly enhanced robustness.
5. Integrating the tubular nanoobjects by incorporating them into larger hierarchical structures and connecting them to solid substrate surfaces.
6. Developing beneficial applications in nanofabrication, smart materials, and biotechnology.

The last point seems to be of particular importance, since a continuously dynamic development of the field will only occur if advantageous properties of the nanotubes are revealed and emerging technological applications provide the necessary driving force. Hopefully, this review has – at least to some extent – risen to the expectations of the expert and – perhaps more importantly – generated interest in entering and contributing to this fascinating field at the interface of chemistry, physics, and biology.

Acknowledgments Generous support by the Sofja Kovalevskaja Award of the Alexander von Humboldt Foundation, endowed by the Federal Ministry of Education and Research (BMBF) within the Program for Investment in the Future (ZIP) of the German Government, is gratefully acknowledged. MABB thanks the Studienstiftung des Deutschen Volkes for providing a doctoral fellowship.

References

1. For a recent comment on Moore's law, see: Marsh G (2003) Mater Today 6:28
2. Special Issue "Nanotechnology" (2001) Sci Am 285:32
3. Love JC, Whitesides GM (2001) Sci Am 285:39
4. Lehn J-M (1995) Supramolecular chemistry: concepts and perspectives. VCH, Weinheim
5. Hecht S (2003) Angew Chem Int Ed 42:24
6. For rod-like modules, see: Schwab PFH, Levin MD, Michl J (1999) Chem Rev 99:1863
7. Special Issue "Supramolecular Chemistry & Self-Assembly" (2002) Science 295:2395 (and references therein)
8. For example: Yazdani A, Lieber CM (1999) Nature 401:227
9. For an excellent and inspiring review, consult: Bong DT, Clark TD, Granja JR, Ghadiri MR (2001) Angew Chem Int Ed 40:988
10. Clark CGJ, Wooley KL (2001) Regioselectively-crosslinked nanostructures. In: Fréchet JMJ, Tomalia DA (eds) Dendrimers and other Dendritic Polymers. Wiley, p 148
11. Special Issue "Carbon Nanotubes" (2001) Top Appl Phys 80:1
12. Special Issue "Carbon Nanotubes" (2002) Acc Chem Res 35:997
13. Xia Y, Yang P, Sun Y, Wu Y, Mayers B, Gates B, Yin Y, Kim F, Yan H (2003) Adv Mater 15:353
14. Steinhart M, Wendorff JH, Greiner A, Wehrspohn RB, Nielsch K, Schilling J, Choi J, Gösele U (2002) Science 296:1997
15. Bognitzki M, Hou H, Ishaque M, Frese T, Hellwig M, Schwarte C, Schaper A, Wendorff JH, Greiner A (2000) Adv Mater 12:637
16. Collins PG, Avouris P (2000) Sci Am 283:62
17. Hirsch A (2002) Angew Chem Int Ed 41:1853
18. See also chapter 5 of this volume
19. Sattler M, Liang H, Nettesheim D, Meadows RP, Harlan JE, Eberstadt M, Yoon HS, Shuker SB, Chang BS, Minn AJ, Thompson CB, Fesik SW (1997) Science 275:983
20. Ketcham RR, Hu W, Cross TA (1993) Science 261:1457
21. Jackson DS, Grant ME (1974) Nature 249:406
22. McDermott G, Prince SM, Freer AA, Hawthornthwaite-Lawless AM, Papiz MZ, Cogdell RJ, Isaacs NW (1995) Nature 374:517
23. Nakano T, Okamoto Y (2001) Chem Rev 101:4013
24. Hill DJ, Mio MJ, Prince RB, Hughes TS, Moore JS (2001) Chem Rev 101:3893
25. Urbano A (2003) Angew Chem Int Ed 42:3986
26. Meurer KP, Voegtle F (1985) Top Curr Chem 127:1
27. Here, we define the inner cavity by the *empty* space within a tubular structure. The inner diameter has been calculated from X-ray crystallographic or modeling data by measuring the distance between opposing atoms at the inner rim, correcting for the helical pitch, and subtracting the corresponding van der Waals radii
28. Han S, Bond AD, Disch RL, Holmes D, Schulman JM, Teat SJ, Vollhardt KPC, Whitener GD (2002) Angew Chem Int Ed 41:3223
29. Han S, Anderson DR, Bond AD, Chu HV, Disch RL, Holmes D, Schulman JM, Teat SJ, Vollhardt KPC, Whitener GD (2002) Angew Chem Int Ed 41:3227
30. De Santis P, Morosetti S, Rizzo R (1974) Macromolecules 7:52
31. Huc I (2004) Eur J Org Chem 1:17
32. Hamuro Y, Geib SJ, Hamilton AD (1994) Angew Chem Int Ed Engl 33:446
33. Hamuro Y, Geib SJ, Hamilton AD (1996) J Am Chem Soc 118:7529
34. Hamuro Y, Geib SJ, Hamilton AD (1997) J Am Chem Soc 119:10587
35. Zhu J, Parra RD, Zeng H, Skrzypczak-Jankun E, Zeng XC, Gong B (2000) J Am Chem Soc 122:4219

36. Gong B, Zeng H, Zhu J, Yuan L, Han Y, Cheng S, Furukawa M, Parra RD, Kovalevsky AY, Mills JL, Skrzypczak-Jankun E, Martinovic S, Smith RD, Zheng C, Szyperski T, Zeng XC (2002) Proc Natl Acad Sci 99:11583
37. Gong B (2001) Chem Eur J 7:4336
38. Cary JM, Moore JS (2002) Org Lett 4:4663
39. Yang X, Brown AL, Furukawa M, Li S, Gardinier WE, Bukowski EJ, Bright FV, Zheng C, Zeng XC, Gong B (2003) Chem Commun 56
40. Berl V, Huc I, Khoury RG, Krische MJ, Lehn J-M (2000) Nature 407:720
41. Berl V, Huc I, Khoury RG, Lehn J-M (2001) Chem Eur J 7:2798
42. Berl V, Huc I, Khoury RG, Lehn J-M (2001) Chem Eur J 7:2810
43. Dolain C, Maurizot V, Huc I (2003) Angew Chem Int Ed 42:2738
44. Kolomiets E, Berl V, Odriozola I, Stadler A-M, Kyritsakas N, Lehn J-M (2003) Chem Commun 2868
45. Jiang H, Leger J-M, Huc I (2003) J Am Chem Soc 125:3448
46. Berl V, Krische MJ, Huc I, Lehn J-M, Schmutz M (2000) Chem Eur J 6:1938
47. van Gorp JJ, Vekemans JAJM, Meijer EW (2004) Chem Commun 60
48. Hanan GS, Lehn J-M, Kyritsakas N, Fischer J (1995) J Chem Soc, Chem Commun 765
49. Hanan GS, Schubert US, Volkmer D, Riviere E, Lehn J-M, Kyritsakas N, Fischer J (1997) Can J Chem 75:169
50. Bassani DM, Lehn J-M (1997) Bull Soc Chim Fr 134:897
51. Bassani DM, Lehn J-M, Baum G, Fenske D (1997) Angew Chem Int Ed Engl 36:1845
52. Ohkita M, Lehn J-M, Baum G, Fenske D (1999) Chem Eur J 5:3471
53. Barboiu M, Lehn J-M (2002) Proc Natl Acad Sci 99:5201
54. Barboiu M, Vaughan G, Kyritsakas N, Lehn J-M (2003) Chem Eur J 9:763
55. Gardinier KM, Khoury RG, Lehn J-M (2000) Chem Eur J 6:4124
56. Schmitt J-L, Stadler A-M, Kyritsakas N, Lehn J-M (2003) Helv Chim Acta 86:1598
57. Cuccia LA, Lehn J-M, Homo J-C, Schmutz M (2000) Angew Chem Int Ed 39:233
58. Petitjean A, Cuccia LA, Lehn J-M, Nierengarten H, Schmutz M (2002) Angew Chem Int Ed 41:1195
59. Nelson JC, Saven JG, Moore JS, Wolynes PG (1997) Science 277:1793
60. Matsuda K, Stone MT, Moore JS (2002) J Am Chem Soc 124:11836
61. Hunter CA, Lawson KR, Perkins J, Urch CJ (2001) J Chem Soc Perkin Trans 2:651
62. Meyer EA, Castellano RK, Diederich F (2003) Angew Chem Int Ed 42:1210
63. Lahiri S, Thompson JL, Moore JS (2000) J Am Chem Soc 122:11315
64. Brunsveld L, Prince RB, Meijer EW, Moore JS (2000) Org Lett 2:1525
65. Hill DJ, Moore JS (2002) Proc Natl Acad Sci 99:5053
66. Prince RB, Saven JG, Wolynes PG, Moore JS (1999) J Am Chem Soc 121:3114
67. Zimm BH, Bragg JK (1959) J Chem Phys 31:526
68. Gin MS, Yokozawa T, Prince RB, Moore JS (1999) J Am Chem Soc 121:2643
69. Gin MS, Moore JS (2000) Org Lett 2:135
70. Prince RB, Brunsveld L, Meijer EW, Moore JS (2000) Angew Chem Int Ed 39:228
71. Prince RB, Moore JS, Brunsveld L, Meijer EW (2001) Chem Eur J 7:4150
72. Prince RB, Okada T, Moore JS (1999) Angew Chem Int Ed 38:233
73. Prince RB, Barnes SA, Moore JS (2000) J Am Chem Soc 122:2758
74. Tanatani A, Mio MJ, Moore JS (2001) J Am Chem Soc 123:1792
75. Tanatani A, Hughes TS, Moore JS (2002) Angew Chem Int Ed 41:325
76. Rowan SJ, Cantrill SJ, Cousins GRL, Sanders JKM, Stoddart JF (2002) Angew Chem Int Ed 41:899
77. Nishinaga T, Tanatani A, Oh K, Moore JS (2002) J Am Chem Soc 124:5934
78. Zhao D, Moore JS (2003) Org Biomol Chem 1:3471
79. Oh K, Jeong K-S, Moore JS (2001) Nature 414:889

80. Oh K, Jeong K-S, Moore JS (2003) J Org Chem 68:8397
81. Zhao D, Moore JS (2003) Macromolecules 36:2712
82. Zhao D, Moore JS (2003) J Am Chem Soc 125:16294
83. Wooley KL (2000) J Polym Sci, Part A: Polym Chem 38:1397
84. Blackwell HE, Grubbs RH (1998) Angew Chem Int Ed 37:3281
85. Blackwell HE, Sadowsky JD, Howard RJ, Sampson JN, Chao JA, Steinmetz WE, O'Leary DJ, Grubbs RH (2001) J Org Chem 66:5291
86. Kumita JR, Smart OS, Woolley GA (2000) Proc Natl Acad Sci 97:3803
87. Flint DG, Kumita JR, Smart OS, Woolley GA (2002) Chem Biol 9:391
88. Hecht S, Khan A (2003) Angew Chem Int Ed 42:6021
89. Khan A, Hecht S (2004) Chem Commun 300
90. Cornelissen JJLM, Rowan AE, Nolte RJM, Sommerdijk NAJM (2001) Chem Rev 101:4039
91. Langs DA (1988) Science 241:188
92. Piguet C, Bernardinelli G, Hopfgartner G (1997) Chem Rev 97:2005
93. Williams A (1997) Chem Eur J 3:15
94. Albrecht M (2001) Chem Rev 101:3457
95. Prest P-J, Prince RB, Moore JS (1999) J Am Chem Soc 121:5933
96. Mio MJ, Prince RB, Moore JS, Kübel C, Martin DC (2000) J Am Chem Soc 122:6134
97. Kübel C, Mio MJ, Moore JS, Martin DC (2002) J Am Chem Soc 124:8605
98. Presnell SR, Cohen FE (1989) Proc Natl Acad Sci 86:6592
99. Doyle DA, Morais Cabral J, Pfuetzner RA, Kuo A, Gulbis JM, Cohen SL, Chait BT, MacKinnon R (1998) Science 280:69
100. Schafmeister CE, Miercke LJW, Stroud RM (1993) Science 262:734
101. Wyman TB, Nicol F, Zelphati O, Scaria PV, Plank C, Szoka FC Jr (1997) Biochemistry 36:3008
102. Matile S (2001) Chem Soc Rev 30:158
103. Flower DR (1996) Biochem J 318:1
104. Nagano N, Hutchinson EG, Thornton JM (1999) Protein Sci 8:2072
105. Song L, Hobaugh MR, Shustak C, Cheley S, Bayley H, Gouaux JE (1996) Science 274:1859
106. Sakai N, Matile S (2003) Chem Commun 2514
107. Baumeister B, Sakai N, Matile S (2000) Angew Chem Int Ed 39:1955
108. Das G, Talukdar P, Matile S (2002) Science 298:1600
109. Höger S (1999) J Polym Sci, Part A: Polym Chem 37:2685
110. Zhao D, Moore JS (2003) Chem Commun 807
111. Grave C, Schlüter AD (2002) Eur J Org Chem 3075
112. Ghadiri MR, Granja JR, Milligan RA, McRee DE, Khazanovich N (1993) Nature 366:324
113. Hartgerink JD, Clark TD, Ghadiri MR (1998) Chem Eur J 4:1367
114. Granja JR, Ghadiri MR (1994) J Am Chem Soc 116:10785
115. Khazanovich N, Granja JR, McRee DE, Milligan RA, Ghadiri MR (1994) J Am Chem Soc 116:6011
116. Ghadiri MR, Granja JR, Buehler LK (1994) Nature 369:301
117. Horne WS, Stout CD, Ghadiri MR (2003) J Am Chem Soc 125:9372
118. Seebach D, Matthews JL, Meden A, Wessels T, Baerlocher C, McCusker LB (1997) Helv Chim Acta 80:173
119. Clark TD, Buehler LK, Ghadiri MR (1998) J Am Chem Soc 120:651
120. Clark TD, Buriak JM, Kobayashi K, Isler MP, McRee DE, Ghadiri MR (1998) J Am Chem Soc 120:8949
121. Ranganathan D, Samant MP, Karle IL (2001) J Am Chem Soc 123:5619
122. Ranganathan D, Haridas V, Gilardi R, Karle IL (1998) J Am Chem Soc 120:10793
123. Gauthier D, Baillargeon P, Drouin M, Dory YL (2001) Angew Chem Int Ed 40:4635
124. Ranganathan D (2001) Acc Chem Res 34:919

125. Ashton PR, Brown CL, Menzer S, Nepogodiev SA, Stoddart JF, Williams DJ (1996) Chem Eur J 2:580
126. Ashton PR, Cantrill SJ, Gattuso G, Menzer S, Nepogodiev SA, Shipway AN, Stoddart JF, Williams DJ (1997) Chem Eur J 3:1299
127. Gattuso G, Menzer S, Nepogodiev SA, Stoddart JF, Williams DJ (1997) Angew Chem Int Ed Engl 36:1451
128. Grave C, Lentz D, Schäfer A, Samori P, Rabe JP, Franke P, Schlüter AD (2003) J Am Chem Soc 125:6907
129. Müller P, Uson I, Hensel V, Schlüter AD, Sheldrick GM (2001) Helv Chim Acta 84:778
130. Young JK, Moore JS (1995) In: Stang PJ, Diederich F (eds) Modern Acetylene Chemistry. VCH, Weinheim, Germany, p 415
131. Moore JS (1997) Acc Chem Res 30:402
132. Venkataraman D, Lee S, Zhang J, Moore JS (1994) Nature 371:591
133. Ge P-H, Fu W, Herrmann WA, Herdtweck E, Campana C, Adams RD, Bunz UHF (2000) Angew Chem Int Ed 39:3607
134. Hosokawa Y, Kawase T, Oda M (2001) Chem Commun 1948
135. Höger S, Morrison DL, Enkelmann V (2002) J Am Chem Soc 124:6734
136. Mindyuk OY, Stetzer MR, Gidalevitz D, Heiney PA, Nelson JC, Moore JS (1999) Langmuir 15:6897
137. Rapaport H, Kim HS, Kjaer K, Howes PB, Cohen S, Als-Nielsen J, Ghadiri MR, Leiserowitz L, Lahav M (1999) J Am Chem Soc 121:1186
138. Zhang J, Moore JS (1994) J Am Chem Soc 116:2655
139. Mindyuk O, Stetzer MR, Heiney PA, Nelson JC, Moore JS (1998) Adv Mater 10:1363
140. Leclair S, Baillargeon P, Skouta R, Gauthier D, Zhao Y, Dory YL (2004) Angew Chem Int Ed 43:349
141. Zhang J, Moore JS (1992) J Am Chem Soc 114:9701
142. Hunter CA, Sanders JKM (1990) J Am Chem Soc 112:5525
143. Prince RB (2000) Phenylene ethynylene foldamers: cooperative conformational transition, twist sense bias, molecular recognition properties, and solid-state organization. PhD thesis, University of Illinois
144. Tobe Y, Utsumi N, Kawabata K, Nagano A, Adachi K, Araki S, Sonoda M, Hirose K, Naemura K (2002) J Am Chem Soc 124:5350
145. Tobe Y, Utsumi N, Nagano A, Naemura K (1998) Angew Chem Int Ed 37:1285
146. Tobe Y, Nagano A, Kawabata K, Sonoda M, Naemura K (2000) Org Lett 2:3265
147. Höger S, Bonrad K, Mourran A, Beginn U, Möller M (2001) J Am Chem Soc 123:5651
148. Rosselli S, Ramminger A-D, Wagner T, Silier B, Wiegand S, Haussler W, Lieser G, Scheumann V, Höger S (2001) Angew Chem Int Ed 40:3138
149. Harada A, Li J, Kamachi M (1992) Nature 356:325
150. Harada A, Li J, Kamachi M (1993) Nature 364:516
151. Liu Y, You C-C, Zhang H-Y, Kang S-Z, Zhu C-F, Wang C (2001) Nano Lett 1:613
152. Zimmerman SC, Wendland MS, Rakow NA, Zharov I, Suslick KS (2002) Nature 418:399
153. Wendland MS, Zimmerman SC (1999) J Am Chem Soc 121:1389
154. Kim Y, Mayer MF, Zimmerman SC (2003) Angew Chem Int Ed 42:1121
155. Clark TD, Ghadiri MR (1995) J Am Chem Soc 117:12364
156. Clark TD, Kobayashi K, Ghadiri MR (1999) Chem Eur J 5:782
157. Venkataraman D, Gardner GB, Lee S, Moore JS (1995) J Am Chem Soc 117:11600
158. Orr GW, Barbour LJ, Atwood JL (1999) Science 285:1049
159. Williamson JR (1994) Annu Rev Biophys Biomol Struct 23:703
160. Marlow AL, Mezzina E, Spada GP, Masiero S, Davis JT, Gottarelli G (1999) J Org Chem 64:5116

161. Forman SL, Fettinger JC, Pieraccini S, Gottarelli G, Davis JT (2000) J Am Chem Soc 122:4060
162. Mezzina E, Mariani P, Itri R, Masiero S, Pieraccini S, Spada GP, Spinozzi F, Davis JT, Gottarelli G (2001) Chem Eur J 7:388
163. Sessler JL, Sathiosatham M, Doerr K, Lynch V, Abboud KA (2000) Angew Chem Int Ed 39:1300
164. Giorgi T, Lena S, Mariani P, Cremonini MA, Masiero S, Pieraccini S, Rabe JP, Samori P, Spada GP, Gottarelli G (2003) J Am Chem Soc 125:14741
165. Bonazzi S, Capobianco M, De Morais MM, Garbesi A, Gottarelli G, Mariani P, Ponzi Bossi MG, Spada GP, Tondelli L (1991) J Am Chem Soc 113:5809
166. Sidorov V, Kotch FW, Davis JT, El-Kouedi M (2000) Chem Commun 2369
167. Tirumala S, Davis JT (1997) J Am Chem Soc 119:2769
168. Cai M, Marlow AL, Fettinger JC, Fabris D, Haverlock TJ, Moyer BA, Davis JT (2000) Angew Chem Int Ed 39:1283
169. Chaput JC, Switzer C (1999) Proc Natl Acad Sci 96:10614
170. Bonazzi S, De Morais MM, Gottarelli G, Mariani P, Spada GP (1993) Angew Chem Int Ed Engl 32:248
171. Ciuchi F, Di Nicola G, Franz H, Gottarelli G, Mariani P, Ponzi Bossi MG, Spada GP (1994) J Am Chem Soc 116:7064
172. Zimmerman SC, Duerr BF (1992) J Org Chem 57:2215
173. Kolotuchin SV, Zimmerman SC (1998) J Am Chem Soc 120:9092
174. Whitesides GM, Simanek EE, Mathias JP, Seto CT, Chin D, Mammen M, Gordon DM (1995) Acc Chem Res 28:37
175. Choi IS, Li X, Simanek EE, Akaba R, Whitesides GM (1999) Chem Mater 11:684
176. Klok H-A, Jolliffe KA, Schauer CL, Prins LJ, Spatz JP, Möller M, Timmerman P, Reinhoudt DN (1999) J Am Chem Soc 121:7154
177. Kimizuka N, Kawasaki T, Hirata K, Kunitake T (1995) J Am Chem Soc 117:6360
178. Fenniri H, Mathivanan P, Vidale KL, Sherman DM, Hallenga K, Wood KV, Stowell JG (2001) J Am Chem Soc 123:3854
179. Fenniri H, Deng B-L, Ribbe AE (2002) J Am Chem Soc 124:11064
180. Jonkheijm P, Miura A, Zdanowska M, Hoeben FJM, De Feyter S, Schenning APHJ, De Schryver FC, Meijer EW (2004) Angew Chem Int Ed 43:74
181. Yang J, Marendaz J-L, Geib SJ, Hamilton AD (1994) Tetrahedron Lett 35:3665
182. Israelachvili JN (1985) Intermolecular and surface forces: with applications to colloidal and biological systems. Academic, London
183. Klug A (1983) Angew Chem Int Ed Engl 22:565
184. Fuhrhop JH, Helfrich W (1993) Chem Rev 93:1565
185. Bates FS (1991) Science 251:898
186. Lee M, Cho B-K, Zin W-C (2001) Chem Rev 101:3869
187. Percec V, Ahn CH, Ungar G, Yeardley DJP, Moller M, Sheiko SS (1998) Nature 391:161
188. Percec V, Johansson G, Heck J, Ungar G, Batty SV (1993) J Chem Soc Perkin Trans 1: 1411
189. Percec V, Heck J, Tomazos D, Falkenberg F, Blackwell H, Ungar G (1993) J Chem Soc Perkin Trans 1:2799
190. Percec V, Johansson G, Ungar G, Zhou J (1996) J Am Chem Soc 118:9855
191. Hudson SD, Jung HT, Percec V, Cho WD, Johansson G, Ungar G, Balagurusamy VSK (1997) Science 278:449
192. Percec V, Heck JA, Tomazos D, Ungar G (1993) J Chem Soc Perkin Trans 2:2381
193. Percec V, Glodde M, Bera TK, Miura Y, Shiyanovskaya I, Singer KD, Balagurusamy VSK, Heiney PA, Schnell I, Rapp A, Spiess HW, Hudson SD, Duan H (2002) Nature 419: 384

194. Percec V, Ahn C-H, Cho W-D, Jamieson AM, Kim J, Leman T, Schmidt M, Gerle M, Möller M, Prokhorova SA, Sheiko SS, Cheng SZD, Zhang A, Ungar G, Yeardley DJP (1998) J Am Chem Soc 120:8619
195. Gin DL, Gu W, Pindzola BA, Zhou WJ (2001) Acc Chem Res 34:973
196. Zhou W, Gu W, Xu Y, Pecinovsky CS, Gin DL (2003) Langmuir 19:6346
197. Kim C, Lee SJ, Lee IH, Kim KT, Song HH, Jeon H-J (2003) Chem Mater 15:3638
198. Ringsdorf H, Schlarb B, Venzmer J (1988) Angew Chem Int Ed Engl 28:113
199. Mueller A, O'Brien DF (2002) Chem Rev 102:727
200. Schnur JM (1993) Science 262:1669
201. Wilson-Kubalek EM, Brown RE, Celia H, Milligan RA (1998) Proc Natl Acad Sci 95:8040
202. Fuhrhop JH, Fritsch D (1986) Acc Chem Res 19:130
203. Imae T, Takahashi Y, Muramatsu H (1992) J Am Chem Soc 114:3414
204. Itojima Y, Ogawa Y, Tsuno K, Handa N, Yanagawa H (1992) Biochemistry 31:4757
205. Nakashima N, Asakuma S, Kunitake T (1985) J Am Chem Soc 107:509
206. Srisiri W, Sisson TM, O'Brien DF, McGrath KM, Han Y, Gruner SM (1997) J Am Chem Soc 119:4866
207. Georger JH, Singh A, Price RR, Schnur JM, Yager P, Schoen PE (1987) J Am Chem Soc 109:6169
208. Giulieri F, Guillod F, Greiner J, Krafft M-P, Riess JG (1996) Chem Eur J 2:1335
209. Fuhrhop J-H, Schnieder P, Boekema E, Helfricho W (1988) J Am Chem Soc 110:2861
210. Frankel DA, O'Brien DF (1994) J Am Chem Soc 116:10057
211. Fuhrhop JH, Blumtritt P, Lehmann C, Luger P (1991) J Am Chem Soc 113:7437
212. Frankel DA, O'Brien DF (1991) J Am Chem Soc 113:7436
213. Singh A, Markowitz MA (1994) New J Chem 18:377
214. Fuhrhop JH, Spiroski D, Boettcher C (1993) J Am Chem Soc 115:1600
215. John G, Jung JH, Minamikawa H, Yoshida K, Shimizu T (2002) Chem Eur J 8:5494
216. Thomas BN, Safinya CR, Plano RJ, Clark NA (1995) Science 267:1635
217. Wong GCL, Tang JX, Lin A, Li Y, Janmey PA, Safinya CR (2000) Science 288:2035
218. Antonietti M, Förster S (2003) Adv Mater 15:1323
219. Stupp SI, Son S, Lin HC, Li LS (1993) Science 259:59
220. Zubarev ER, Pralle MU, Li L, Stupp SI (1999) Science 283:523
221. Hartgerink JD, Beniash E, Stupp SI (2001) Science 294:1684
222. Sone ED, Zubarev ER, Stupp SI (2002) Angew Chem Int Ed 41:1705
223. Asakuma S, Okada H, Kunitake T (1991) J Am Chem Soc 113:1749
224. Yu K, Eisenberg A (1998) Macromolecules 31:3509
225. Discher BM, Won Y-Y, Ege DS, Lee JCM, Bates FS, Discher DE, Hammer DA (1999) Science 284:1143
226. Won Y-Y, Davis HT, Bates FS (1999) Science 283:960
227. Stewart S, Liu G (2000) Angew Chem Int Ed 39:340
228. Yan D, Zhou Y, Hou J (2004) Science 303:65

A Covalent Chemistry Approach to Giant Macromolecules with Cylindrical Shape and an Engineerable Interior and Surface

A. Dieter Schlüter (✉)

Chair of Polymer Chemistry, Swiss Federal Institute of Technology, Department of Materials, Institute of Polymers, ETH-Hönggerberg, HCI J 541, 8093 Zürich, Switzerland
dieter.schluter@mat.ethz.ch

1	Introduction	152
2	**Synthetic Routes**	155
2.1	Strategies	155
2.2	Divergent Route	158
2.3	Macromonomer Route	163
2.3.1	By Step-Growth Mechanisms	163
2.3.2	By Chain-Growth Mechanisms	167
2.4	Towards Dendronized Polymers with Narrow Length Distributions	170
2.5	How to Make the Cylinders Thicker and to Provide them with a Broadly Usable Surface	171
2.6	Comments on Solution and Bulk Characterization	175
3	Towards Applications	178
4	Miscellaneous	185
5	Summary and Outlook	187
	References	188

Abstract This article divides the synthetic routes leading to dendronized polymers into two main categories (attach-to and macromonomer) and compares their respective advantages and disadvantages. It gives an overview on the kind of structures presently available and also spans the bridge to both polymers and molecular objects decorated with only a few dendrons and to the structurally related, but less defined polymers with hyperbranched side chains. The presently observed shift from mere synthetic issues to more application-driven research on dendronized polymers is reflected in corresponding chapters which highlight some important developments in this direction.

Keywords Dendrimers · Nanoobjects · Polymer synthesis · Shape control

1
Introduction

Mimicking the size and – at a later stage – the function of biological functional units has always been a dream of chemists. Such an aim requires not only giant molecular structures to be generated, whose dimensions are on the order of tens and even hundreds of nanometers, but also that these man-made objects can be given a useful, predeterminable shape and be equipped, at both the periphery and the interior with functionalities such as recognition or catalytically active sites. It is evident that successful projects in this direction will have considerable impact on both the biological and materials sciences. Given the comparatively tiny size of building blocks available to the chemist for repetitive construction [1], it is obvious that hundreds and thousands of them would have to be connected in a more or less controlled fashion in order to arrive at large objects. Impressive progress along these lines has been achieved by self-assembly of appropriately designed building blocks into three-dimensional objects such as micelles and vesicles [2–4], ribbons [5], mushroom-type structures [6], stacks of all sorts [7–9], and the like. Some concrete examples and access to pertinent literature can be found in chapters 1, 2 and 3 of this volume. Related approaches to (non-confined) three-dimensionally ordered macroscopic matter have also been reported by several research groups [10–15]. A key element in all these self-assembly processes to ordered matter is healing. Since only weak intermolecular forces are used for the assembly of the individual constituents [16] all growth steps are reversible and wrong steps are healed until the energetic minimum structure is reached.

Compared to the self-assembly's state of the art, the alternative covalent chemistry approach to functional objects of the size of viruses, for example, is still in its infancy. There is not even one controlled three-dimensional covalent growth process, either to laterally confined products (molecular nanoobjects) or to extended macroscopic materials. An obvious intrinsic difficulty with the use of covalent chemistry is the lack of a healing option. Once a mistake occurs, a permanent structural or, even worse, a functional defect is created. This may have discouraged research in this direction. On the other hand, the clear advantage of the covalent approach is that once an object with all its more or, hopefully, less defects has been achieved, it does not disintegrate easily anymore and can thus be exposed to and used in different environments and conditions. Additionally, its mechanical stability is higher than that of a self-assembled congener. This stability is relevant, e.g., for individualization and manipulation sequences in connection with the bottom-up approach to functional molecular units (see paragraph 3). It should be mentioned that there have also been attempts to increase the mechanical stability of self-assembled [17–20] or conformationally preoriented structures [21] (see chapter 3). This required incorporation of cross-linkable units at designated positions and execution of an additional, difficult-to-achieve cross-linking of these units in the solid state after the self-assembly or conformational preorientation has taken place. The ultimate test for the mechanical stability of a structure stabilized by this method, however, is still missing. No such object has yet been, for example, dragged across a surface by a scanning force microscope's tip.

Covalent chemistry can easily produce long molecules by linear polymerization. Polymers commonly lack, however, any sizable extension perpendicular to the chain axis (thickness) and are best described as more or less flexible threads. In solution they attain random globular structures whose extension and segmental mobility depend on interaction with the surrounding medium. Because of this conformational mobility, terms like "interior/exterior" or "surface" cannot be reasonably applied, and linear polymerization is therefore not normally a means for the creation of huge and structurally defined molecular objects. Dendrimers consist of a few dendrons (tree-like branched fragments) attached to a dot-like core [22, 23]. They are prototypes for a controlled three-dimensional covalent growth, and thus potentially interesting for the above objective. For intrinsic synthetic and structural reasons, however, dendrimers cannot be obtained with diameters exceeding 5–10 nm. About a decade or so ago, the first steps toward overcoming this size limitation were taken, by merging the polymer and the dendrimer concepts. A new subclass of comb polymers, the so-called dendronized polymers (denpols) was created [24, 25]. Denpols consist of a polymeric backbone with the dendrons being attached to all or almost all repeat units. By attaching dendrons of increasing size (generation) to a polymer rather than to a small core, the steric repulsion between consecutive dendrons was implemented as a shape-creating element. The dendrons needed to have the largest realizable steric demand (size, generation) and be attached with the tightest possible spacings. In the extreme case, their mutual repulsion would stretch the backbone into a basically linear chain, tightly wrapped by a dendritic layer, thus creating a (filled) cylindrical object. Figure 1

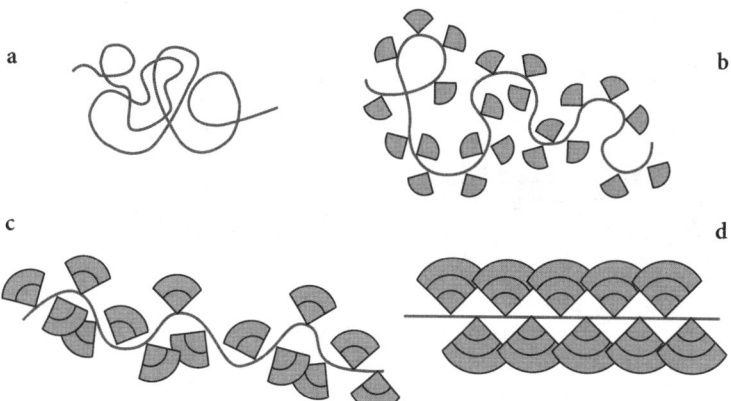

Fig. 1a–d Cartoon representation of a coiled polymer backbone's stretching through the attachment of increasingly sterically demanding dendrons. Polymer backbone with no dendrons (**a**) and dendrons of the first (**b**), second (**c**), and third generations (**d**)

illustrates this transition of a coiled polymer into a cylindrical object by the attachment of increasingly more demanding dendrons. Covalent cylindrical objects based on denpols have been realized with lengths of several hundred nanometers and widths of up to 5–7 nm and, in this regard, are amongst the largest molecules ever prepared. Denpols have been developed to the point that one now can choose between various representatives which differ in chemical constitution, stiffness, surface decoration, backbone properties, applicational functions, and the like [26–28]. As a result of a considerable effort in straightening and simplifying synthetic procedures, some of these cylinders can be additionally provided on the 5–10 g scale with high and quantifiable structural perfection [29].

Higher generation denpols have a set of features which make them unique macromolecules and interesting candidates for a variety of applications, some of which were unforeseeable at the beginning of this research. They are not only huge molecular objects of near-cylindrical shape, but their surface and backbone's chemical nature are engineerable as well. Figure 2 gives an overview of

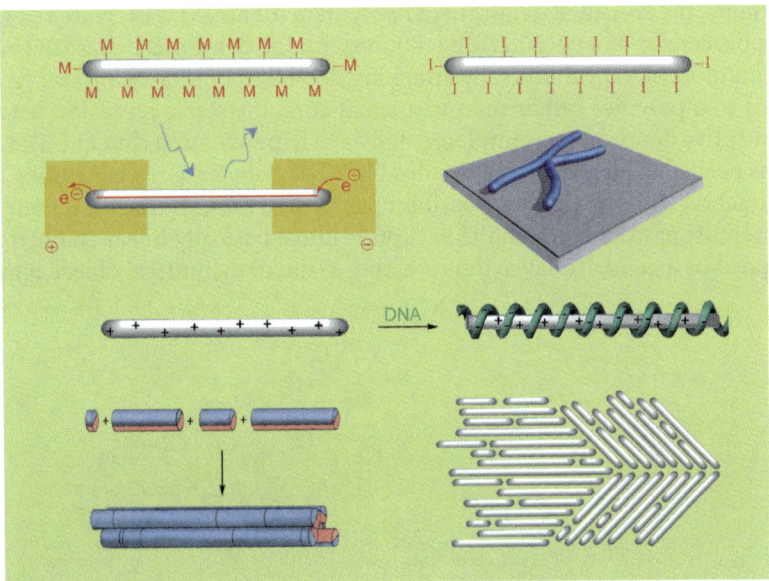

Fig. 2 Dendronized polymers, represented as *white/gray* (or *blue/red*) cylinders, put to work as catalyst (*M*) supports in nanodimensions (*top left*), polyinitiator for the synthesis of "hairy" functional derivatives (*top right*), energy transfer, light harvesting, and/or electrically conducting materials (*left, second down*), objects for covalent attachment by the move-connect-prove strategy between individualized molecules on solid surfaces (*right, second down*), novel, ultra highly charged polyelectrolytes for, e.g., wrapping with DNA and subsequent gene transfection studies (*third down*), lengthwise segregated polar/non-polar constituents of novel "supercylinders" (*bottom left*), nanoobjects for surface patterning and to induce periodicity changes from the Angstrom to nanometer scale (*bottom right*)

areas in which denpols were and are being put to work. This helps illustrate their versatility and helps one understand why a steadily increasing number of research groups worldwide are now entering this fascinating field of research.

This chapter presents synthetic issues of denpols, provides and compares different methods, tries to draw an overall feasibility picture, and addresses some pitfalls and occasional difficulty with an unambiguous structural characterization. Since synthetic matters have meanwhile reached some maturity, research has shifted more towards characterizational and applicational matters of this class of polymers. This is reflected in this chapter's organization, which has its own paragraphs on characterization (2.6) and applications (3). The chapter relies on earlier reviews [26–29] and mainly concentrates on what has been reported since then. A few of the less recent cases will also only be discussed in cases where it is important to draw general conclusions or to contrast certain findings. Otherwise the main attempt here is to comprehensively cover recent findings in the rapidly developing denpol area. Most new denpols are introduced in the synthetic part, though some of them are first mentioned in the application section. Those with low-generation dendrons have also been incorporated for the sake of completeness. One must be aware, of course, that only the higher generation ones will exhibit rigidity and can be reasonably considered cylinders. Most of the work involved in the synthesis of denpols has to be invested in their corresponding dendrons and macromonomers, respectively. For reasons of space, the present chapter concentrates on matters dealing directly with the polymers rather than their precursors, and describes only a few aspects regarding the latter. Finally, there is no uniformly accepted definition of a what a dendron generation is. This chapter does not make an attempt to standardize this and rather uses the nomenclature suggested by the respective authors.

2
Synthetic Routes

2.1
Strategies

All syntheses of denpols can be divided into two main categories, designated here routes I and II. The main difference is that route I starts from a linear polymer, to which the dendritic layer is attached by some occasionally rather complex chemistry, whereas route II starts from a monomer already carrying the final dendron as substituent, which is polymerized (Scheme 1). In the following, these routes will be compared, at first in general and then in more specific terms, in order to provide an understanding for the respective synthetic problems associated with them.

It is useful to further subdivide route I into routes Ia and Ib, although the closer one looks at them, the more artificial this division may appear to be at times. Both need the same starting polymer which carries anchor groups for

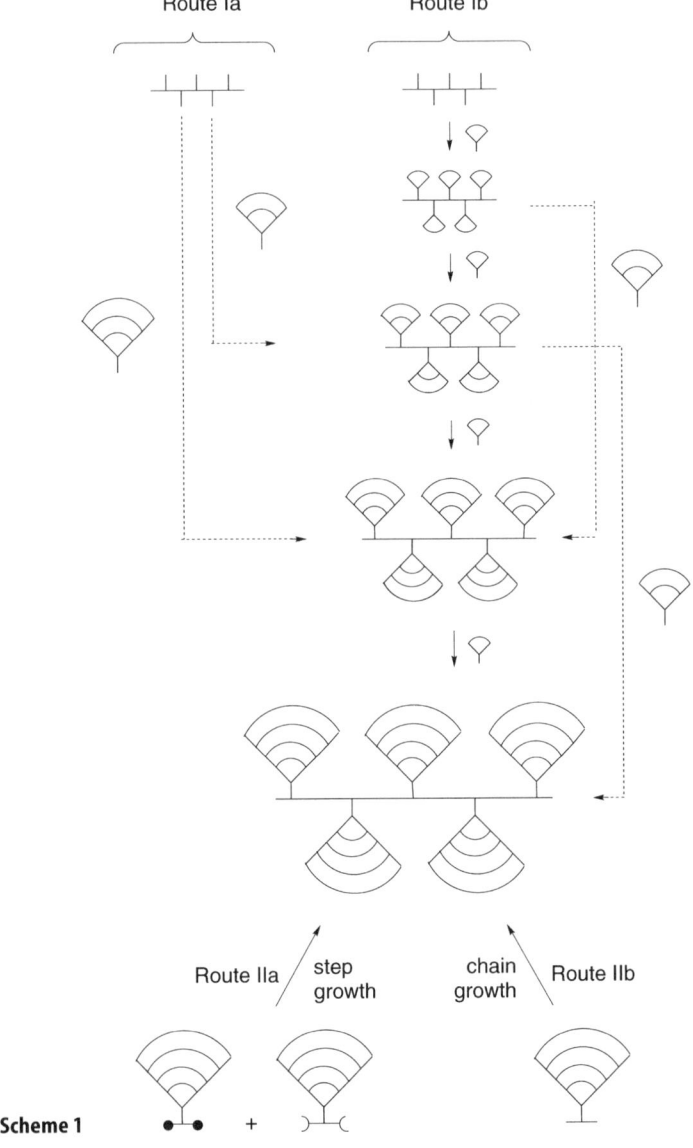

Scheme 1

dendron attachment. Route Ia uses dendrons which are already preformed and of the size required for the product [larger than first generation (G1)]. They normally do not carry peripheral functional groups for further growth and the final polymer is obtained in one step [30]. This sounds attractive effort-wise. The reality, however, shows that the achievement of a complete coverage is a major problem, at least for the systems tested. So far, this has only been accomplished on the G2 level. For many applications, specifically those where the compactness and rigidity of the structure play a role, a dendritic layer of G2 dendrons around a backbone is not fully sufficient. It is too early to conclude that route Ia is intrinsically problematical, but it is noteworthy in this context that only one further report on this route has appeared yet, which by and large also supports the above observations. Except for this one example, route Ia will, therefore, not be treated any further in this chapter.

Route Ib is a step-wise synthetic procedure. It employs preformed G1 dendrons, which are attached in a series of subsequent reactions to the starting polymer and its resulting homologs (PG1, PG2, etc.), until the final dendronized polymer is reached. This way, successive generations are added onto the ones already existing. Alternatively, the individual generations themselves may also be assembled on the polymer, typically in two steps each (not shown in Scheme 1, for clarity). In both these cases and specifically in the latter, numerous reactions on polymers are required, which may be considered unattractive regarding the overall synthetic effort involved and the achievable degree of structure control. As will be seen in the following, route Ib is nevertheless feasible under certain circumstances and is more attractive than Ia, judging by the respective number of publications. Route Ib can additionally be performed in an accelerated fashion whereby, e.g., a G1 or a G2 polymer is directly converted to a G3 and G4 polymer by reaction with G2 dendrons. Route Ib is referred to in the literature as either iterative, attach-to, divergent, or grafting from. All these terms mean that the final denpols dendritic layer is built step by step. Important synthesis goals of route Ib are to find the optimum reactions and conditions for all these modifications, to quantify the coupling efficiencies for each step, and to find the maximum generation achievable this way. Tomalia and co-workers first claimed the divergent growth of dendronized polymers some 15 years ago [31].

Route II starts from monomers already carrying the final dendron (Scheme 1, shown for G4 dendrons). They are polymerized either according to step- (IIa) or chain-growth mechanisms (IIb) with all the enormous consequences this difference in mechanism has on the achievable molar mass and molar mass distribution, the applicability of controlled procedures, and the importance of monomer purity and stoichiometric balance, etc. Route II has the advantage that, intrinsically, the dendronized products must be structurally perfect regarding their dendrons, which is sometimes a critical aspect, especially for route Ib. On the other hand, the price paid for this advantage is steric hindrance, which can make polymerization, especially of monomers with high generation dendrons, furnish only low molar mass products or even no polymer at all. Because of the considerable molar mass of dendronized monomers, route II is normally referred to as the macromonomer approach. Synthetic aspects regarding route II include finding out the macromonomer's compatibility with the polymerization conditions (no chain transfer, no loss of end groups, etc.) and the mechanism tolerating large dendron sizes. It is also interesting to investigate the applicability of the so-called controlled methods to macromonomers in order to achieve more narrowly distributed dendronized polymers (see 2.4).

Both routes I and II have in common that the steric repulsion intentionally implemented into many (but not all) of the final products sooner or later causes problems during synthesis, in the sense that reaction centers cannot reach each other anymore or that anticipated reactions turn too slowly and side channels come into play. Thus, synthesis of densely packed dendronized polymers always has an element of the quadrature of a circle and seeming subtleties in the reaction conditions can have an enormous effect on the outcome. Besides the general aspects discussed so far there are more specific, structure-related ones which also need to be considered whenever the synthesis of a dendronized polymer with a certain function or an application is the goal. The choice of reactions and conditions then depends on their compatibility with their respective requirement. This may pose considerable limitations to the tools available to a synthetic chemist and cause almost insurmountable hurdles.

2.2
Divergent Route

There is one recent example for route Ia. Hawker et al. reported an extended study on the impact of both dendron size and distance of anchor groups along a backbone upon the achievable degree of attachment [32]. They used the narrowly distributed copolymer 1 as starting material whose content of active ester anchor group containing repeat units (x) to parent ones (y) could be easily modified (Scheme 2). These copolymers were reacted with G1–G5 Fréchet-type dendrons 2 and the efficiency of amidation to polymer 3 was quantitatively determined with regard to the x/y ratio. Grafting of dendrons to polymer 1 went smoothly for all generations with loadings of 10–20 mol% of active ester in the backbone. For example, for polymer 1 with 20% active ester repeat units, 2(G5) underwent amidation with 91% of the active esters. On going to 1 with 30% anchor groups, 2(G4) could not be attached completely anymore, and for 1 with 40% anchor groups, 2(G2) was the largest dendron to be attached completely. To sum up, there is a clear interplay of the two opposing effects mentioned which renders the attachment of dendrons higher than G2 to all repeat units of a linear polymer complicated if not impossible. This finding seems to be independent of backbone rigidity: the present example used flexible vinyl-type polymers, whereas the previous one employed rigid rod type poly(*para*-phenylene) scaffolds [24, 30].

The majority of the recent examples for route Ib deal with siloxane polymers (Fig. 3). In a thorough study, Mery et al. grew branches off a poly(methylhydrosiloxane) with approximately 78 repeat units by standard platinum-catalyzed hydrosilylation chemistry with an excess of allyltrichlorosilane, followed by quenching of the pendant chlorosilane functions with an excess of allyl Grignard reagent [33]. The corresponding G1 polymer 4(G1) (not shown) with its threefold-branched dendron was obtained in a 60% yield and carefully analyzed. The coverage achieved was virtually 100% by ^{1}H-NMR integration, elemental analysis, and titration of the double bonds. Application of the same

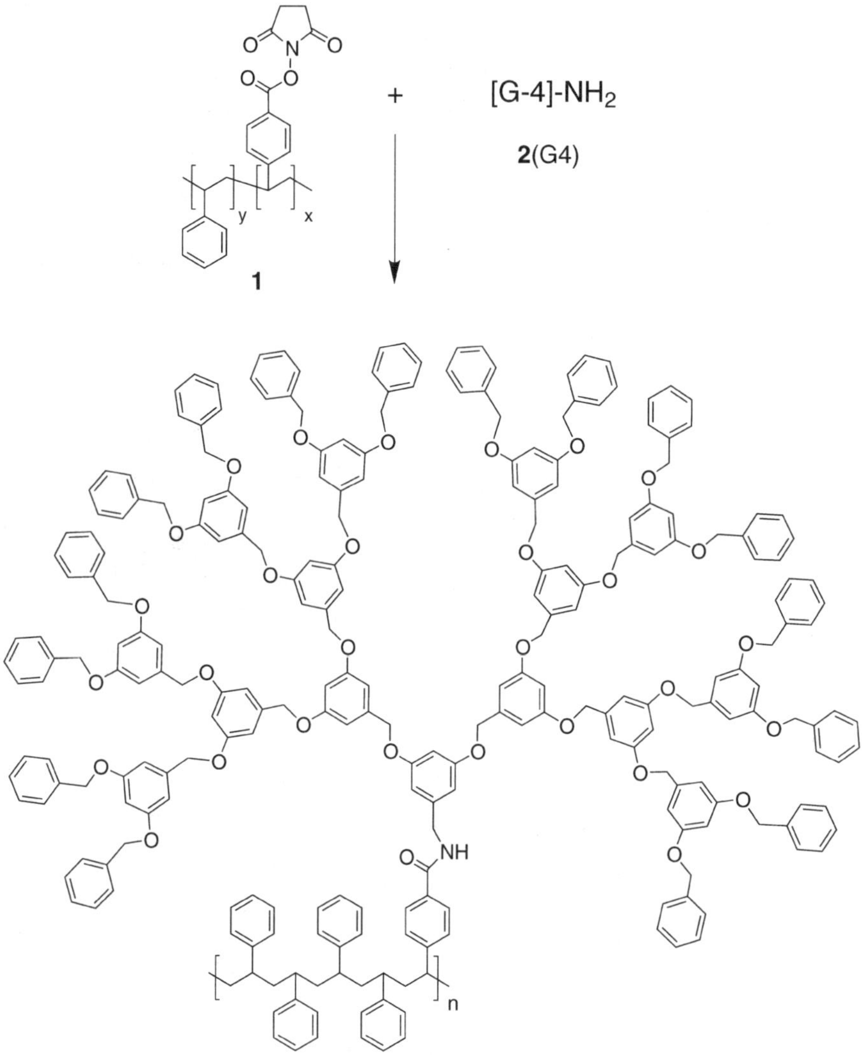

Scheme 2

sequence to **4(G1)** under significantly more rigorous conditions afforded **4(G2)** in a yield of 50%. Despite larger excesses of catalysts and reagents and prolonged reaction times, the coverage of the terminal silyls with allyl groups could not be driven beyond 80–85%. This was explained in terms of steric crowding and surface congestion as de Gennes proposed [34]. Small angle neutron scattering investigations showed that the dendronization of the starting polysiloxane to **4(G1)** and **4(G2)** resulted in a considerable backbone stiffening, as one may have intuitively expected. Almost simultaneously, Kim et al. reported the allyl-terminated polysiloxane-based dendronized polymers **5(G1)**, **5(G2)**, **6(G1)**, and

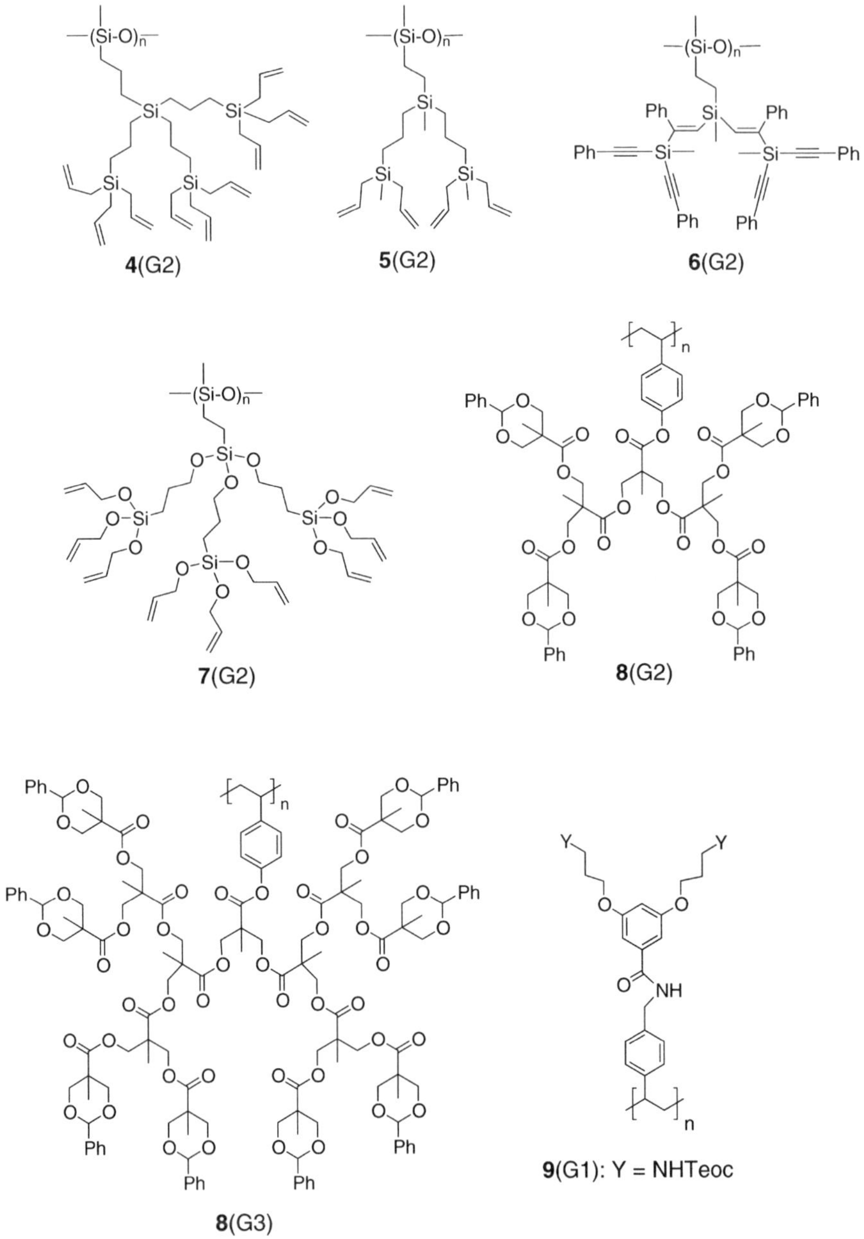

Fig. 3 Chemical structures of denpols 4(G2)–10(G4)

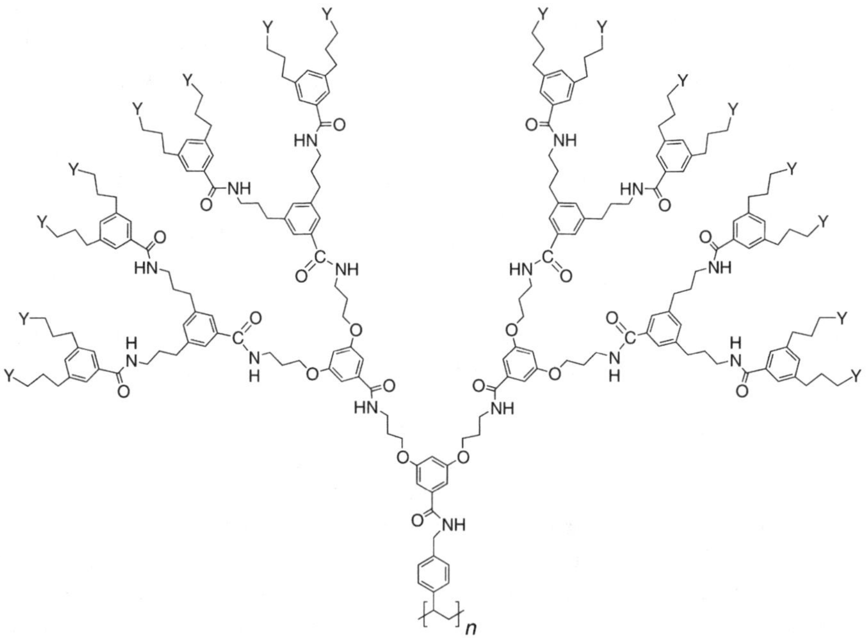

Fig. 3 (continued)   **10**(G4): Y = NHTeoc

6(G2), which differ from the Mery polymers in that they have only twofold branching units [35]. Nevertheless, attempts to grow a third generation also failed here. Unfortunately the polymers were incompletely characterized and it was not clear what molar mass and distribution the starting polymer had. More recently, the same laboratory briefly reported the synthesis of **7**(G1) and **7**(G2) [36]. Hydrosilylation reactions by which vinyl-functionalized, silicon-bridged biferrocene G1 dendrons were added to poly(methylhydrosiloxane) and related copolymers were also reported [37].

The subtleties occasionally encountered in the synthesis of dendronized polymers are nicely illustrated by an interesting example reported from Fréchet's laboratory [38]. They started their sequence with two fractions of the commercially available, narrowly distributed poly(4-hydroxystyrene). These fractions had degrees of polymerization (*DP*) of 40 (sample 1) and 90, respectively (sample 2). The sequence involves the coupling of the polymer's hydroxy groups with a benzylidene-protected bis(hydroxymethyl)propanoic acid followed by deprotection, resulting in the liberation of two hydroxy groups available for repetition of this sequence. For the lower molar mass sample 1, this sequence could be carried out three times, thus furnishing **8**(G1), **8**(G2), and **8**(G3), which were characterized by a variety of methods including MALDI-TOF MS. For sample 2, however, realization of **8**(G3) was complicated either by covalent side reactions or aggregation phenomena. At this point, careful development and fine-tuning of reaction conditions turned out to be very fruitful. If the polar solvent mixture DMF/pyridine was used, **8**(G3) could also be reached for this higher molar mass sample, and sample 1 carried even further to **8**(G4) (not shown). Besides the impact that reaction conditions can have, the lesson to be learned from this example is the importance of the molar mass of the starting polymer. It controls the total number of conversions to be executed per molecule, and thus the overall complexity of the synthesis process. It also has an impact on solubility and aggregation behavior. As will be seen below, it is reasonable to optimize reaction conditions for all growth steps on a lower molar mass sample and only subsequently apply these conditions to one of higher molar mass. In a more recent study, Fréchet's laboratory transferred the same methodology to biodegradable polyester backbones [39].

Dendronization was also proven to be feasible for a high molar mass sample of the G1 polymer **9**(G1), which itself was prepared according to route IIb [40, 41]. According to gel permeation chromatography (GPC) calibrated with polystyrene standards it had a DP=430. It has been consistently shown for a number of structurally rather different examples that GPC considerably underestimates the actual molar mass of dendronized polymers [42]. Factors of 2 or 3 are not uncommon. Polymer **9**(G1) was reacted at its deprotected amine functions with an active ester G1 building block, which also carried protected amines at its periphery. Deprotection of these amines on the thus-formed G2 polymer followed by repetition of the same procedure was carried all the way through to **10**(G4) (intermediates not shown). Scanning force microscopy (SFM) investigations of individualized **10**(G4)s on highly oriented pyrrolytic graphite (HOPG) proved that some of the chains were several hundred nanometers long. One may ask why **10**(G4) has two different repeat units and, therefore, why two obviously different G1 building blocks were used during its synthesis. There is, again, a very simple synthetic reason behind this: solubility. When the sequence was attempted with the building block first used, the corresponding G3 polymer (not shown) was only sparingly soluble and the G4 homolog could not be reasonably and reproducibly prepared. Taking two phenolic ether oxygens per repeat unit out, however, furnished fully soluble intermediates and **10**(G4) could be easily and fully reproducibly reached. All steps were quantified by applying methods from biochemistry and found to be virtually complete (quantification of unreacted amines by UV absorbance measurements of reaction with the Sanger reagent) [41]. The disadvantage of this series of experiments is the relatively broad molar mass distribution of the starting polymer, which was prepared by standard radical processes. This was, of course, carried through the entire series of dendronization experiments. For a recent application of controlled radical techniques to dendritic macromonomers, see paragraph 2.4. The quasi-homologous products were used for several applications including wrapping of the corresponding polyelectrolytes by DNA and connecting individualized chains on a solid surface by SFM (see paragraph 3).

2.3
Macromonomer Route

2.3.1
By Step-Growth Mechanisms

Denpols offer the unique opportunity to surround an electro-optically active (conjugated) backbone by its own solubilizing and/or shielding/insulating (dendritic) layer. It was thus natural for the corresponding representatives to be synthesized and tested for energy transfer and conductivity properties. Most work with conjugated backbones used step-growth mechanisms. For a few recent reports from Percec's laboratory using chain-growth procedures, see below.

Aida et al. and Bao et al. were the first to report work along these lines [43–45]. Recently Müllen et al. also contributed interesting examples based on the polyfluorenes **11**($n=0$) and **11**($n=1$) (Fig. 4) [46, 47]. Specifically, the former turned out to be a useful material for organic light emitting diodes (LEDs) with blue emission. Their G1 dendrons do not interfere with the charge transport or the material's emission properties and help the color stability of the devices. Both polymers were synthesized according to the Yamamoto procedure [48], which is basically a Ni-catalyzed reductive homocoupling of aromatic dibromides (here: dendronized 4,4'-dibromofluorene macromonomers). The GPC molar masses obtained are approximately in the order of 50,000–60,000 Da, and thus within the normal limits for this method, even if one takes into account the molar masses of rod-like polymers being normally overestimated by GPC versus polystyrene standards. This overestimation will to some unknown degree be compensated for by the dendritic substitution's opposite effect. Along the same lines, Shu et al. reported on the dendronized polyfluorenes **12**(G1-G3), the highest homolog of which also showed promising optoelectronic properties for an application in a blue-emitting polymer LED [49].

Fréchet et al. reported first results on the synthesis of the dendronized polythiophenes **13**(G2), **13**(G3), **14**(G2), and **14**(G3) using the Stille cross-coupling procedure of AA- and BB-type monomers [50]. They employed 2,5-bis(trimethylstannyl)thiophene as AA monomer in all cases and coupled it with the dendronized dibromothiophene and dibromopentathiophene as the corresponding BB-type macromonomers. For **14**(G2) the dendron size was not sufficient to mediate solubility in organic solvents. All other products were soluble. Stille cross-coupling is normally accompanied by homocoupling and methyl-group transfer reactions and is thus a problematical reaction for polymerization purposes [51]. This can be seen from the GPC elution curve of polymer **14**(G3), which is rather complex. Nevertheless, high molar mass fractions could be obtained by preparative GPC and the electrical properties successfully tested.

Both these works are intentionally exploratory and preliminary. Synthesis was purposely only developed to the point where one can see whether the product actually shows the expected property. Since the properties in both cases

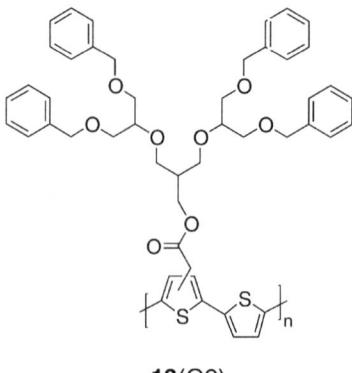

Fig. 4 Chemical structures of denpols 11 (*n*=0,1)–16(G2, p=1,R=H)

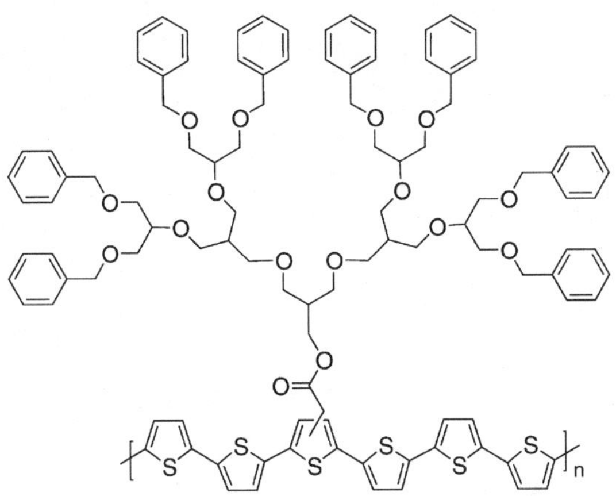

14(G3)

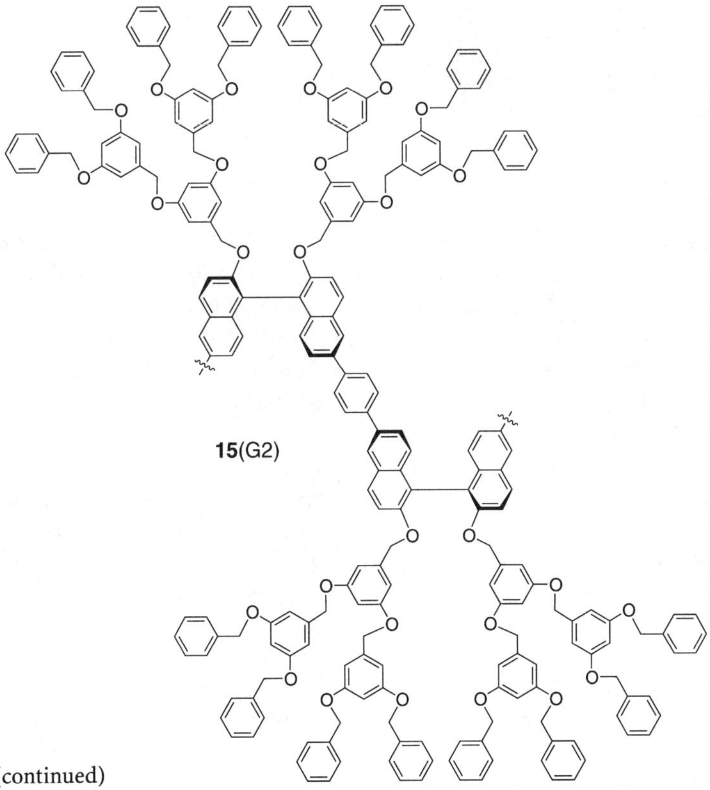

15(G2)

Fig. 4 (continued)

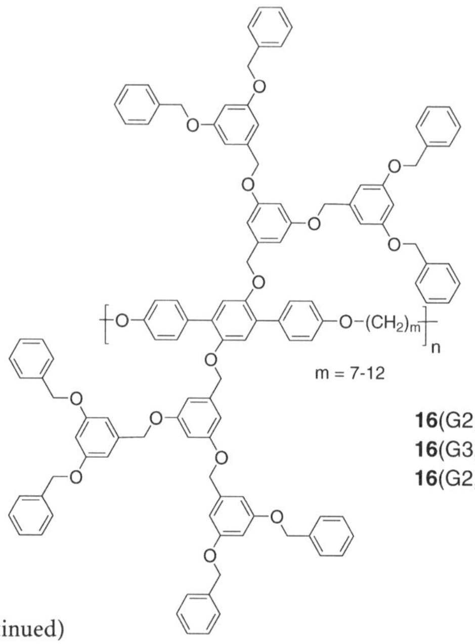

Fig. 4 (continued)

16(G2,p=0,R=H)
16(G3,p=0,R=OC$_{12}$H$_{25}$)
16(G2,p=1,R=H)

m = 7-12

look promising, this is now a justification to further optimize the synthetic procedures.

Suzuki polycondensation (SPC) is perhaps the most powerful tool for synthesizing polyarylenes [52]. It was shown that even for macromonomers with enormous steric load (G4 Fréchet-type dendrons) high molar mass polymers can be obtained after some optimization work [53]. It is, therefore, not surprising that this method has recently also been used for the synthesis of novel dendronized polyarylenes. Xi et al. reported the chiral main chain representatives 15(G1) and 15(G2), whose chirality is based on the known atropisomeric dinaphthyl units [54]. 1,4 Benzene bisboronic acid served as the AA monomer and the G1- and G2-dendronized 6,6'-dibromo-1,1'-dinaphthyls, respectively, as the BB macromonomer counterparts. For 15(G1) a weight average molar mass was stated as 240,000 (corresponding to a weight average DP_w=250), and for 15(G2) as 11,000 (corresponding to DP_w=6). Although these values have to be considered with care, it is not easy to understand why the molar mass of the higher generation polymer seems to be drastically lower than for 15(G1). A problem here may have been the difficulty in precisely matching the required 1:1 stoichiometry. Boronic acids tend to contain water and the weighed quantities have to be corrected accordingly, which is not always easy to do. The above example 12 used the same synthetic method.

Route IIa has not only been followed by transition-metal-mediated stepgrowth procedures. Kallitsis et al. reported a successful polyether formation between G1 and G2 dendronized terphenyls with terminal phenolic groups and α,ω-dibromoalkanes as coupling partners to give **16**(G1, p=0,R=H) (structure not shown) and **16**(G2, p=0,R=H) [55]. The polymerizations were carried out in the two-phase system *ortho*-dichlorobenzene/NaOH (10 N) in the presence of a phase transfer catalyst. Depending on reaction time respectable molar masses as high as, e.g., M_w=280,000 (GPC calibrated by light-scattering) for **16**(G2) were obtained. It is helpful in this regard that the steric repulsion of consecutive dendrons is small. The polymers' thermal and dynamic mechanical behaviors reveal a tendency towards phase separation of the dendritic fragments and the main chain. This work was recently considerably extended to polymers with more mesogenic dendrons, **16**(G2, p=0, R=OC$_{12}$H$_{25}$) and those with longer rigid units in the main chain, **16**(G2, p=1, R=H) [56].

2.3.2
By Chain-Growth Mechanisms

As pointed out above and in the earlier reviews [28, 29], a main concern with route II is the achievement of high molar mass polymers for macromonomers with generation three and higher. In most cases, irrespective of the polymerization mechanism, only relatively low molar mass products were obtained. This general observation was further substantiated by a work of Scrivante, Chessa et al. [57]. They tried to synthesize polymers **17**(G1–G3) which carry rather interesting pyridine-containing dendrons for complexation studies (Fig. 5). Reasonable molar masses could only be obtained for **17**(G1), by standard radical techniques. For the next higher homolog, **17**(G2), the product already contains oligomers, and dimers and trimers form the majority of the material for **17**(G3). A drop in the number of repeat units going from the G1 to G2 macromonomer was also observed for the recently reported polyacrylamides **18**(G1) and **18**(G2) [58, 59]. Whereas the former has DP_w=1,800 (GPC), the latter has only DP_w=107.

In an intense effort to see whether higher generation vinyl-type macromonomers can be polymerized to high molar mass product, the two homologous series of dendronized methacrylate-based macromonomers were prepared, obtained on the several-gram-scale as analytically pure materials, and polymerized to **19**(G1-G4) and **20**(G1-G4) [60]. The only difference between these two series is the ethyleneoxy spacer of **20**. This spacer was introduced because it was not clear from the literature whether or not a spacer facilitates polymerization, as one intuitively may be inclined to think. Both series of macromonomers, **19** and **20**, were polymerized under thermally induced radical polymerization (TRP) conditions which means that no initiator precursor was deliberately added. The molar masses were determined by GPC referenced to both a poly(methylmethacrylate) (PMMA) and a G1 denpol standard. TRP of all monomers of series **19** gave polymers whose molar masses decrease from several million for **19**(G1), to several hundred thousand for **19**(G2) and **19**(G3), and to approximately 100,000 for **19**(G4). The molar masses of series **20** were practically identical for all generations. In all experiments it was of the

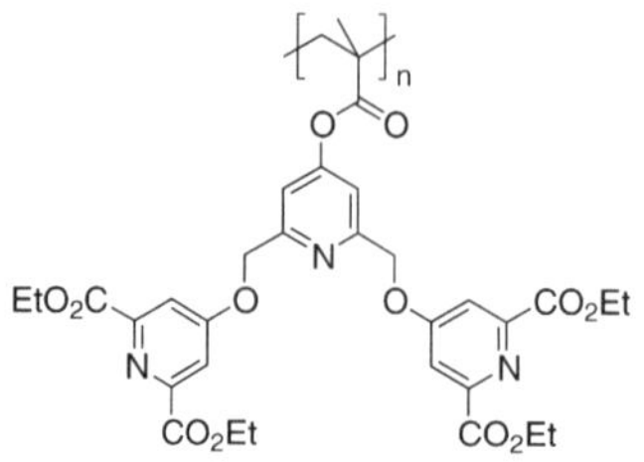

Fig. 5 Chemical structures of denpols 17(G2)–21(G2)

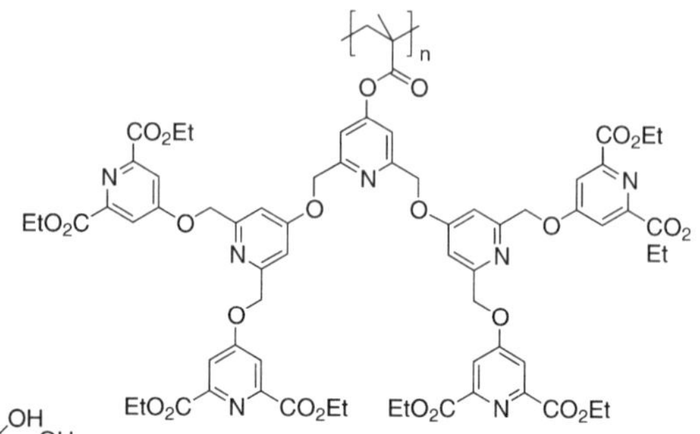

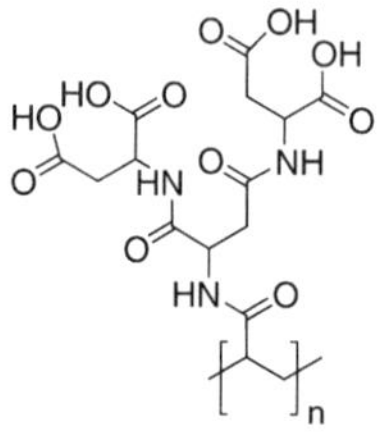

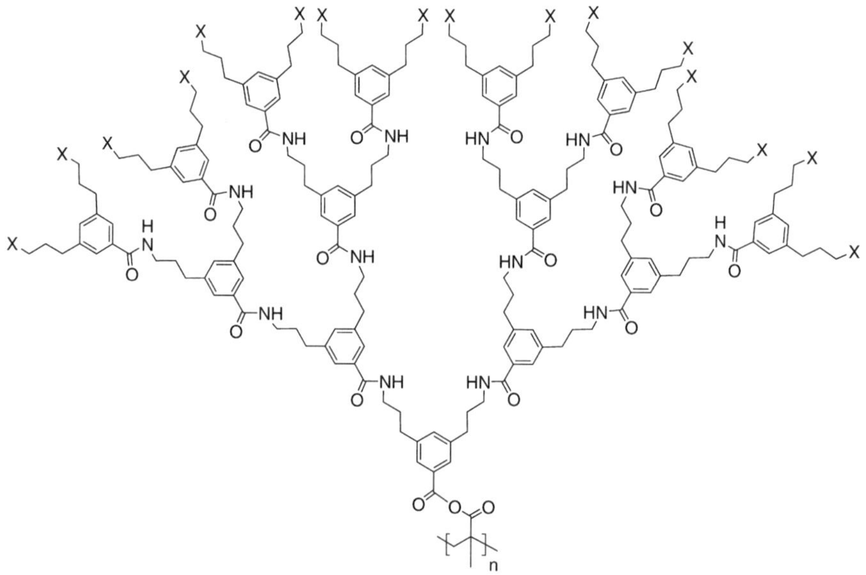

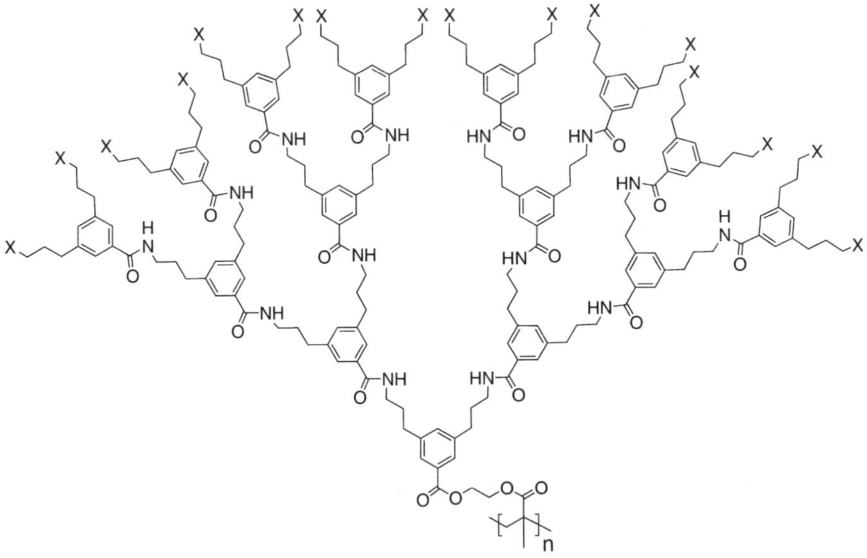

20(G4): X = NHBoc

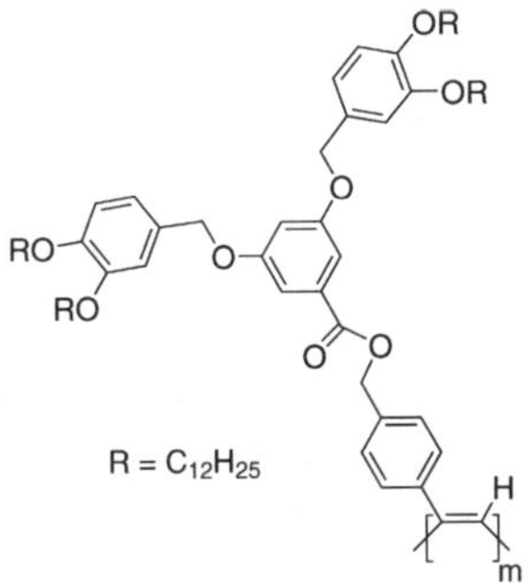

R = C₁₂H₂₅

Fig. 5 (continued)   **21 (G2)**

utmost importance to reach the highest synthetically achievable concentration. If fluidity was sufficient, they were carried out in bulk. Otherwise the molar masses dropped sharply, sometimes to a point where no polymer was obtained at all [61].

What conclusions can be drawn from these surprising findings?:

a. High molar mass denpols can be obtained at least from some vinyl-type macromonomers if high concentrations are applied. Even fourth generation macromonomers can be polymerized.
b. Qualitatively there is an inverse dependence of molar mass on generation number. In certain cases and under certain conditions it is possible to com-

pensate for the detrimental effect of increasing dendron size on obtainable *DP*s to the point that very long chains with even G3 dendrons can be obtained.

c. At least for the short ethyleneoxy group, there is no recognizable spacer effect. Monomer purity and concentration of the polymerizable unit seem to play the major role.

Recently, Percec's laboratory reported on the Ziegler-Natta-type polymerization of dendronized phenylactylenes like **21**(G2) [62], which is based on related earlier reports in the Japanese literature [63]. The molar masses were relatively high and the distribution narrow. Within the same framework, ethynyl carbazoles carrying a branched, but not strictly dendritic substituent, were also successfully polymerized (structures not shown) [64].

2.4
Towards Dendronized Polymers with Narrow Length Distributions

The diameter of denpols in the stretched conformation is determined by the generation number, and thus – by conformity with organic chemistry laws – monodisperse. Their lengths obey the fundamentals of polymer chemistry, however, and are therefore inevitably polydisperse. The polydispersity values (*PDI*) of the denpols discussed so far lie in the range *PDI*=2–3. Sometimes even higher values were reported. Though serious attempts were not normally made to accurately determine the PDIs, it is reasonable to assume that all samples have a rather broad length distribution. As determined, e.g., by SFM of spin-coated samples, dendronized polymers obtained by radical polymerization contain macromolecules with contour lengths from a few tens of nanometers all the way up to several hundreds of nanometers [65].

Control of polymer molar masses and PDIs allows precise adjustment of properties to an intended application. Different techniques, including the controlled radical procedures, have been successfully utilized in polymer chemistry for such a purpose. Among the latter, atom transfer radical polymerization (ATRP) [66–68] and reversible addition-fragmentation chain transfer (RAFT) polymerization have recently been intensively explored [69–72]. Both methods were tried for the corresponding methacrylate monomers of polymers **19**(G1) and **19**(G2) [60,73]. ATRPs proceeded surprisingly fast and the denpols obtained had low *PDI*s of 1.1–1.2. The molar masses, however, could not be driven beyond roughly 100,000, which translates into 200 repeat units for **19**(G1) and 90 repeat units for **19**(G2). If higher molar masses were attempted, the *PDI* values increased sharply. Though it was interesting and a bit surprising to see that ATRP can actually be applied to such sterically demanding macromonomers, the observed limitation in regard to samples with both high molar mass and narrow distribution was not fully satisfying. Fortunately, with RAFT the results looked better [73]. In the presence of AIBN as radical initiator and, e.g., 2-(2-cyanopropyl)dithiobenzoate (CPDB) as mediator and at mediator-to-monomer ratios of 2:200 for the monomers of **19**(G1) and **19**(G2), the following data were obtained: **19**(G1), number average molar mass (M_n) =320,000, *PDI*=1.24; **19**(G2), M_n=178,000, *PDI*=1.20. The more sterically demanding G2 monomer required a higher polymerization temperature (90 °C instead of 60 °C) but could nevertheless be put to work. It does not seem likely that any of the controlled radical techniques can be reasonably applied to higher than G2 macromonomers. Nevertheless, the above RAFT result has considerable importance for the achievement of low *PDI* denpols with generations G3 or G4. Both **19**(G1) and **19**(G2) carry peripheral functional groups for further divergent growth and are therefore ideal candidates for a route Ib growth protocol.

Recently Cheng and Xi also reported an ATRP of a G1 and G2 methacrylic macromonomer (not shown). Though the molar masses were low, the dispersities were narrow and it was shown that the chains ends can be used for block copolymer formation [74]. Another ATRP study appeared practically simultaneously with Xi's work, from Malmström's laboratory. Here the focus was on both control of molar mass distribution and generation of higher generation denpols starting from G1 or G2 polymers (not shown) [75]. Again the molar masses were relatively low. This work is a case of route Ib (paragraph 2.2) but is nevertheless mentioned here because the aspect of its controlledness was considered especially important.

2.5
How to Make the Cylinders Thicker and to Provide them with a Broadly Usable Surface

The rod-like tobacco mosaic virus (TMV) has a length and a diameter of approximately 300 and 18 nm, respectively, and the rod part of the λ bacteriophage is 135 nm long and 15 nm thick. Lengthwise, denpols can easily compete with these biologically relevant objects; diameterwise, however, they cannot. Their diameters do not exceed 6–7 nm. In light of the above discussion on the reverse impact of dendron size on polymerizability of the corresponding dendronized macromonomers, it is clear that a macromonomer strategy of any sort will not provide access to denpols with doubled or tripled diameter. Likewise, the attach-to route is bound to reach its limits if, say, a G10 denpol is attempted. Both the structural aspects and the required synthetic effort would be totally unacceptable. Recently, molecular brushes were developed [76]. They consist, as their rather descriptive name implies, of a polymer backbone with appendent linear polymeric side chains at each repeat unit [77, 78], and thus represent the class of comb polymers. They can be obtained by polymerizing short polymer chains which will later be the brushes' hairs and which carry a terminal polymerizable unit [79]. Alternatively, the side chains can also be grafted to a polymer chain [80, 81] or grown from it [82–84]. This latter strategy is of interest in regard to increasing the denpols' thickness. In the first orienting experiments, the "surface" of G2 denpols was decorated with initiating sites (like in **22**) [85] or with polymerizable units at which a grafting process can be started (like in **23a**) and then hairs grown off it (Fig. 6) [86]. Both of these starting polymers carry four active termini per repeat unit and can thus, in principle, start four hairs. They were grown by ATRP with methylmethacrylate in the case of **22** ($m=0$) and a ring-opening metathesis polymerization (ROMP) with norbornene in the case of **23a**. When these grafting processes were carried out under carefully optimized conditions, cross-linking reactions could be largely suppressed and the "hairy" denpols **22**($m>0$) and **23b** obtained.

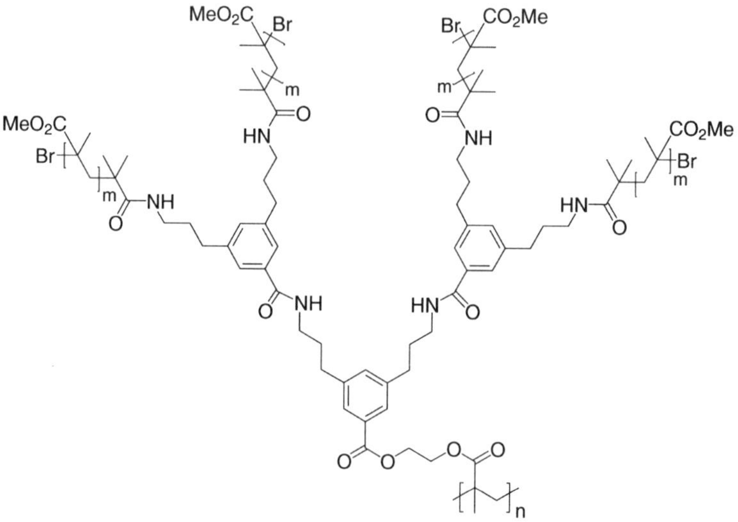

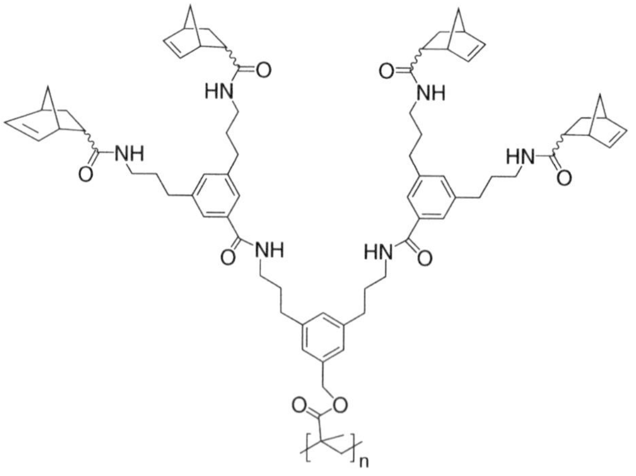

Fig. 6 Chemical structures of denpols **22(G2)–23b**

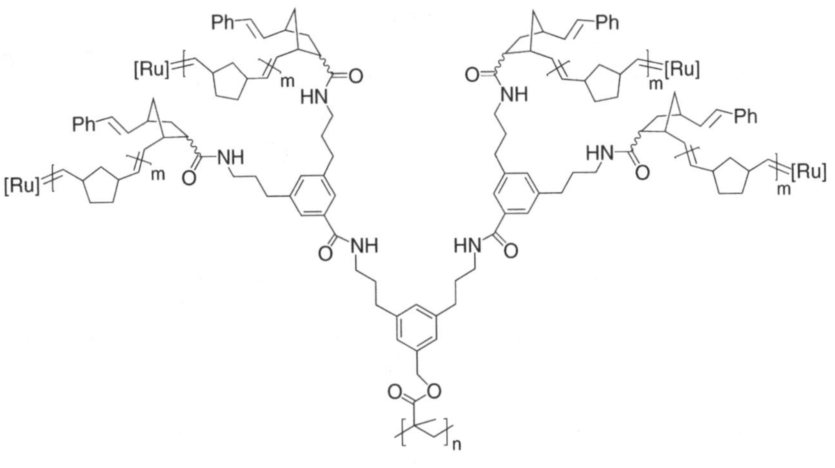

A Covalent Chemistry Approach to Giant Macromolecules with Cylindrical Shape 173

Fig. 6 (continued)   23b

What about the thickness now? This is where the scanning force microscope comes into play. Figure 7 shows the SFM height image of polymer **22**($m > 0$), for example, after these molecules were spin-coated from diluted solutions onto mica. As one can see, the individualized features consist of a lighter (higher) part surrounded by a darker (thinner) corona-like one. The lighter part is caused by the denpol backbone and the darker by the hairs. The corona's lateral dimension in principle allows one to determine the hairs' minimum length and its

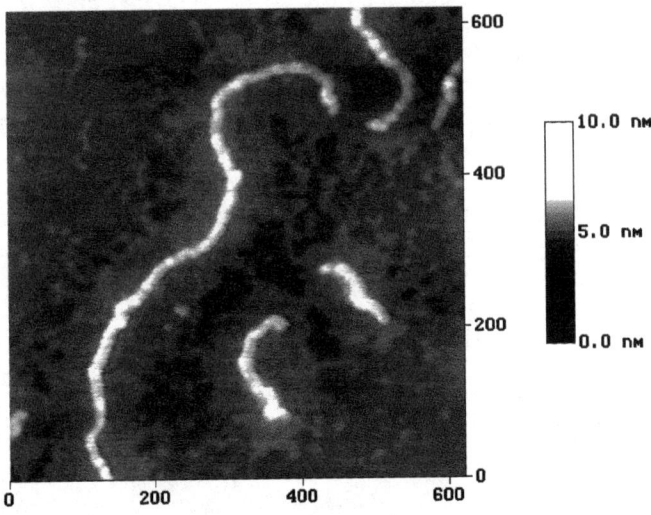

Fig. 7 Scanning force microscopy (SFM) height image of **22**(G2) spin-coated on mica (scan size: 622 nm, scan rate: 1.485 Hz). Corona diameter approximately 100 nm

thickness must reasonably agree with the number of hairs per repeat unit, i.e., the initiation efficiency. Unfortunately, tapping mode SFM does not allow per se a precise determination of the heights of objects which interact with the tip differently than the substrate does. Nevertheless, the mere existence of a corona is convincing experimental evidence for hairy denpols. The polar PMMA hairs spread out on a polar surface rather than fold back. It should be mentioned that the corona may also contain some hairs which happen not to be connected to the denpol. Having thus proven the existence of such a thing as hairy denpols, the next step is to force the hairs to fold back and produce sausage-like objects whose thickness is basically determined by the hairs' mass per repeat unit and the diameter of the starting denpol. This is presently under investigation for the same G2 denpol [**22**(m>0)] by individualizing it on the non-polar HOPG or on monolayer modified MoS_2. Initial results indicate that conditions can be found where the reduced attractive interaction between hairs and the surface makes the hairs actually fold back and the desired sausage-type objects be created. The apparent heights and widths of these objects are presently under investigation. This result defines the present state of the art. The obvious next steps are to use the thicker G4 denpol as polyinitiator, grow even longer hairs off it, and select hairs which contain polymerizable units so that the sausage-type structures not only get thicker but can also be covalently stabilized by some intramolecular polymerization reaction.

The impact denpol nanocylinders will have in the future depends not only on their sheer size and their shape persistency, but also on the versatility and ease with which their "surface" can be engineered. Of specific interest is whether certain functional, catalytical, or targeting units, and fluorescence tags or modifiers of all sorts can be anchored to it in predeterminable absolute and relative concentrations. Translated into chemical needs, such a requirement means that denpols have to have selectively addressable functional groups at their surface in controllable absolute and relative concentrations. Attempts in this direction have been undertaken and should be briefly described here. Figure 8a shows a set of G2 polymers all carrying terminal amine functional groups, which are ideal candidates for subsequent attachments. They are individually addressable because – at least in low molar mass chemistry – the *tert*-butyloxycarbonyl (Boc) or benzyloxycarbonyl (Cbz) protecting groups are orthogonal to one another. This means that the Boc-protected amines can be deprotected leaving the Cbz-protected ones untouched and vice versa. For two of these polymers this orthogonality has already been shown [87]. According to their mode of construction, the proportions between the orthogonally protected amines can be varied almost at will. Five different ratios have been realized so far (Fig. 8a). These structures will open access to cylinders of the kind shown in Fig. 8b where the differently colored dots represent different surface decorations.

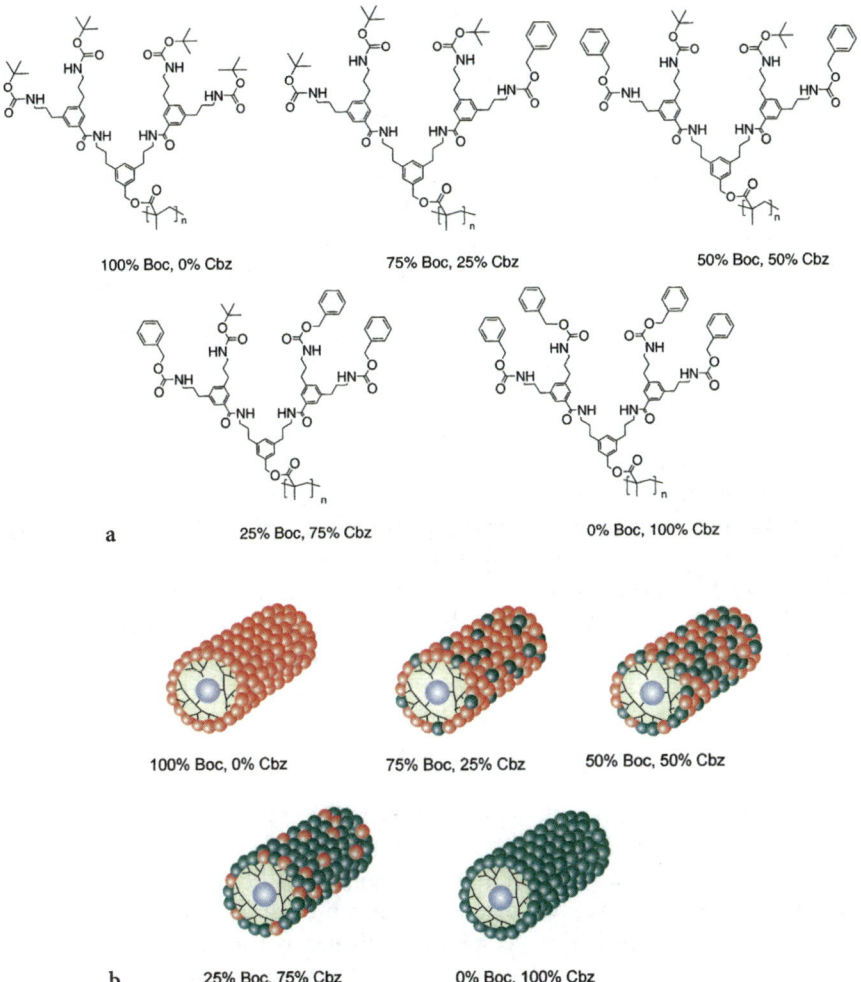

Fig. 8a, b Denpol **19**(G2) with different ratios of orthogonally *tert*-butyloxycarbonyl (*Boc*)- and benzyloxycarbonyl (*Cbz*)-protected peripheral amine groups as chemical structure (**a**) and cartoon representation (**b**)

2.6
Comments on Solution and Bulk Characterization

Characterization of dendronized polymers is a bit like taming a wild beast. Specifically for the high molar mass cases, which can easily be in the order of a few millions, NMR spectroscopy, the synthetic chemist's favorite toy for characterization in solution, cannot be reasonably used anymore. Lines can become so broad that they virtually disappear in the baseline, not to mention any fine structure. Determinations of coverage, as briefly discussed in paragraph 2.1, therefore have to be done with quantitative UV absorbance or fluorescence measurements after appropriate labeling. NMR spectroscopy has, however, been usefully applied in several cases where the molar mass did not exceed several hundred thousands. Another problem of high molar mass denpols is their work-up. Specifically for those with relatively low generation dendrons, the products should not be completely dried because this sharply reduces their solubility. This can go so far as to render them completely insoluble, even when treated with high-boiling organic solvents for extended periods of time. This effect seems to be general; its seriousness depends, of course, on the concrete structure. Denpols can only be re-dissolved in freeze-dried form. For polymers with higher generation dendrons, solubility is not such a critical problem. This may indicate that the lower generation representatives have a higher tendency to entangle and/or interdigitate, whereas the presumed higher stiffness of the higher generation ones hinders them from doing exactly this. A broad investigation of the molar mass determination by light-scattering and small angle neutron scattering technics was started a while ago [42] and is still being continued. An interesting tool for molar mass determination is SFM. Except for when the aggregation is too strong, polymer samples can be spin-coated on a solid surface such as HOPG, mica, or molybdenum disulfide in a form such that basically all chains can be individually imaged. Under certain circumstances this allows measurement of contour lengths and length distributions and the combining of these data into a histographic analysis. Such analyses have, for example, been done for polymers 9(G1,G2) and 10(G3,G4) [41], as well as several others and represent an interesting and relatively easy complementary tool for scattering techniques. Obtaining correct data from elemental analysis is another problematical aspect. Because of the sheer dimensions of these highly branched structures, solvents are often entrapped, even after prolonged treatment of freeze-dried material in high vacuum. Repetition of the same procedure at elevated temperatures can sometimes be helpful. Solvent-free NMR spectra can be obtained, if necessary, by repeated dissolution/drying cycles with deuterated solvents. That way, non-deuterated solvent molecules are exchanged. This does not, of course, help to achieve correct data from elemental analysis.

The first bulk characterization of denpols was done back in 1995 by Blackwell using Percec's polymers with so-called tapered side chains (structure not shown) [88]. These denpols have low generation dendrons (typically G1) which are attached mostly to PMMA main chains by flexible spacers of varying length and carry substituents at the dendrons. This helps to improve their mesogenic properties. The idea behind this design was to induce the denpols' dendrons to self-assemble into cylindrical, columnar moieties whereby the tapered side chains the played role of the oligopeptide shell in TMV, more or less, and the denpols' backbone the role of TMV's inner RNA. The materials showed liquid crystalline properties. This work was then carried further, to the point where the shape of the denpols in bulk could be varied from spherical to cylindrical depending on chain length and dendron decoration [89]. Spherical structures

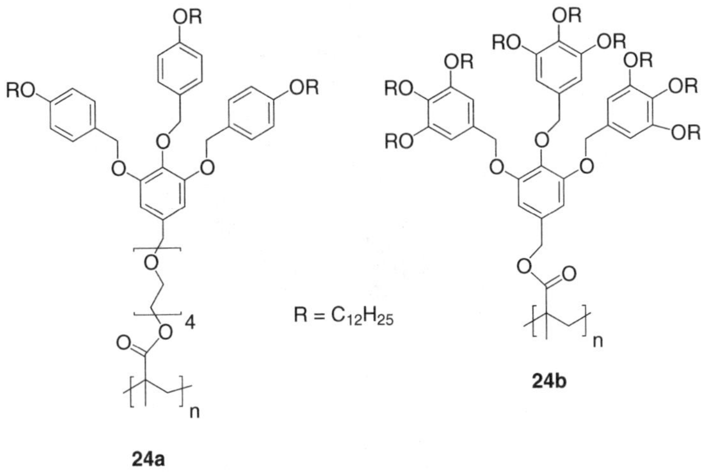

Fig. 9 Chemical structures of denpols **24a** and **24b**. **24a** has a tetraethyleneoxy spacer between the dendron and the backbone

were obtained when the dendrons caused the growing chain to self-terminate at an early stage of growth. Recently the same family of denpols was investigated by ^{1}H and ^{13}C solid-state magic angle spinning (MAS) NMR spectroscopy in order to identify structure-directing elements in the self-assembly process, and to understand the order phenomena [90]. Two representatives used for these studies are **24a** and **24b** (Fig. 9). From the high local order among the aromatic subunits it was concluded that they play a structure-directing role in the creation of the cylindrical overall shape and the helical arrangement of the backbone in the cylinder's most inner part. The flexible ethyleneoxy linker between backbone and mesogenic dendron in denpol **24a** was found to mechanically decouple both these parts from another. This allows the dendrons' aromatic units to attain the most favorable relative conformation. Not unexpectedly, the dodecyl chains in the periphery of this family's representatives are mobile.

Pakula et al. recently did systematic studies on the bulk properties of two homologous series of denpols **19** and **20** [91]. The complete G1–G4 series of both denpols were investigated by DSC, X-ray diffraction, and dynamic mechanical measurements. The X-ray diffractograms showed distinct small angle peaks which were attributed to backbone-backbone distances ranging from roughly 2.7 to 5.6 nm for both series. Figure 10 shows the diffractograms of **19**. Signatures of a long range order as suggested for the above family [90] were not found. As one might expect, increasing the dendron generation led to increased backbone stretching. It also showed a pronounced effect on the glass transition temperature. For the first three generations these temperatures increased by approximately 10° per generation. Increasing the dendron generation also involved a considerable broadening of the segmental relaxation process, which

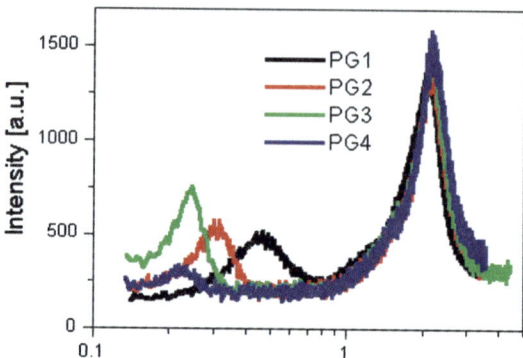

Fig. 10 X-ray diffractograms of bulk samples of the homologous series **19(G1)–19(G4)** assigned as *PG1–PG4* at room temperature. All diffractograms show an amorphous halo at wide angles and relatively sharp low angle peaks with positions, intensities, and widths depending on the generation. The positions correspond to backbone–backbone spacings (diameters of denpols in bulk) of 2.69, 4.06, 5.05, and 5.56 nm, respectively

indicates an increasing dynamic heterogeneity of the systems. Mechanistic suggestions were made for this.

Lezov et al. studied denpols with polystyrene backbone and G1 through G4 Fréchet-type dendrons (not shown) for their conformational, electrooptical, and dynamic properties by using light-scattering, viscometry, isothermal diffusion, and electric and flow birefringence methods [92, 93]. In solution these macromolecules showed a coil-like conformation. The equilibrium rigidity increased with generation number and was found to be higher than that of parent polystyrene by factors of 2.7 and 11.3 for G1 and G4 polymer, respectively. The optical anisotropy of the repeat units had negative signs and were larger in absolute values than the parent polystyrene. Electrical birefringence showed that the orientation of the denpols takes place on a scale comparable to their end-to-end distance.

3
Towards Applications

Figure 2 highlights some of the potential applications of denpols. They have been investigated so far (a) as catalyst supports, (b) as starting materials for further growth towards huge molecular objects, (c) for electroluminescence or electrical conductivity applications, (d) as constituents for elementary steps towards the bottom-up to the nanosciences, (e) for wrapping studies of these novel cationic polyelectrolytes with DNA as the negatively charged counterpart, (f) to assemble amphiphilically equipped representatives into huge cylinders, and (g) for coatings of surfaces with monolayers which render an ångström periodicity into a nanometer one. Applications (a), (c), (d), and (e) will be explained in the following, (b) has already been addressed above, and regarding (f) and (g), the reader is referred to an earlier review [28] and the pertinent literature.

Two reports appeared in 2003 which dealt with catalysis in the interior and on the "surface", respectively, of denpols (Fig. 11). Polymer 25 carries roughly 10% of non-dendronized pyrrolidino repeat units [94]. They have a C-4 connected pyridyl unit instead, which renders the properties of the repeat unit similar to *p*-dimethylaminopyridine, a well known catalyst for esterification reactions. The remaining 90% are dendronized with G4 polyester-type dendrons. The complete structure was designed so that there is a polarity gradient, with increasing polarity from the periphery to the backbone. This design could lead to interesting microenvironmental effects of transport and catalysis. The first reaction, which was tested under the influence of 25 as a catalyst, was the acylation reaction of a tertiary alcohol with pivalic anhydride. A pronounced catalytic effect was observed and ascribed to a microenvironmental effect created by the denpol.

The second example aims at the issue of whether denpols could possibly play a role between homogeneous and heterogeneous catalysis [95]. Depending on the number of repeat units and generation number, hundreds and thousands of catalytically active sites could be attached to their "surface." This reminds one of heterogeneous surfaces, yet they are still soluble. The first polymers tried for such an objective was the G1–G3 series of 26. They were decorated with van Koten's NCN-palladium and platinum pincers, which were proven successful in the case of surface-coated spherical dendrimers [96]. The test reaction was the aldol condensation of benzaldehyde and methyl isocyanoacetate. Interestingly the catalytic activity per metal center did not depend on the denpol's generation but was somewhat lower than for single pincers. It seems that the pincers are easily available on the "surface" and not hidden by backfolding processes. Because of the diversity of metals, which can be accommodated by pincer-type ligands [96], further development of the catalytic properties looks promising.

Catalyst supports are amply available. Regarding overall synthetic effort and costs involved, denpols cannot compete at all. Why, then, catalytically active materials based on denpols? Is there more to them than just a (bad) selling argument for a new class of polymers? The somewhat plain answer is: the future will show. Denpol catalysts combine features like high structure perfection and concrete and predeterminable areal density of catalytically active sites with their potential to gradually change the surface activity and to manipulate and integrate them into molecular functional units on surfaces (see below). More intense research will be required before we will be able to differentiate between a dead and a gold mine.

Regarding electro-optical applications, the initial works have already been mentioned in an earlier review [28]. For LED applications the most important issues are color stability and device lifetime. For a while it was thought that shielding of the emitting backbone by dendrons would improve these aspects because of reduced aggregation (and thus quenching) and excimer formation.

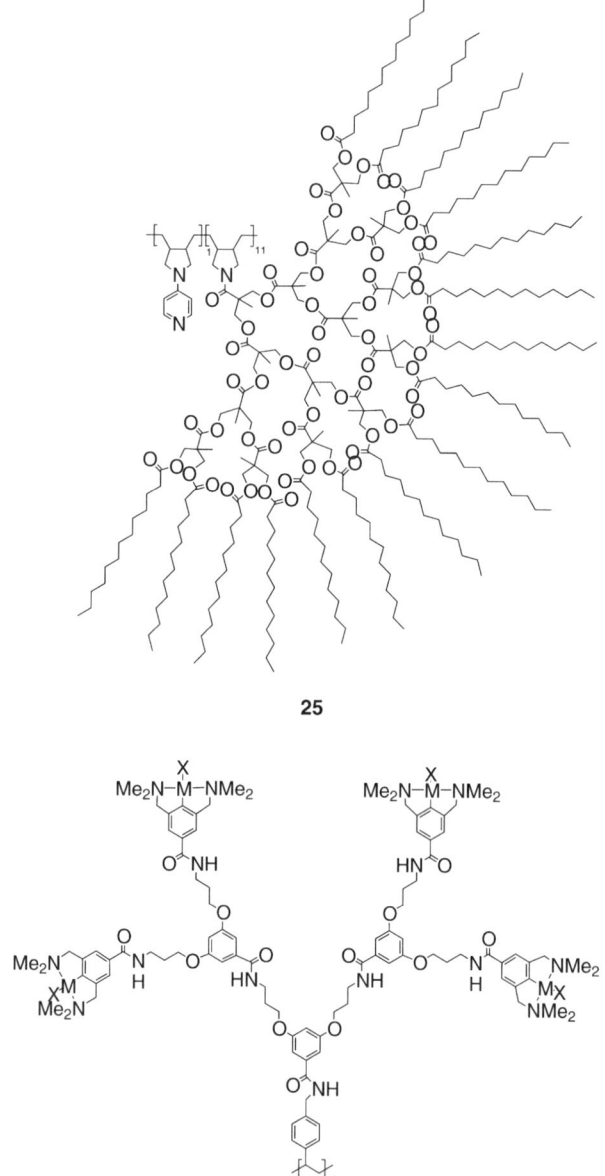

Fig. 11 Chemical structures of the two denpols which have so far been used for catalysis **25**(G4) and **26**(G2). The former has the catalytically active site (the aminopyridine) located in the denpols interior at the backbone whereas the latter carries them (the metallo pincers) – in the fully unfolded conformation – in the periphery

Due to the work of List et al. [47, 97] and Meijer et al. [98] it meanwhile became increasingly accepted that rigorous exclusion of traces of fluorenone-type defects in the backbones is more important for achieving high-performance devices than the dendronization. They should neither form during device operation nor be incorporated through synthesis from the very beginning. Fluorenones emit green–orange, which together with the bluely emitting fluorene units causes the observed devices' white emission. Both effects can be dealt with, at least to some degree. The oxidative sensitivity and thus the bleaching can be reduced, for example, by an appropriate structure design. Consequently, the fluorene C-9 position should carry two organic substituents (and not only one) and aryls seems to better than alkyls. This is what the color stabilities of **11**($n=1$) [46] and **11**($n=0$) showed [47]. Additionally, serious precautions must be taken during the synthesis in order to prevent the incorporation of fluorenone impurities [98]. It seems that many published data were obtained from non-optimally synthesized materials, and it is thus difficult to seriously compare different polyfluorene derivatives and even batches of one and the same structure in a quantitative manner. There are too many factors which have an influence on color stability. For **11**($n=0$) under presently optimal conditions, a blue color stability can be achieved for up to approximately 30 min. Though this is a considerable improvement compared to many other published systems, it is not sufficient to be industrially relevant. This also holds true for **12** (Fig. 4). It may turn out in the forthcoming years that the very enticing concept of dendronization does not prove to not be the best for such applications, but rather that a perfectly synthesized polymer with a more simple substitution pattern will lead to even better devices.

Dalton's group has been developing organic materials for second order nonlinear optical (NLO) applications [99, 100]. They achieved electro-optic coefficients of greater than 100 pm/V at telecommunication wavelength. These values are a factor of 3 larger than those of the best inorganic material, lithium niobate. The macroscopic electro-optic activity of organic materials depends on the molecular hyperpolarizability, β, of individual organic chromophores, and on the product of number density, N, and noncentrosymmetric order, $<\cos^3\Theta>$, of the chromophores in a hardened polymer matrix. Based on attempts for a rational material's design, this group recently prepared the dendronized copolymer **27** and related structures (Fig. 12) and tested them for their NLO behavior. The best representative of this series exhibited electro-optic coefficients of 182 pm/V at 1.3 µm communication wavelength and, upon further optimization, values of 300 pm/V are assumed realizable within the near future. Very recently there appeared another interesting work in the same direction, which describes a dendronized polyimide showing both high thermal stability and considerable electro-optic activity [101].

Denpols and SFM fit together like a high speed car and a German Autobahn. The car can only develop its full potential when driven at high speed which the Autobahn allows it to do. On the one hand, SFM is an invaluable tool for both characterization and full exploitation of potential applications of denpols and,

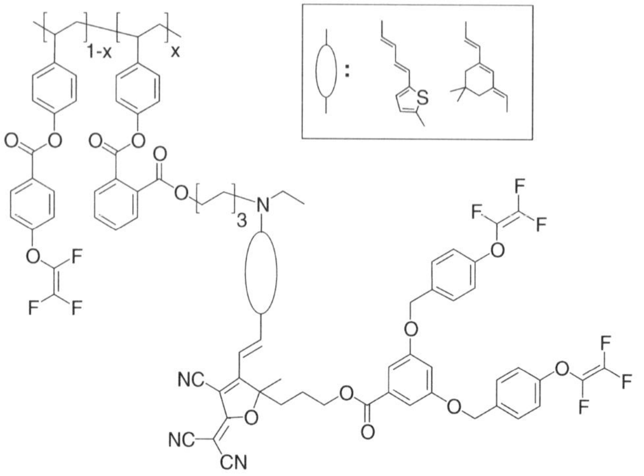

Fig. 12 Chemical structure of the rather complex first generation denpol 27 which shows a remarkably high second order non-linear optical effect

on the other, high-generation and long denpols are an ideal study object for the SFM. Denpols meet all the requirements of SFM to allow this method to play to its full strength for soft organic matter. Far beyond mere imaging of individualized denpols, which can be used for histographic contour lengths determinations, for example, these macromolecules are ideal candidates for manipulation experiments as they can be dragged in a predetermined way across atomically flat surfaces such as HOPG [40]. By combining this controlled dragging of denpols with a chemical reaction within the SFM experimental set-up a so-called move-connect-prove sequence was recently developed which allows covalent chemistry between two individual molecules [65]. This can thus be regarded as a basic step towards the bottom-up approach to the nanosciences. This experiment shall be briefly described in the following.

The idea is outlined in Fig. 13a; the real experiment is shown in Fig. 13b. The blue worms resemble denpols whose periphery is decorated with thousands of photosensitive azide groups. Azides, upon irradiation with UV-light, decompose into the highly reactive nitrene intermediates. Nitrenes exhibit a high and non-selective reactivity towards formation of covalent chemical bonds. They "bite" into almost everything they can find. If one irradiates two strands of such denpols after they have been brought into close contact by the SFM, an intermolecular nitrene reaction will occur at the site where the two strands are in intimate contact. In order to prove that this connection has actually taken place, the presumed linkage was mechanically challenged (Fig. 13b). All arms of the X-shaped arrangement were moved individually leaving the connection intact. Even if the arms were cut off and the whole arrangement of the formerly

A Covalent Chemistry Approach to Giant Macromolecules with Cylindrical Shape 183

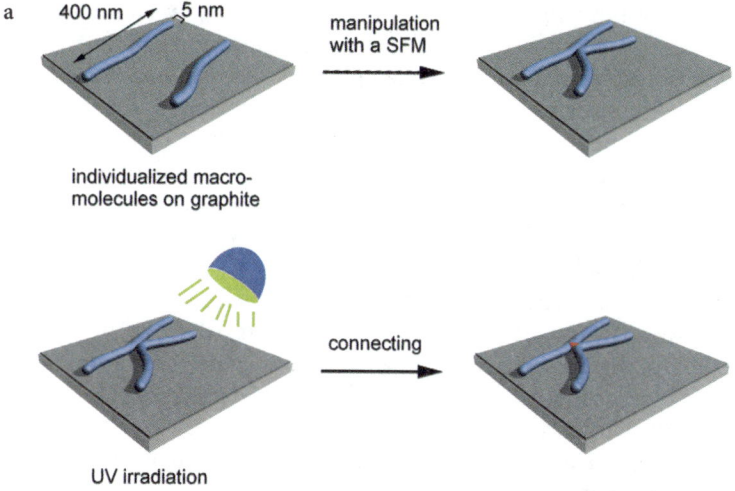

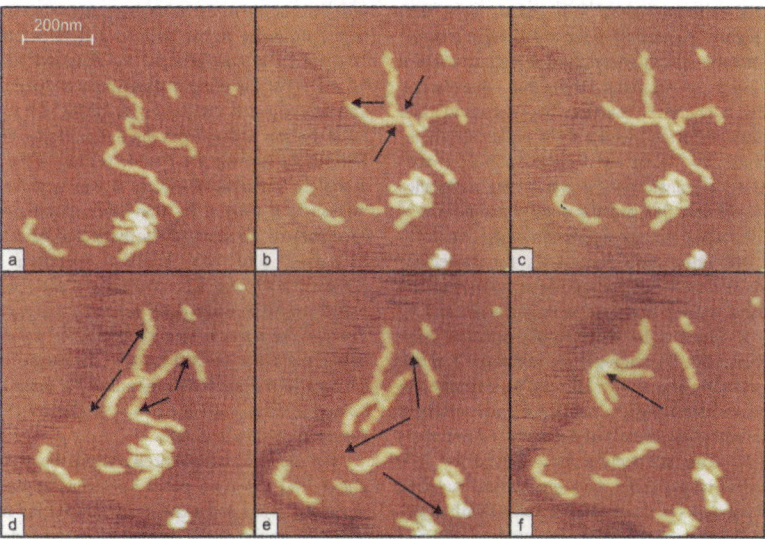

Fig. 13 a Cartoon representation and **b** real experiment SFM image series of a move-connect-prove sequence in whose course two individual denpols are covalently connected by a photochemically induced cross-linking reaction. The *arrows* in **b** indicate the SFM tip direction

two molecules was moved 100 nm or so across the surface, no disassembly into two individual denpols was observed. Thus, the connection had been achieved. In a control experiment in which two denpols were brought into the same tight contact but without irradiation the two strands remained individual species and could easily be separated from one another. With this experiment, the systematic generation of functional assemblies between denpols and other large molecules/objects like DNA or buckytubes is the next important step. Before any application could be envisaged, however, it would be necessary to run such processes serially.

The last potential application of denpols which should be mentioned here is directed towards compactization of DNA and gene delivery [102]. It uses the fact that many denpols are novel polyelectrolytes and, depending on their charge, should give rise to the formation of ordered aggregates with DNA which is a negatively charged polyelectrolyte. Many denpols carry protected amine functionalities in the periphery which can be converted into the corresponding positively charged ammoniums. Spherically shaped, cationic dendrimers have been successfully employed for DNA compactization and transfection studies [103–105]. A molecular model for the formed polyelectrolyte aggregates, however, has not yet been obtained. This is where denpols with their enormous size and defined cylindrical shape come into play. Rabe and his coworkers mixed quaternized denpols 9(G2) and 10(G4) with double-stranded, linearized plasmid DNA in aqueous buffer and spin-coated the obtained homogenous solution onto a polyornithine-modified mica surface. Besides complex aggregates between denpols and DNA, defined features were also observed by SFM. This allowed one to determine how much of a given DNA strand is involved in the

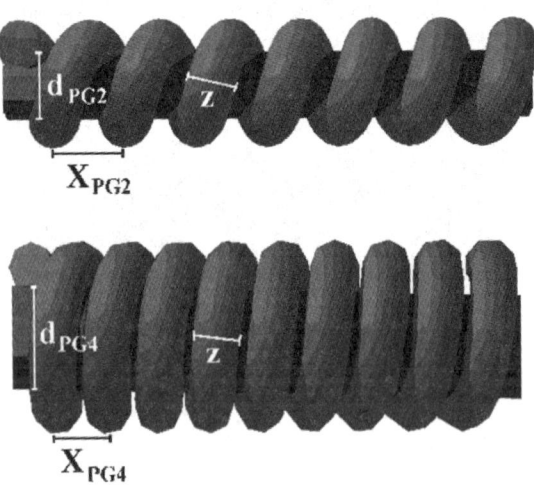

Fig. 14 Molecular scale model based on SFM data of complexes between the deprotected and thus positively charged 9(G2) (*upper*) and 10(G4) (*lower*) with linearized double stranded plasmid DNA ($d_{PG2}=3.2+0.3$, $X_{PG2}=2.30+0.27$, $z=2$; $d_{PG4}=6.6+0.7$, $X_{PG4}=2.16+0.27$, $z=2$)

aggregate and also how many DNA strands contribute to the aggregate. The complete analysis of all SFM data indicated that DNA wraps around the denpols (and not the other way around) and that they form defined helical aggregates with the structural characteristics given in Fig. 14. This is the first molecular model for such a polyelectrolyte aggregate.

4
Miscellaneous

The polymers discussed so far (ideally) have structurally perfect dendritic side chains and normally one such chain per repeat unit, even though there are exceptions such as polymers **15** and **16** which carry two. Another exception is **25**, only roughly 90% of whose repeat units carry a dendron. There are a number of works on macromolecules combining linear and branched structural elements that do not have these features, but have arisen from similar thinking patterns and will, therefore, be mentioned here. The first work to refer to, although now relatively old, is a 1992 publication by Fréchet in which he describes a copolymerization between dendronized styrene macromonomers and styrene. Even though only a few macromonomers could actually be incorporated into the backbone and, thus, only a few repeat units carry dendrons, this was the first fully proven case of a dendronized polymer [106]. Reduced coverages by dendrons were recently also realized by Roy, who decorated chitosan with G1 and G2 sialic acid dendrons, employing reductive amination as the coupling chemistry [107]. Considering the complexity of carbohydrate chemistry, the coverages of 13% (G1) and 6% (G2) achieved are respectable. Along the same line, attempts were made to decorate single and multi-walled nanotubes with dendrons. The first approach by Hirsch et al. [108] and Sun et al. [109] utilized defect sites and the functional groups located there (mostly carboxylic acids) as anchor groups for dendron attachment. Very recently, Hirsch et al. succeeded with the more elegant and more widely applicable approach in which dendrons were directly attached to the intact buckytube surface [110, 111]. He reacted single-walled carbon nanotubes (SWCNTs) dispersed in 1,1,2,2-tetrachloroethane with Frechet-type dendrons carrying alkoxycarbonyl azides at the focal point, under thermal treatment. As mentioned above, the azides decompose to nitrenes which bite into the tube's surface and, thus, generate structure **28** with covalently bound dendrons (Fig. 15). The main advantage of dendronized nanotubes is their good solubility in organic solvents under ambient conditions, which is an essential feature for processing and characterization. For the latter, a variety of techniques such as NMR, Raman, UV/Vis/NIR, and XPS spectroscopies, thermal analysis, and SFM were used. The dendronization of nanotubes also helps with another important problem, which is the high tendency of nanotubes to form stable and ill-defined clusters. This way the chance to access isolated nanotubes is greatly improved.

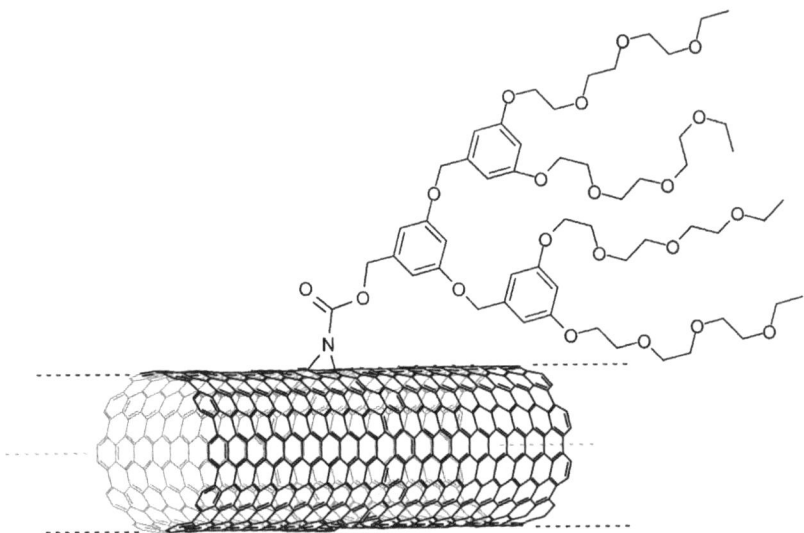

Fig. 15 Structure of a dendronized carbon nanotube

Dendrimers and hyperbranched polymers [112–114] are often compared regarding their respective advantages and disadvantages. On the positive side of dendrimers is their (near-) perfect structure, important for all matters where the definedness of an interior and monodispersity play a role. In contrast, hyperbranched polymers are clearly superior as far as synthetic effort is concerned. They can normally be obtained in a simple one-pot procedure rather than multi-step sequences, which for certain dendrimers can reach the dimension of a complex natural product synthesis. Nevertheless, some hyperbranched polymers reach dendrimer properties to a degree which is sufficient for some applications. It is not surprising, therefore, that efforts were undertaken to also make polymers with hyperbranched side chains. Credit is due to Frey et al. for developing a hyperbranched carbosilane macromonomer with an oxazoline polymerizable group and successfully subjecting it to a cationic polymerization [114]. Through a careful optimization of the polymerization conditions, DP values of 15–20 could be reached. As mentioned above, the achievement of as high as possible a concentration of the polymerizable group, and yet with a stirrable reaction mixture was the critical issue here. Copolymers with a reduced dendron density along the backbone were also prepared by the same procedure. Meanwhile, Frey et al. have extended this work to polybutadienes, to which methyldi(undecenyl)silane was grafted by hydrosilylation [115, 116]. By the nature of this attempt, whether the silane adds to the backbone olefins or to the ones it carries itself at the termini cannot be completely controlled.

Meanwhile, the first theoretical study on denpols has appeared. Stimson, Photinos, and coworkers performed molecular simulation studies [117] on a homologous series of dendronized poly(*para*-phenylene)s, for which very similar derivatives had been synthesized earlier [53, 118]. They found that for generations up to G3 the conformations were basically unhindered and therefore the chains remained flexible. At G4 the relative orientation of the repeat units becomes more restricted but almost rigid conformations were not observed until G5. Interestingly enough, it has been proposed that the backbones for these rigid structures start to meander. Meander formation has also been observed experimentally [119].

Finally, three publications may help underline how lively research in the denpol area has been lately. Zimmerman's group reported a fascinating approach to nanotubes based on denpols [120]. He equipped the denpols with both a cross-linkable surface and a hydrolyzable connection between backbone and dendrons. This way the interior of cross-linked denpols can be hydrolytically removed, thus creating a hollow denpol, which is nothing other than a molecular tube, and whose diameter should be similar to the starting denpol. Though not a denpol in a strict sense, Chow et al. introduced so-called outer sphere–outer sphere connected dendronized organoplatinum polymers [121]. They can be considered a nice development for both the dendrimer and polymer areas. Over the past few years, the Diederich group enriched the whole area by preparing valuable oligomeric denpols. They are interesting for synthetic reasons because new grounds for possible growth reactions like oxidative acetylene dimerization and Hagihara-type cross-couplings were explored, and additionally deserve attention as model compounds for electrochemical and fluorescence studies [122].

5
Summary and Outlook

What direction will the journey take, now that the main synthetic problems to high molar mass denpols with high generation dendrons at every repeat unit have been solved, at least for a few cases with different structures and polymerization methods and for the most part by following route II? Two factors immediately come to mind under synthetic aspects: chain length control and surface decoration. Promising attempts have already been undertaken for both, but quite some more time will be required before these matters can be considered solved and sufficiently exploited. Presently, it still is a non-trivial task to reach narrowly distributed material over a molar mass of 100,000. From a more speculative angle, there is the question of whether there are other targets for the dendronization approach. Would it be feasible, for example, to merge this concept with the biological world of viruses, proteins, membranes, and the like, or the colloid sciences? One may also ask whether certain denpols could be designed so as to show a biological function. As mentioned, research aiming at mimicking the size and shape of TMV is ongoing. Considerable potential is also seen in the use of denpols both with appropriate surface decoration and internal structural conditions as constituents for the bottom-up approach to functional units on the molecular level. They are amongst the largest manmade molecules ever, which enables application of physical tools like SFM for manipulating, connecting, and oxidizing/reducing. Finally, it would be desirable to initiate more exchange between people doing denpols and others working with the so-called poly(macromonomers) [123–132]. The latter polymers have linear chains of considerable molar mass attached to their backbones and in a sense complement the ones with branched chains discussed here. The hairy denpols could serve as a vehicle here.

Acknowledgements I would like to cordially thank the following coworkers for their considerable involvement, creative input, and occasionally high frustration threshold: Drs. Afang Zhang, Lijin Shu, Zhishan Bo, and M. Sc. Rabie el-Hellani. Of equal importance are our cooperation partners, who helped us enormously in discovering the denpols' surprising potential: Profs. Jürgen P. Rabe, Berlin, Manfred Schmidt, Mainz, and Tadeusz Pakula, Mainz. I also wish to thank Prof. Emil List, Graz, Dr. Andrew Grimsdale, Mainz, and Prof. Larry Dalton, Seattle, for giving advice and/or making results available to us prior to publication. Generous financial support by the Deutsche Forschungsgemeinschaft (Sfb 448, TP 1) is gratefully acknowledged. Finally, I would like to thank Dr. Pamela Winchester for her continuous help with the drawings and her competent language check.

References

1. Michl J (ed) (1997) Modular chemistry. Kluwer, Dordrecht
2. Fuhrhop JH, Endisch C (2000) Molecular and supramolecular chemistry of natural products and their model compounds. Dekker, New York
3. Fuhrhop JH, Köning J (1994) Membranes and molecular assemblies: the synkinetic approach. Royal Society of Chemistry, Cambridge
4. Antonietti M, Förster S (2003) Adv Mater 15:1323
5. Claussen RC, Rabatic BM, Stupp SI (2003) J Am Chem Soc 125:12680
6. Zubarev ER, Pralle MU, Li L, Stupp SI (1999) Science 283:523
7. Hartgerink JD, Clark TD, Ghadriri MR (1998) Chem Eur J 4:1367
8. Rosselli S, Ramminger AD, Wagner T, Lieser G, Höger S (2003) Chem Eur J 9:3482
9. Simpson CD, Wu J, Watson MD, Müllen K (2004) J Mater Chem 14:494
10. Tschierske C (2002) Nature 419:681
11. Cheng XH, Das MK, Siegmar D, Tschierske C (2002) Angew Chem Int Ed 41:4031
12. Ungar G, Lin Y, Zeng X, Percec V (2003) Science 299:1208
13. Antonietti M (2003) Nature Mater 2:9
14. Decher G (1997) Science 277:1232
15. Goldacker T, Abetz V, Stadler R, Erukhimovich I, Leibler L (1999) Nature 398:137
16. Lehn JM (1995) Supramolecular chemistry, concepts and perspectives. VCH, Weinheim
17. Clark CG, Wooley KL (2001) Regioselectively crosslinked nanostructures. In: Fréchet JMJ, Tomalia DA (eds) Dendrimers and other dendritic polymers. Wiley, Chichester, p 147
18. Mertesdorf C, Ringsdorf H, Stumpe J (1991) Liq Cryst 9:337
19. Kim Y, Mayer MF, Zimmerman SC (2003) Angew Chem Int Ed 42:1121
20. Ghadiri MR, Granja JR, Milligan RA, McRee DE, Khazanovich N (1993) Nature 366:324
21. Hecht S, Khan A (2003) Angew Chem Int Ed 42:6021

22. Newkome GR, Moorefield CN, Vögtle F (2001) Dendrimers and dendrons – concepts, syntheses, and applications. Wiley-VCH, Weinheim
23. Fréchet JMJ, Tomalia DA (2001) Dendrimers and other dendritic polymers. Wiley-VCH, Chichester
24. Freudenberger R, Claussen W, Schlüter AD (1994) Polymer 35:4496
25. Percec V, Heck J, Tomazos D, Falkenberg F, Blackwell H, Ungar G (1993) J Chem Soc Perkin Trans 2799
26. Schlüter AD (1998) Top Curr Chem 197:165
27. Frey H (1998) Angew Chem Int Ed 37:2193
28. Schlüter AD, Rabe JP (2000) Angew Chem Int Ed 39:864
29. Zhang A, Shu L, Bo Z, Schlüter AD (2003) Macromol Chem Phys 204:328
30. Karakaya B, Claussen W, Schäfer A, Lehmann A, Schlüter AD (1996) Acta Polymer 47:79
31. Tomalia DA, Stahlbush JR (1988) US Patent 4857599
32. Desal A, Atkinson N, Rivera F, Devonport W, Ress I, Branz SE, Hawker CJ (2000) J Polym Sci Polym Chem 38:1033
33. Quali N, Méry S, Skoulios A, Noirez L (2000) Macromolecules 33:6185
34. de Gennes PG, Hervét H (1983) J Phys (Paris) 44:L351
35. Kim C, Kang S (2000) J Polym Sci Polym Chem 38:724
36. Kim C, Kwark K (2002) J Polym Sci Polym Chem 40:976
37. Alonso B, González B, García B, Ramírez-Oliva E, Zamora M, Casado CM, Cuadrado I (2001) J Organometal Chem 637–639:642
38. Grayson SM, Fréchet JMJ (2001) Macromolecules 34:6542
39. Grayson SM, Lee CC, Fréchet JMJ (2002) Am Chem Soc Polym Div Polym Prepr 43:775
40. Shu L, Schlüter AD, Ecker C, Severin N, Rabe JP (2001) Angew Chem Int Ed 40:4666
41. Shu L, Goessl I, Rabe JP, Schlüter AD (2002) Macromol Chem Phys 203:2540
42. Förster S, Neubert I, Schlüter AD, Lindner P (1999) Macromolecules 32:4043
43. Sato T, Jiang DL, Aida T (1999) J Am Chem Soc 121:10568
44. Bao Z, Amundson KR, Lovinger AJ (1998) Macromolecules 31:8647
45. Jakubiak R, Bao Z, Rothberg L (2000) Synth Met 61:114
46. Setayesh S, Grimsdale A, Weil T, Enkelmann V, Müllen K, Meghdadi F, List EJW, Leising G (2001) J Am Chem Soc 123:946
47. Pogantsch A, Wenzl, FP, List EJW, Leising G, Grimsdale AC, Müllen K (2002) Adv Mater 14:1061
48. Yamamoto T (1992) Progr Polym Sci 17:1153
49. Chou CH, Shu CF (2002) Macromolecules 35:9673
50. Malenfant PRL, Fréchet JMJ (2000) Macromolecules 33:3634
51. Geissler H (1998) Transition metal-catalyzed cross coupling reactions. In: Beller M, Bolm C (eds) Transition metals for organic synthesis. Wiley VCH, Weinheim, p 165
52. Schlüter AD (2001) J Polym Sci Polym Chem 39:1533
53. Bo Z, Schlüter AD (2000) Chem Eur J 6:3235
54. Jiang J, Liu HW, Zhao YL, Chen CF, Xi F (2002) J Polym Sci Polym Chem 40:1167
55. Andreopoulou AK, Kallitsis JK (2002) Macromolecules 35:5808
56. Andreopoulou AK, Carbonnier B, Kallitsis JK, Pakula T (in press) Macromolecules
57. Scrivante A, Fasan S, Matteoli U, Seraglia R, Chessa G (2000) Macromol Chem Phys 201:326
58. Girbasova N, Aseyev V, Saratovsky S, Moukhina I, Tenhu H, Bilibin A (2003) Macromol Chem Phys 204:2258
59. Bilibin A, Zorin I, Saratovsky S, Moukhina I, Egovora G, Girbasova N (2003) Macromol Symp 199:197
60. Zhang A, Zhang B, Wächtersbach E, Schmidt M, Schlüter AD (2003) Chem Eur J 9:6083
61. Shu L, Schlüter AD (2000) Macromol Chem Phys 201:239

62. Percec V, Rudick JG, Wagner M, Maganov S, Obata M, Balagurusamy VSK, Heiney PJ (2002) Am Chem Soc Polym Div Polym Prepr 43:428
63. Kaneko T, Horie T, Aoki T, Oikawa E (2000) Kobunshi Ronbunshu 57:672
64. Percec V, Obata M, Rudick JG, De BB, Glodde M, Bera TK, Magonov SN, Balagurusamy VSK, Heiney PA (2002) J Polym Sci Polym Chem 40:3509
65. Barner J, Mallwitz F, Shu L, Schlüter AD, Rabe JP (2003) Angew Chem Int Ed 42:1932
66. Matyjaszewski K, Xia J (2001) Chem Rev 101:2921
67. Kamigato M, Ando T, Sawamoto M (2001) Chem Rev 101:3689
68. Hawker CJ, Bosman AW, Harth E (2001) Chem Rev 101:3661
69. Moad G, Mayadunne RTA, Rizzardo E, Skidmore M, Thang SH (2003) Macromol Symp 192:1
70. Barner-Kowollik, Davis TP, Heuts JPA, Stenzel MH, Vana P, Whittaker M (2003) J Polym Sci Polym Chem 41:365
71. Chiefari J, Mayadunne RTA, Moad CL, Moag G, Rizzardo E, Postma A, Skidmore MA, Thang SH (2003) Macromolecules 36:2273
72. Chong YK, Krstina J, Le TPT, Moad G, Postma A, Rizzardo E, Thang SH (2003) Macromolecules 36:2256
73. Zhang A, Wei L, Schlüter AD (in press) Macromol Rapid Commun
74. Cheng CX, Tang RP, Zhao YL, Xi F (2004) J Appl Polym Sci 91:2733
75. Malkoch M, Carlmark A, Woldegiorgis A, Hult A, Malmström EE (2004) Macromolecules 37:322
76. Hadjichristidis N, Pitsikalis M, Iatrou H, Pispas S (2003) Macromol Rapid Commun 24:979
77. Tsukahara Y, Mizuno K, Segawa A, Yamashita Y (1989) Macromolecules 22:1546
78. Wintermantel M, Schmidt M, Tsukahara Y, Kajiwara K, Kohjiya S (1994) Macromol Rapid Commun 18:279
79. Dziezok P, Sheilo SS, Fischer K, Schmidt M, Möller M (1997) Angew Chem Int Ed Engl 36:2812
80. Schappacher M, Billaud C, Paulo C, Deffieux A (1999) Macromol Chem Phys 200:2377
81. Ryu SW, Hirao A (2000) Macromolecules 33:4765
82. Kotani Y, Kato M, Kamigaito M, Sawamoto M (1996) Macromolecules 29:6979
83. Qin S, Matyjaszewski K, Xu H, Sheiko SS (2003) Macromolecules 36:605
84. Zhang M, Breiner T, Mori H, Müller AHE (2003) Polymer 44:1449
85. Zhang A, Barner J, Goessl I, Rabe RJ, Schlüter AD (2004) Angew Chem Int Ed 43:5185
86. Zhang A, Schlüter AD (2004) Am Chem Soc Polym Div Polym Prepr 44:524
87. El-Hellani R, Schlüter AD (2004) Polym Mater Sci Engin 91:387
88. Kwon YK, Chvalun SN, Blackwell J, Percec V, Heck JA (1995) Macromolecules 28:1552
89. Percec V, Ahn CH, Ungar G, Yeardley DJP, Möller M, Sheiko SS (1998) Nature 391:161
90. Rapp A, Schnell I, Sebastiani D, Brown SP, Percec V, Spiess HW (2003) J Am Chem Soc 125:13284
91. Zhang A, Okrasa L, Pakula T, Schlüter AD (2004) J Am Chem Soc 126:6658
92. Mel'nikov AB, Polushina GE, Atonov EA, Ryumtsev EI, Lezov AV (2000) Polymer Sci Ser A 42:760
93. Lezov AV, Rjumtsev EE, Mel'nikov AB, Filippov SK, Polushina GE, Mikhailova ME, Antonov EA. Private communication; to be published
94. Liang CO, Helms B, Hawker CJ, Fréchet JMJ (2003) Chem Commun 2524
95. Suijkerbuijk BMJM, Shu L, KleinGebbink RJM, Schlüter AD, van Koten G (2003) Organomet 22:4175
96. Albrecht M, van Koten G (2001) Angew Chem Int Ed 40:3750
97. Scherf U, List E (2002) Adv Mater 14:477

98. Craig MR, de Kok MM, Hofstaat JW, Schenning APHJ, Meijer EW (2003) J Mater Chem 13:2861
99. Dalton LR (2004) Pure Appl Chem 76:1421
100. Dalton LR (2003) J Phys Cond Matter 15:R897
101. Luo J, Haller M, Li H, Tang HZ, Jen AKY (2004) Macromolecules 37:248
102. Goessl I, Shu L, Schlüter AD, Rabe JP (2002) J Am Chem Soc 124:6860
103. Roessler BJ, Bielinski AU, Janczak K, Lee I, Baker Jr JR (2001) Biochem Biophys Res Commun 283:124
104. Zinselmayer BH, Mackay SP, Schatzlein AG, Uchegbu IF (2002) Pharm Res 19:960
105. Eliyahu H, Servel N, Domb AJ, Barenholz Y (2002) Gene Ther 9:850
106. Hawker CJ, Fréchet JMJ, (1992) Polymer 33:1507
107. Sashiwa H, Shigemasa Y, Roy R (2001) Macromolecules 34:3905
108. Holzinger M, Hirsch A, Bernier P, Duesberg GS, Burghard M (2000) Appl Phys A70:599
109. Sun YP, Huang W, Lin Y, Fu K, Kitaygorodskiy A, Riddle LA, Yu YJ, Carroll DL (2001) Chem Mater 13:2864
110. Holzinger M, Abraham J, Whelan P, Graupner R, Ley L, Hennrich F, Kappes M, Hirsch A (2003) J Am Chem Soc 125:8566
111. Holzinger M, Vostrowsky O, Hirsch A, Hennrich F, Kappes M, Weiss R, Jellen F (2001) Angew Chem Int Ed 40:4002
112. Voit B (2000) J Polym Sci Polym Chem 38:2505
113. Mitsutoshi J, Masa-aki K (2001) Progr Polym Sci 26:1233
114. Lach C, Hanselmann H, Frey H, Mülhaupt R (1998) Macromol Rapid Commun 19:461
115. Pusel T, Frey H, Lee YU, Jo WH (2001) Polym Mater Sci Eng 84:730
116. Pusel T (2002) PhD thesis, University of Mainz
117. Christopoulos DK, Photinos DJ, Stimson LM, Terzis AF, Vanakaras G (2003) J Mater Chem 13:2756
118. Karakaya B, Claussen W, Gessler K, Saenger W, Schlüter AD (1997) J Am Chem Soc 119:3296
119. Ecker, C, Shu L, Schlüter AD, Rabe JP (2004) Macromolecules 37:2484
120. Kim Y, Mayer MF, Zimmerman SC (2003) Angew Chem Int Ed 42:1121
121. Chow HF, Leung CF, Li W, Wong KW, Xi L (2003) Angew Chem Int Ed 42:4919
122. Schenning APHJ, Arndt JD, Ito M, Stoddart A, Schreiber M, Siemsen P, Martin RE, Boudon C, Gisselbrecht JP, Gross M, Gramlich V, Diederich F (2001) Helv Chim Acta 84:296
123. Yoshimoto N, Wakiyama S, Tsukahara Y, Kaeriyama K (2001) Design Monom Polym 4:203
124. Nomura K, Takahashi S, Imanishi Y (2001) Macromolecules 34:4712
125. Tsukahara Y, Namba SI, Iwasa J, Nakano Y, Kaeriyama K, Takahashi M (2001) Macromolecules 34:2624
126. Müller AHE, Cheng G, Böker A, Krausch G (2001) Polym Mater Sci Eng 84:91
127. Fischer K, Schmidt M (2001) Macromol Rapid Commun 22:787
128. Shinoda H, Miller PJ, Matyjaszewski (2001) Macromolecules 34:3186
129. Gnanou Y, Héroguez V, Six JL, Grande D, Fontanille M (1999) Am Chem Soc Polym Chem Div Polym Prepr 40:161
130. Dziecok P, Sheiko SS, Fischer K, Schmidt M, Möller M (1998) Angew Chem Int Ed 36:2812
131. Zhang M, Breiner T, Mori H, Müller AHE (2003) Polymer 44:1449
132. Li C, Gunari N, Fischer K, Janshoff A, Schmidt M (2004) Angew Chem Int Ed 43:1101

Functionalization of Carbon Nanotubes

Andreas Hirsch · Otto Vostrowsky (✉)

Institut für Organische Chemie, Universität Erlangen-Nürnberg, Henkestrasse 42, 91054 Erlangen, Germany
andreas.hirsch@chemie.uni-erlangen.de, otto.vostrowsky@chemie.uni-erlangen.de

1	Carbon Nanotubes – A New Carbon Allotrope	194
2	Functionalization of Carbon Nanotubes	196
3	The Reactivity of Carbon Nanotubes	197
4	Covalent Functionalization	199
4.1	Oxidative Purification – Carboxylation of CNTs	199
4.2	Defect Functionalization – Transformation and Modification of Carboxylic Functions	201
4.2.1	Amidation – Formation of Carbon Nanotube-Acyl Amides	201
4.2.2	Formation of Carbon Nanotube-Esters	207
4.2.3	Thiolation – Formation of SWCNT-CH_2-SH by Modification of Oxidized SWCNTs	208
4.3	Halogenation of Carbon Nanotubes	209
4.3.1	Fluorination of Nanotubes	209
4.3.2	Chlorination of Carbon Nanotubes	211
4.3.3	Bromination of MWCNTs	212
4.4	Hydrogenation of Carbon Nanotubes	212
4.5	Addition of Radicals	213
4.6	Addition of Nucleophilic Carbenes	214
4.7	Sidewall Functionalization Through Electrophilic Addition	214
4.8	Cycloadditions	215
4.8.1	Carbene Addition	215
4.8.2	Addition of Nitrenes	215
4.8.3	Nucleophilic Cyclopropanation	217
4.8.4	Azomethine Ylides	217
4.8.5	Diels-Alder Reaction	218
4.8.6	Sidewall Osmylation of Individual SWCNTs	219
4.9	Aryl Diazonium Chemistry – Electrochemical Functionalization – Cathodic and Anodic Coupling	219
5	Noncovalent Exohedral Functionalization	221
6	Endohedral Functionalization	230
7	Conclusions	230
References		231

Abstract Through functionalization of carbon nanotubes, i.e., the attachment of appropriate chemical functionalities onto their conjugated sp^2 carbon scaffold, the prerequisites for faciliating the production of possible applications of such nanostructures are established. The derivatized tubes exhibit improved properties with respect to solubility and ease of dispersion, manipulation and processibility, and can be considered a true subdivision of organic molecules. In the present contribution, the current state of functionalization is reviewed, distinguishing between covalent and noncovalent functionalization, defect and sidewall functionalization, exohedral and endohedral functionalization and supramolecular complexation. The covalent functionalizations are classified according to their chemistry and type of reaction.

Keywords Carbon nanotubes · Functionalization · Supramolecular chemistry · Nanochemistry

1
Carbon Nanotubes – A New Carbon Allotrope

When the Japanese scientist Sumio Iijima in 1991 observed hollow carbon tubules by high-resolution electron microscopy formed by a high current arc discharge process to evaporate graphite [1], a novel carbon allotrope was discovered. The small tubules consisted of concentrically nested tube-like graphene structures, each successive outer shell having a larger diameter. Because of their nanometer-scale diameters and their consisting of up to dozens of concentric tubes, these tubules were called multi-walled carbon nanotubes (MWCNTs). Two years later, with the same arc process and in the presence of catalytic particles, single-shelled nanotubes, made up of just one single layer of carbon atoms, were synthesized and consequently called single-walled carbon nanotubes (SWCNTs) (Fig. 1) [2, 3]. Both MWCNTs and SWCNTs can be grown

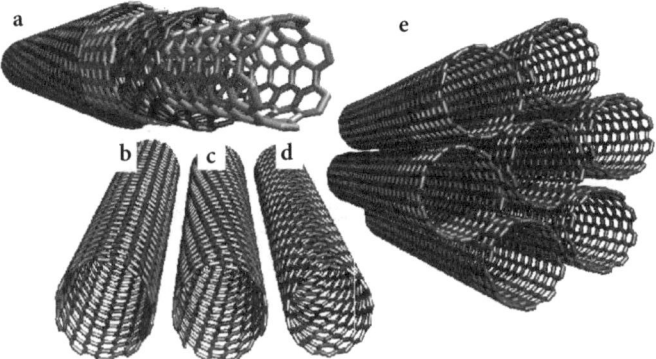

Fig. 1a–e Idealized representations of different structures of defect-free and opened carbon nanotubes. **a** Concentrical multi-walled carbon nanotubes (MWCNT). **b** "Metallic" armchair (10,10) single-walled carbon nanotubes (SWCNT). **c** Helical chiral semiconducting SWCNT. **d** Zigzag (15,0) SWCNT. **e** SWCNT bundle. The armchair (**b**) and zigzag tubes (**d**) are achiral

in tangled structures or ordered close-packed structures. A typical SWCNT will have a cylindrical structure with a diameter in the order of nanometers and can be up to a few micrometers long. Within the group of SWCNTs, because of different "rollings" of a graphene sheet into a cylinder, achiral zigzag and armchair tubes and helical chiral nanotubes are to be distinguished [4]. SWCNTs typically exist as ropes or bundles 10–30 nm in diameter and a few micrometers in length. The SWCNT ropes are entangled together in the solid state to form a highly dense, complex network structure. Despite their simple chemical composition and atomic bonding configuration, carbon nanotubes (CNTs) exhibit an extreme diversity in structures and structure–property relations [4].

CNTs exhibit rather outstanding and unique mechanical and physical properties, and immediately after their discovery were regarded as new materials for future technologies. Today they are widely recognized as *the* essential contributors to nanotechnology, a technology concerned with the control and usage of matter on the nanometer to molecular scale and with the creation of nanoscale building blocks with fundamentally new physical, chemical, biological properties and functions. Depending on their structure, diameter and chirality (and thus on their production), CNTs may be either metallic or semiconductors and represent ballistic electron conductors [5–7], carrying current with essentially no resistance, the electrons streaming through the tube without scattering off defects or atoms [8]. Consequently, nanotubes have also been isolated that had both a metallic region and a semiconducting region, resulting from a specific defect [9]. They have been shown to possess outstanding field emission properties and may be used as electronic elements in future nanoelectronics [10, 11]. Individual tubes as well as ensembles of nanotubes are potential candidates as components in the area of electron field emission sources [10, 12], scanning probes [11, 13], nanoelectrical devices [12, 14–17], components of electrochemical energy storage systems [18, 19] or of hydrogen storage systems [20, 21]. They represent the ultimate carbon fiber, with the highest thermal conductivity [22], exceptional mechanical properties [23–26] and the highest tensile strength of any material [27], having Young's moduli of around 1 Tpa [28] and being up to 100 times stronger than steel.

Research on CNTs has, in the past, mainly concentrated on the investigation of their physical properties [29–32]. Continuing developments in the fields of microscopy and electron spectroscopy has made it possible to conduct precise investigations on individual molecules. However, to apply the property profile of CNTs and to profit from their promising technological applications, it is also necessary that large ensembles and bulk samples of pure and uniform nanotube molecules can be investigated and manipulated. CNTs provided by conventional syntheses and with mixed chiralities represent the bottleneck in any application to electronic devices which require purely semiconducting or metallic CNTs [12, 33, 34].

2
Functionalization of Carbon Nanotubes

CNTs in all their forms are difficult to disperse and dissolve in water and in organic media, and they are extremely resistant to wetting. For a long time they defied chemical characterization, synthetic chemical treatment and any solution chemistry. Difficulties also arise making composites of insoluble nanotubes with other materials, and in generating, for example, aligned assemblies, which is important for the construction of photonic and electronic devices. A suitable functionalization of the nanotubes, i.e., the attachment of "chemical functionalities" represents a strategy for overcoming these barriers and has thus become an attractive target for synthetic chemists and materials scientists. Functionalization can improve solubility [35] and processability, and will allow combinination of the unique properties of nanotubes with those of other types of materials. Chemical bonds might be used to tailor the interaction of the nanotube with other entities, such as a solvent, polymer and biopolymer matrices and other nanotubes. A functionalized nanotube might have mechanical or electrical properties that are different from those of the unfunctionalized nanotube, and thus may be utilized for fine-tuning the chemistry and physics of CNTs.

When dealing with functionalization of CNTs, a distinction must be made between covalent and noncovalent functionalization, the functionalization of

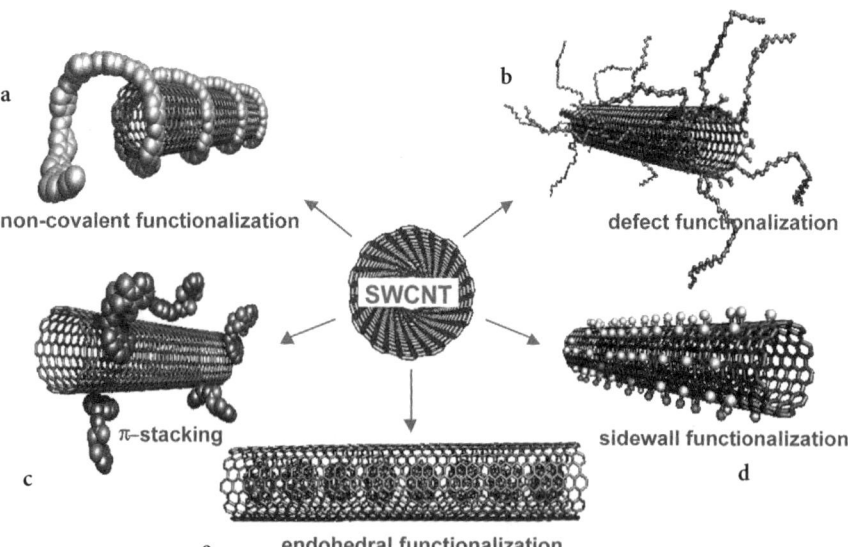

Fig. 2a–e Different possibilities of the functionalization of *SWCNTs*. **a** Noncovalent exohedral functionalization with polymers. **b** Defect-group functionalization. **c** Noncovalent exohedral functionalization with molecules through π-stacking. **d** Sidewall functionalization. **e** Endohedral functionalization, in this case C_{60}@SWCNT

SWCNTs and MWCNTs, and that of individual tubes and tube bundles. Covalent functionalization is based on covalent linkage of functional entities onto the nanotube's carbon scaffold. It can be performed at the termini of the tubes or at their sidewalls. Direct covalent sidewall functionalization is associated with a change of hybridization from sp^2 to sp^3 and a simultaneous loss of conjugation. Defect functionalization takes advantage of chemical transformations of defect sites already present. Defect sites can be the open ends and holes in the sidewalls, terminated, for example, by carboxylic groups, and pentagon and heptagon irregularities in the hexagon graphene framework. Oxygenated sites, formed through oxidative purification, have also to be considered as defects. A non-covalent functionalization is mainly based on supramolecular complexation using various adsorption forces, such as van der Waals' and π-stacking interactions. All these functionalizations are exohedral derivatizations. A special case is the endohedral functionalization of CNTs, i.e., the filling of the tubes with atoms or small molecules (Fig. 2). The chemistry of CNTs has been reviewed in detail over the last two years in several papers [36–40].

3
The Reactivity of Carbon Nanotubes

A perfect SWCNT without functional groups is a quasi-1D cylindrical aromatic macromolecule and is thus expected to be fairly chemically inert. However, in nonplanar conjugated organic molecules, two principal sources, namely curvature-induced pyramidalization of the conjugated carbon atoms and π-orbital misalignment between adjacent pairs of conjugated carbon atoms, induce local strain [36] Although the sidewalls of CNTs and the soccerball-like surfaces of fullerenes are both examples of curved carbon, there are significant structural differences reflected in their chemistry. In particular, fullerenes are curved in two dimensions, 2D, whereas nanotubes are curved in one dimension, 1D only. Thus for a curved carbon structure of given radius, the carbon atom network in a fullerene is more distorted than that in the corresponding CNT. Among the stable fullerenes, C_{60} is the smallest with the highest curvature, and because of the pronounced pyramidalization of the C atoms within the carbon framework, the convex surface is very susceptible towards addition reactions [41]. To curve a graphene sheet into a (10,10) SWCNT requires a pyramidalization angle Θ_P for the carbon atoms of about $\Theta_P=3.0°$. The fullerene of equivalent radius, C_{240} [a (10,10) SWCNT can be capped by a hemisphere of C_{240}], has a $\Theta_P^{max}=9.7°$ (Fig. 3). The strain energy of pyramidalization is roughly proportional to Θ_P^2, so in this case the fullerene stores about ten times the strain energy of pyramidalization per carbon atom, compared to the "equivalent" CNT [36]. Only the smallest (and probably unstable) tube with (2,2) dimensions has a higher degree of pyramidalization than C_{60}, although the C_{60} diameter (7.10 Å) is much larger than that of a (2,2) SWCNT (2.75 Å). The first armchair

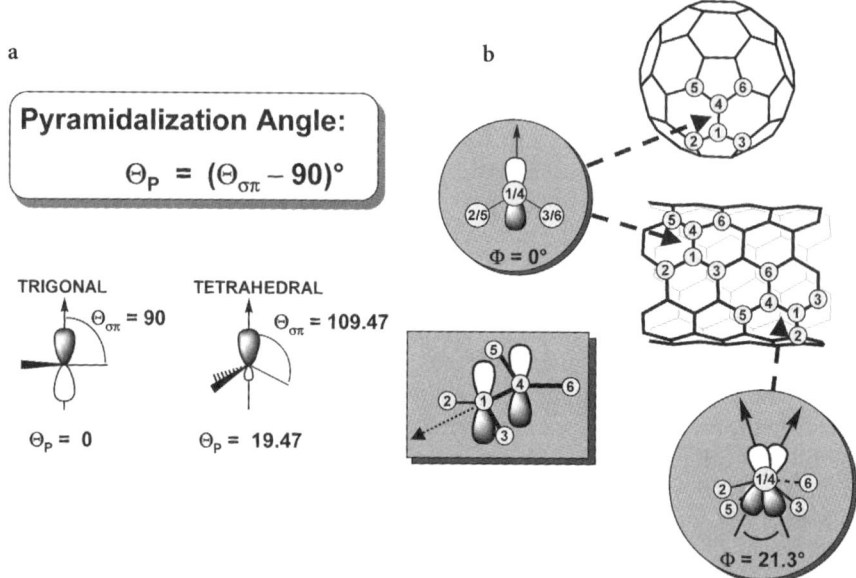

Fig. 3a, b Diagrams of **a** the pyramidalization angle (Θ_P), and **b** the π-orbital misalignment angles (Φ) along the C1-C4 bond in a (5,5)-SWCNT and C_{60}

tube with a larger diameter than C_{60} is a (6,6) SWCNT. CNTs, even the thin highpressure CO conversion (HiPco)-tubes with a typical diameter of 7 Å, comparable to that of C_{60}, are expected to be less reactive. The more pyramidalized C-atoms of the end caps of closed CNTs resemble a hemispherical fullerene and ensure that they are always more reactive than the sidewalls, irrespective of the diameter of the CNT [36].

However, while the π-orbital alignment in the fullerenes is almost perfect, this is not the case for all the bonds in the CNTs. This can be seen in the illustration of an arm-chair (5,5) SWCNT in Fig. 3 (pyramidalization angle, $\Theta_P^{max}=6°$). Although all carbon atoms are equivalent, there are two types of bonds: those that run parallel to the circumference (or perpendicular to the nanotube axis) and those at an angle to the circumference with π-orbital misalignment angles (Φ, deg) of $\Phi=0°$ and 21.3°, respectively. The analogous values for the (10,10) SWCNT are $\Phi=0°$ and 10.4 °. On the basis of previous calculations of torsional strain energies in conjugated organic molecules, π-orbital misalignment is likely to be the main source of strain in the CNTs. This represents a clear contrast with fullerene chemistry, although there are parallels with the reactivity of homofullerenes. Just as in the case of the fullerenes, the reactivity of CNTs arises out of their topology but for different reasons. Furthermore, since the pyramidalization angles and the π-orbital misalignment angles of SWCNTs scale inversely with the diameter of the tubes, a differentiation is expected between the reactivity of CNTs of different diameters [36].

4
Covalent Functionalization

Covalent functionalization of SWCNTs using addition chemistry has recently started to be developed, an approach considered to be very promising for nanotube modification and derivatization. However, it is difficult to achieve control over the chemo- and regioselectivity of such additions, which require very "hot" addends such as arynes, carbenes, or halogens, and drastic reaction conditions, to form covalent bonds. Furthermore, it is very difficult to characterize functionalized SWCNTs as such and to determine the precise location and mode of addition [41].

The covalent chemistry of nanotubes is not particularly rich with respect to different chemical reactions to date. This holds for SWCNTs as well as for MWCNTs and SWCNT bundles etc. Most of the examples of chemistry and functionalization discussed in the following will be focussed on SWCNTs.

4.1
Oxidative Purification – Carboxylation of CNTs

One of the most common functionalization techniques is the "oxidative purification" of nanotubes by liquid-phase or gas-phase oxidation, introducing carboxylic groups and some other oxygen-bearing functionalities such as hydroxy, carbonyl, ester and nitro groups into the tubes. Oxidative treatment ("purification") has been achieved using boiling nitric acid [42–47], sulfuric acid [48] or mixtures of both [42], "piranha" (sulfuric acid–hydrogen peroxide) [49] or gasous oxygen [50, 51], ozone [52–55] or air as oxidant [56–58] at elevated temperatures, or combinations of nitric acid and air oxidation [20]. The aim of this oxidative treatment, drawing its inspiration from well-known graphite chemistry [59, 60], is the oxidative removal of the metallic catalyst particles used in the synthesis of the tubes and of the amorphous carbon, a byproduct of most synthetic methods [49].

Upon oxidative treatment the introduction of carboxylic groups and other oxygen-bearing groups at the end of the tubes and at defect sites is promoted, decorating the tubes with a somewhat indeterminate number of oxygenated functionalities. However, mainly because of the large aspect ratio of CNTs, considerable sidewall functionalization also takes place (Fig. 4). HNO_3-oxidation in combination with column chromatography and vacuum filtration led to a new purification method for SWCNTs, developed by Holzinger et al. [47] The oxidative treatment introduces defects on the nanotube surface [61], oxidizes the CNTs ("hole-doping"), and produces impurity states at the Fermi level of the tubes [62]. In addition to purifying the raw material by removal of impurities, oxidation can be used to cut (or "etch"), shorten and open the CNTs [43, 49, 63–65]. Cutting and shortening depends on the rate and extent of the reaction and gives rise to a new length distribution which can be determined by transmission electron microscopy (TEM). The cutting procedure will leave

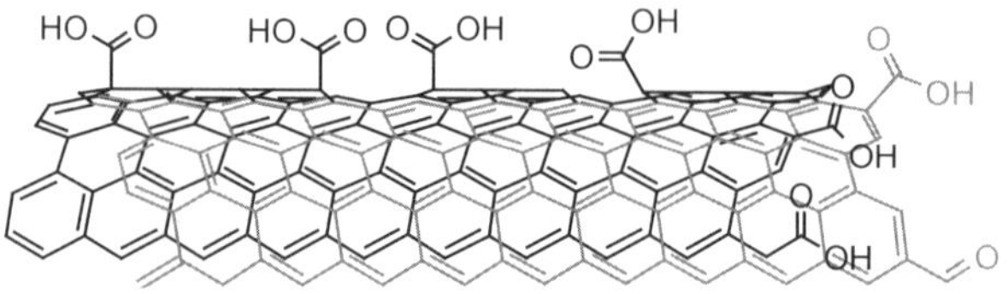

Fig. 4 Section of an oxidized SWCNT, reflecting terminal and sidewall oxidation

SWCNTs with open and oxygenated ends. The occurrence, plurality and degree of oxygenated functionalities have been determined spectroscopically [52, 53]. Since reactivity is a function of curvature [66], the oxidative stability also depends on the tubes' diameter [46, 67] and on the production process responsible for the tubes' dimensions. Yates et al. [52] showed that with 1.4 nm diameter SWCNTs, room temperature oxidation by ozone is confined to the end caps and the "dangling" bonds created by the removal of these caps. A detailed study of diameter-dependent oxidative stability by Fischer et al. [67] recently confirmed a direct relationship between diameter and reactivity. Using the resonance-enhanced Raman radial breathing mode (rbm) the authors clearly showed that smaller diameter tubes are more rapidly air-oxidized than larger diameter tubes [67]. Bahr and Tour cited in their review on the covalent chemistry of SWCNTs [37] that smaller diameter SWCNTs produced by the HiPCO process (Ø ca. 0.7 nm) were more reactive towards ozone than larger diameter SWCNTs produced by laser ablation. This led to a simple way to enrich large-diameter SWCNTs by using a mixed, concentrated H_2SO_4/HNO_3 treatment [68].

Solution phase mid-IR spectroscopy was used to assess the amount of functionality introduced into SWCNTs by oxidation. For a (10,10) SWCNT containing 40 carbon atoms in the unit cell, 20 carboxylic acid groups at each end of the tubes were found. A perfect 100 nm long (10,10) SWCNT-COOH contains $40\times(1000/2.46)$, approximately 16.000, C-atoms and 40 carboxylic acid groups [69]. Zhang et al. [70] claim to have improved the efficiency of nanotube oxidation and functionalization by using $KMnO_4$ as oxidant with the help of a phase transfer catalyst (PTC). A preliminary comparison between the nanotube oxidation with $KMnO_4$ with and without PTC resulted in a yield of about 35–40% of functionalized tubes in terms of the total weight of starting material for the reaction without PTC. For the reaction with PTC, the yield of functionalized nanotubes was about 65–70%. However, it should be mentioned that these functionalized CNTs had a higher concentration of -OH groups (~23%) and a lower concentration of -COOH groups (~3.8%) [70].

SWCNTs oxidized by either acid or ozone treatment have been assembled on a number of surfaces, including silver [71], highly-oriented pyrolitic graphite (HOPG) [72], and silicon [73]. To mitigate the problem of poor matrix–SWCNT connectivity and phase segregation in polymer–SWCNT hybrid materials, acid-treatment-oxidized and negatively charged SWCNTs were assembled layer-by-layer (LBL) according to the LBL technique with positively charged poly(ethyleneimine) (PEI) polyelectrolyte [74]. After subsequent chemical cross-linking, a nanometer-scale composite with SWCNT loadings as high as 50 wt% could be obtained with a tensile strength approaching that of ceramics [74]. Oxidized aligned MWCNT arrays, grown by chemical vapour deposition (CVD) on platinum substrate and acid- or air-treated, were used to immobilize the enzyme glucose oxidase [75]. The enzyme immobilization, achieved by incubation of the oxidized nanotube "sensor", allows for a direct electron transfer from the enzyme to the platinum transducer. The device can thus be used as an amperometric biosensor to record the conversion of glucose to gluconic acid electrically [75].

4.2
Defect Functionalization – Transformation and Modification of Carboxylic Functions

4.2.1
Amidation – Formation of Carbon Nanotube-Acyl Amides

The introduction of carboxylic groups to nanotubes by oxidative procedures gives access to a large number of functional exploitations by transformation of the carboxylic functions, and provides anchor groups for further modification. From carboxylic acids, carboxamides can be formed via carboxylic acid chlorides and allow for the decoration of oxidized tubes with aliphatic amines, aryl amines, amino acid derivatives, peptides, amino-group-substituted dendrimers etc. as nucleophiles. The carboxylic groups can be activated by conversion into acyl chloride groups with thionyl chloride [49, 76], and the acyl chlorides formed can be transformed to carboxamides by amidation (Scheme 1). Similarly, carboxamide nanotubes have been prepared using di-

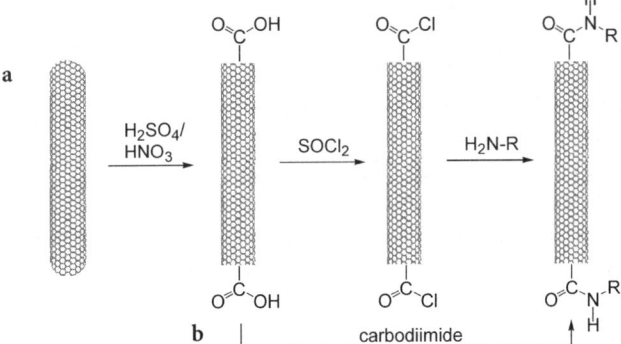

Scheme 1 Schematic representation of oxidative etching ("cutting") of SWCNTs followed by treatment with thionyl chloride and subsequent amidation (**a**). For simplicity, the depiction here as well as the following are not considering the nature and location of functionalities. Only one functional group per tube end is shown. The oxidations also occur at defect sites along the sidewalls, and other functionalities such as esters, quinones, and anhydrides are also formed. Dicyclohexylcarbodiimide is shown as coupling reagent (**b**)

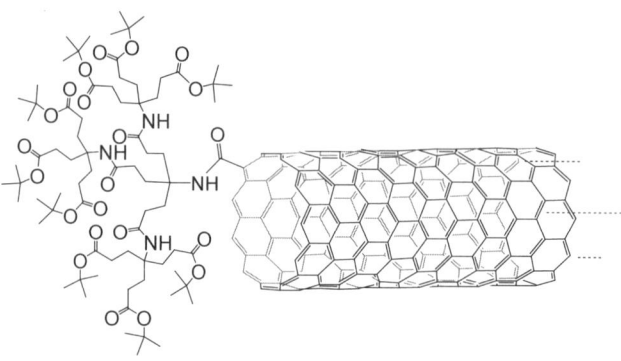

Fig. 5 Modification of oxidized carbon nanotubes (CNTs) with Newkome type amino dendrimers. Second generation dendron and SWCNT core

cyclohexylcarbodiimide (DCC) as dehydrating agent and allowing the direct coupling of amines and carboxylic functions under mild, neutral conditions [77–79].

Newkome-type dendrons were attached to the carbon scaffolds of SWCNTs and MWCNTs by defect group functionalization [80]. First- and second-generation amine dendrons like the ones depicted in Fig. 5 were condensed with the carboxyl groups of purified and opened SWCNTs and MWCNTs according to the carbodiimide technique [80]. CNT derivatives of that kind can be expected to combine the characteristics of CNTs, i.e., electrical, optical and mechanical properties, with those of dendrimers, potential building units for supramolecular, self-assembling and interphase systems.

Haddon et al. were the first to report the functionalization of oxidatively treated SWCNTs with alkyl amines and less nucleophilic aniline derivatives [76, 81]. The conversion of the acid functionality to the amide of octadecyl amine led to the first shortened, soluble SWCNTs [76]. The analysis of (end group and defect site) octadecylamido (ODA)-functionalized SWCNTs by solution phase mid-IR spectroscopy showed that the weight percentage of the acylamido functionality was about 50% [82]. Holzinger [77] obtained ODA SWCNTs by direct condensation of carboxylated SWCNTs with ODA according to the carbodiimide (DCC) coupling technique. Water solubilization of SWCNTs has been achieved by amidation of SWCNT-COCl with glucosamine. The solubility ranged from 0.1 to 0.3 mg/mL [83]. Exposure of CNT-acyl chlorides to H_2N-$(CH_2)_{11}$-SH in toluene at room temperature produced an amide linkage of the nanotube to the alkanethiol [49]. Although the more nucleophilic thiols would be expected to form thioesters predominantly, free thiols were shown to exist by atomic force microscopy (AFM) imaging of attached 10 nm gold nanoparticles [49]. Similarly, Liu et al. [84] achieved the thiolation of SWCNT pipes by reacting cysteamin HS-$(CH_2)_2$-NH_2 with carboxyl-terminated nanotubes under carbodiimide conditions yielding CNT-CONH-$(CH_2)_2$-SH. The functionalized tubes could be assembled as monolayers on a gold surface via Au-S chemical bonding (Scheme 2) [84]. Selective thiolation may be used to make electrical junctions between CNTs and metal electrodes or to position SWCNTs relative to a noble metal surface by taking advantage of the strong thiol-metal interactions.

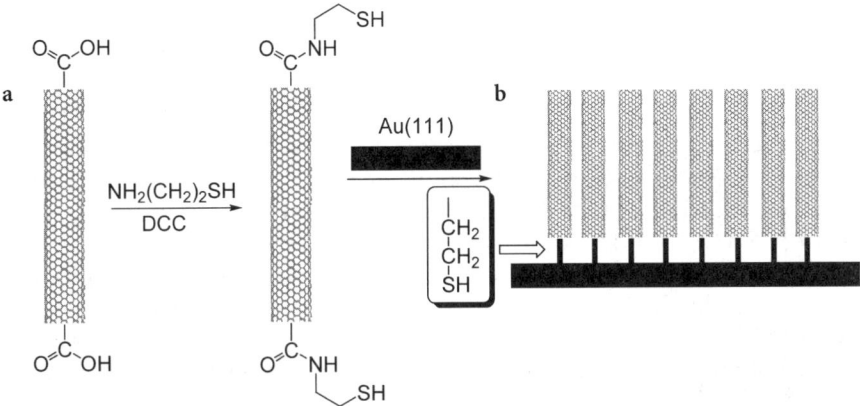

Scheme 2 Schematic diagram of the thiolization reaction of carboxyl-terminated CNTs with cysteamine $NH_2CH_2CH_2SH$ (a) and of the assembling structure of SWCNTs on gold (b)

In a theoretical study by Basiuk et al. it was found that a significant energetic difference between amidation of carboxylic groups in armchair and zigzag SWCNTs exists [85]. Experimental verification of this prediction might enable selective solubilization, and hence separation, of different SWCNT types.

Upon reacting SWCNT-acyl chlorides with α,ω-diamines like, e.g., tripropylenetetramine as molecular linker and subsequent diamide formation Roth et al. [86] and Kiricsi et al. [87] succeeded in the interconnection of tubes and the formation of CNT junctions (Fig. 6). End-to-end and end-to-side nanotube interconnections were formed and observed by AFM. Statistical analyses of the AFM images showed around 30% junctions in functionalized material [88].

A gas-phase derivatization procedure was employed for direct amidization of oxidized SWCNTs with simple aliphatic amines (Scheme 3). In some cases a minor amount of chemically formed amides in addition to a larger portion of physisorbed amines were observed [89]. Full-length oxidatively purified SWCNTs were rendered soluble in common organic solvents by non-covalent (zwitterionic) functionalization in high yield [81, 90]. Of the SWCNT C-atoms, 4–8% can be functionalized by octadecyl amine in this way [81, 90, 91]. This

Scheme 3 Gas phase amidation of oxidatively purified SWCNT with octadecylamin (ODA) to obtain zwitterionic functionalization products

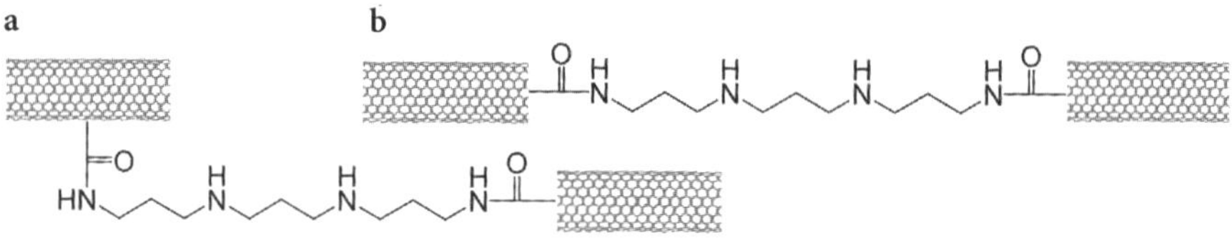

Fig. 6a, b Schematic representation of SWCNT intermolecular junctions connected via two amide linkages, **a** side-to-end and **b** end-to-end junction

acid–base reaction functionalization represents a simple route to solubilizing SWCNTs. The ammonium cation can be readily exchanged by other cations, and the ionic feature of the modified tube may allow electrostatic interactions between SWCNTs and biological molecules.

This procedure led to the exfoliation of the large SWCNTs ropes into small ropes and individual nanotubes. AFM micrographs showed that the majority of the SWCNT ropes were exfoliated into small ropes (∅ 2–5 nm) and individual nanotubes with lengths of several micrometers during the dissolution process. The combination of multiwavelength laser excitation Raman scattering spectroscopy and solution-phase visible and near-IR spectroscopies was used to characterize the library of SWCNTs that are produced using current preparations. The average diameter of nanotubes of metallic nature was found by Raman spectroscopy to be smaller than that of semiconducting nanotubes in the various types of full-length SWCNT preparations. This observation sheds new light on the mechanism of SWCNT formation [90]. Such zwitterion-functionalized SWCNTs in THF solution were length-separated and size-fractioned by GPC chromatography by Papadimitrakopoulos et al. [92]. A total of 40 fractions were collected (1 fraction every 15 s) and AFM was used to obtain the SWCNTs length distribution/fraction.

HiPCO-SWCNTs were oxidized in a UV/O_3 gas-solid interface reaction and subsequently assembled on a rigid oligo(phenylene ethynylene) self-assembled monolayer (SAM) [93]. In a "chemical assembly", based on condensation between the carboxylic acid functionalities of the O_3-SWCNTs and the amine functionalities of the SAMs, SWCNT-amides were formed. A second assembling method was based on physical adsorption via deposition with bridging of metal cations Fe^{3+} on carboxylate-terminated SAMs. High coverage densities and orientation normal to the surface have been shown. The site of functionalization has been suggested to be of higher degree at the nanotube ends [93]. These ordered arrays of SWCNTs may find use in a variety of applications, including field emission devices.

Carboxylic groups at the open ends of SWCNTs were coupled to amines to create AFM probes with basic or hydrophobic functionalities by Lieber et al. (Scheme 4) [94]. Subsequent force titrations recorded between the ends of the SWCNT tips and hydroxy-terminated SAMs confirmed the chemical sensitivity and robustness of these SWCNT tips. Images recorded on patterned SAMs enabled true molecular-resolution imaging [94].

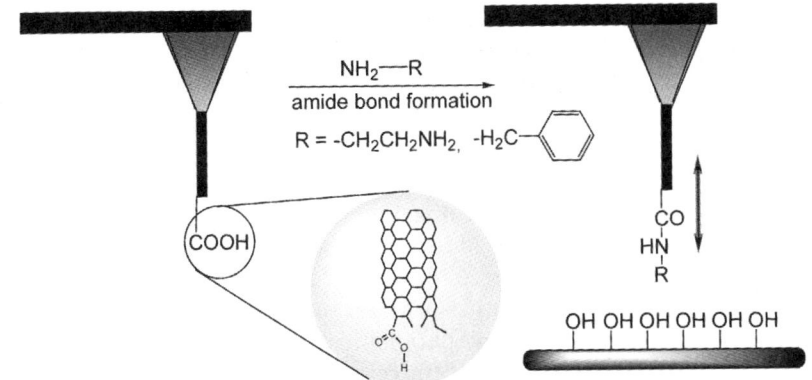

Scheme 4 Schematic illustration of a SWCNT force microscope probe and modification of an oxidized SWCNT tip by coupling an amine RNH$_2$ to a terminal -COOH. The probe is able to sense specific interactions between the functional group R and surface -OH groups

With this amido functionalization, new routes were opened to a covalent linkage of oligomers and polymers, dendrimers, peptides and biopolymers, and to the formation of bioconjugates of CNTs. α,ω-Diaminopolyethylene glycol and long chain ethers of hydroxyaniline were attached to CNTs via amide bonds [95, 96]. Polyethylene oxide (PEO) was grafted onto SWCNTs by amide formation of SWCNT-COCl with monoamine-terminated PEO in DMF and water, and the aggregation behavior of the resulting PEO-*graft*-SWCNTs in solution and in Langmuir-Blodgett (LB) films was investigated [97]. Amino-terminated polystyrene was grafted to oxidatively cut nanotubes via amide formation and nanotubes were obtained which were well dispersible in some organic solvents [98]. Purified MWCNTs were covalently functionalized via attaching aminocopolymer poly(propionylethyleneimine-*co*-ethyleneimine) to the tubes by two different techniques, namely the amidation of CNT-carbonyl chlorides, and heating of the CNT-caboxylic acids in the presence of the aminopolymer to 180 °C for 12 h [99]. First-, second- and third-generation Frechet-type benzylic ether dendrimers and Newkome-type dendrimers with a multifunctional nanotube core were synthesized by Holzinger et al. using DCC and water-soluble 1-ethyl-3-(3-dimethylaminopropyl) carbodiimide (EDC) as condensing reagents in order to improve the solubility of highly oxidized SWCNTs in water [77, 100]. Nanotubes were functionalized by bovine serum albumin (BSA) via diimide-activated amidation; the BSA conjugates obtained were highly water-soluble. Results from characterizations showed the intimate association with the tubes; bioactivity of the nanotube-bound protein was evaluated by a total protein micro-detection assay and showed the majority of the protein in the nanotube-BSA conjugate to be bioactive [101]. Current research is underway to investigate the biocompatibility of chemically inert CNTs by immobilization of biopolymers and proteins on the tubes, to be used for biological applications. The immobilization occurs by covalent amide formation with the protein transferrin, tagged with a fluorescent label, on CNTs in the presence of EDC and sulfo-N-hydroxysuccinimide (sulfo-NHS) [102]. Amino-terminated DNA strands were used to functionalize the open ends and defect sites of oxidatively prepared SWCNTs with ECD as condensing agent [103]. A research group from Israel reports carbodiimide-assisted amidation of SWCNT-COOH with oligonucleotides, and the preparation of a highly water-soluble adduct. Fluorescence imaging of individual nanotube bundles showed that the SWCNT-DNA adducts hybridized selectively with complementary strands [104]. A multistep route to the formation of covalently linked SWCNTs and DNA oligonucleotides was developed by Lasseter et al. and the covalent linkage proven by X-ray photoemission spectroscopy (XPS) [105]. The nanoscience group from Delft University had developed a technique to couple SWCNTs to peptide nucleic acids (PNA). The incorporation of the molecular-recognition features of PNAs into the nanotubes can be used to program selective attachment to a complementary DNA strand or other PNA-SWCNT conjugates by nucleotide hybridization. The method provides a new means of specific deposition of SWCNTs into electronic devices by sequence recognition-based self-assembly and of using SWCNTs as probes in biological systems [106]. From oxidized MWCNT-COOHs with a diameter of 10–30 nm, a glassy carbon electrode (GCE) was fabricated. In the presence of the water-soluble coupling reagent EDC, oligonucleotide probes with an amino group at the 5′-phosphate end were covalently attached, to be used as a DNA-biosensor (Scheme 5). Nucleotide hybridization could be carried out by exposure of the probe-containing electrode to its complementary DNA sequence and detected with the redox intercalator daunomycin as an indicator [107].

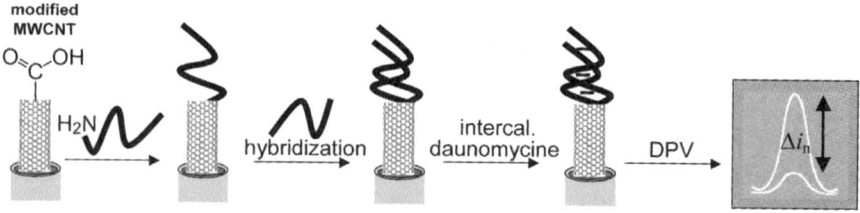

Scheme 5 Schematic representation of the enhanced electrochemical detection of DNA hybridization based on a MWCNT-COOH constructed DNA biosensor. *DPV* Differential pulse voltammetry

By the chemical reduction of amide-solubilized MWCNTs by $LiAlH_4$ the amide groups on the tubes could be reduced to hydroxyl groups, confirmed by FT-IR and XPS studies [108]. The Raman spectroscopic investigation showed that the morphology of the nanotubes did not change after the reduction [108].

4.2.2
Formation of Carbon Nanotube-Esters

Acyl-chloride-functionalized SWCNTs are also susceptible to reaction with other nucleophiles, e.g., alcohols. The Haddon group at Riverside reported the preparation of soluble ester-functionalized SWCNTs: SWCNT-COO$(CH_2)_{17}CH_3$, obtained by reaction with octadecylalcohol [109]. The preparation of soluble polymer-bound and dendritic ester-functionalized SWCNTs has been achieved by Sun and coworkers by attaching poly(vinyl acetate-*co*-vinyl alcohol) [110] and hydrophilic and lipophilic dendron-type benzyl alcohols [96], respectively, to SWCNT-COCl (Fig. 7). These functional groups could be removed under basic and acidic hydrolysis conditions, and thus additional evidence for the nature of the attachment was provided [96, 111].

Labelling of SWCNTs with fluorescence probes was accomplished via esterification by oligomeric species containing derivatized pyrenes, and the products were spectroscopically studied. The fluorescence and fluorescence excitation results showed that the tethered pyrenes form "intramolecular" excimers by π-π-interactions (Scheme 6). The time-resolved fluorescence results show

Fig. 7 **a** SWCNT-ester SWCNT-COO-$(CH_2)_{17}$-CH_3. **b** Ester from SWCNT-COOH and poly(vinyl acetate-*co*-vinyl alcohol). **c** Dendritic type benzyl alcohol ethers

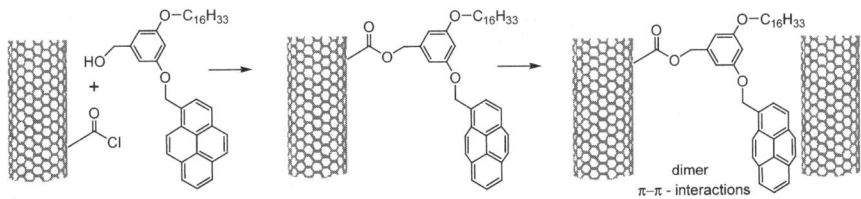

Scheme 6 Labelling of SWCNTs with fluorescence probes via esterification by derivatized pyrenes and formation of "intertubulary" dimers and excimers by π-π-interactions

that the pyrene monomer and excimer emissions are significantly quenched by the attached SWCNTs. The quenching is explained in terms of a mechanism in which CNTs serve as acceptors for excited-state energy transfers from the tethered pyrene moieties [112].

4.2.3
Thiolation – Formation of SWCNT-CH$_2$-SH by Modification of Oxidized SWCNTs

Kim et al. [113] succeeded in 2003 in a direct thiolation by attaching thiol groups to the tubes via successive carboxylation (H$_2$SO$_4$/HNO$_3$; H$_2$O$_2$/H$_2$SO$_4$; sonication), reduction (NaBH$_4$), chlorination (SOCl$_2$) and thiolation (Na$_2$S/NaOH), to the open ends of CNTs (Scheme 7). These open ends were formed by breaking the tubes by sonochemical activation. The intermediate as well as the final products were verified by FT-IR and NMR spectroscopies [113].

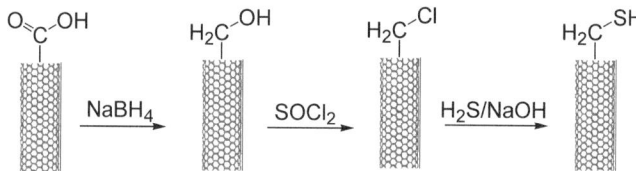

Scheme 7 Thiolation of CNTs by modification of carboxylated tubes

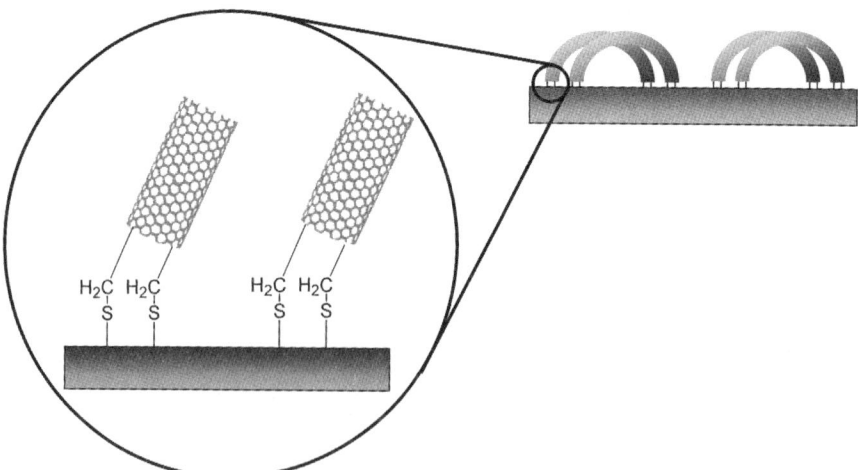

Fig. 8 Schematic representation of the vertical view image of the thiolated CNTs with thiol groups at both ends, binding to a gold surface. The flexible tube body of SWCNTs allows the CNT to conform its geometry for maximum binding energy at the expense of the bending energy. The result is a "bow-type" bundle of the thiolated CNTs

The thiolated CNTs were adsorbed on micrometer-sized silver and gold particles as well as gold surfaces to study the interaction between the thiol groups of the nanotube and the noble metals. The thiol-metal adhesion was studied by scanning electron microscopy (SEM), AFM, wavelength dispersive electron spectroscopy, and Raman spectroscopy. A new type of bonding between the CNT and a noble metal surface was proposed by the authors (Fig. 8), involving a bow-type SWCNT with its two ends strongly attached to the metal surface [113].

4.3 Halogenation of Carbon Nanotubes

4.3.1 Fluorination of Nanotubes

Because of the low reactivity of CNTs' surfaces, fluorination was taken into consideration as one of the first sidewall functionalization reactions in 1996 [114]. Later on, a series of papers dealing with the fluorination of SWCNTs, MWCNTs and CNT-bucky paper (formed by filtration of an SWCNT suspension over a PTFE membrane), were published [115–120]. Most of the papers on the fluorination of CNTs have appeared in the last three years. Extensive controlled and non-destructive sidewall fluorinations of SWCNTs (Fig. 9) using elemental fluorine as the fluorinating agent, and similar to that of the fluorination of fullerenes, [121] were reported by Margrave et al. in 1998 [116]. The authors obtained "fluorocarbon single-walled nanotubes" (F-SWCNT) of approximately C_2F product stoichiometries at temperatures from 150 °C up to 500 °C. Attempts to produce more highly fluorinated material at 500 °C resulted in the destruction of the tubes. Depending on the reaction temperature, the degree of fluorination of nanotubes from different origin varied from $C_{3.9}F$ to $C_{1.9}F$, as determined by XPS [119]. In further work, the authors have demonstrated that F-SWCNTs dissolve well in alcohol solvents and give long-living metastable solutions [117]. Using iodine pentafluoride IF_5, the highest degree of fluorination could be achieved with a composition CF [115]. MWCNTs were fluorinated with elemental fluorine in the temperature range 323–473 K by Touhara et al. [120]. Selective fluorination of the inner surfaces of the CNTs was observed, bringing about a decrease in the surface free energy of the inner tubes' surfaces and an increase in coulombic efficiency of Li/nanotubes rechargeable cells [120].

The fluorinated material exhibited significant changes in its spectroscopic properties (UV/Vis/NIR absorption and Raman scattering), providing evidence

Fig. 9 Schematic depiction of a sidewall-fluorinated SWCNT

for electronic perturbation that was not found in the earlier oxidized materials (see above). The electrical conductivity of the functionalized material differed dramatically from the pristine SWCNTs: untreated SWCNTs were reported to have a resistance of 10–15 Ω, while fluorinated SWCNTs had a resistance of >20 MΩ. Fluorination may result in modifying the electronic structures of the tubes to be either metallic or semiconducting, depending on the coverage and method of fluorine application [117, 122, 123]. For fluorinated SWCNTs, evidence was shown for exfoliation of nanotube-ropes and bundles into individual nanotubes [117].

The major disadvantage of this mode of functionalization must be seen as the high degree of fluorine atoms addition reflecting a great number of tube defects and damage. However, the fluorotubes are very soluble and open new avenues to solution-phase chemistry. SWCNTs, once fluorinated, can serve as a staging point for a wide variety of sidewall chemical functionalizations. By treating fluorinated tubes with strong nucleophiles such as Grignard reagents, alkyl lithium reagents and metal alkoxides, the fluorine substituents were replaced and derivatized products e.g., methyl-, butyl-, hexyl- and methoxyl-nanotubes were formed (Scheme 8, a) [117, 124, 125]. These reactions are thought to proceed via a concerted, allylic displacement mechanism, as backside S_N2 attack is not possible, and the stability of a cationic S_N1 intermediate is questionable. Extension of this displacement capability to functional group-tolerant reagents will further the utility of this chemistry. SWCNTs were found to be "cut" at the fluorinated sites during pyrolysis [126]. F-SWCNTs seem to be a rather promising material, especially for the production of composite materials. Fluorinated SWCNTs have been employed as cathodes in lithium electrochemical cells and are expected to be useful in other applications [118]. By treatment with hydrazine or $LiBH_4/LiAlH_4$, the majority of the covalently bound fluorine could be removed (Scheme 8, b) [117, 119, 124], restoring most of the conductivity and spectroscopic properties of the pristine material.

Korean researchers have investigated the change in atomic and electronic structures of fluorinated SWCNTs using XPS, electrical resistivity measure-

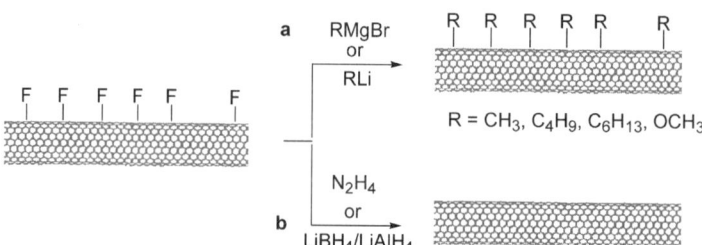

Scheme 8 Treatment of fluorinated tubes with strong nucleophiles and replacement of the fluorine substituents, leading to derivatized methyl-, butyl-, hexyl- and methoxyl-nanotubes as products (**a**). Removal of the covalently bound fluorine by treatment with hydrazine (**b**)

ments and TEM [123]. XPS indicated that the fluorinated SWCNT has an ionic-bonding character at low concentration and covalent-bonding character at high concentration. The resistivity increases with reaction temperatures, resulting from the band-gap enlargement at high fluorine concentration [123].

K.F. Kelly and coworkers explored the rule of addition of fluorine atoms to the walls of armchair SWCNTs using the AM1 and CNDO methods [127]. Scanning tunneling microscopy (STM) images of fluorinated SWCNTs revealed a dramatic banded structure indicating broad continous regions of fluorination terminating abruptly in bands orthogonal to the tube axis. Microscopy images of sparsely butylated SWCNTs, for which fluorinated SWCNTs served as a precursor, were also reported [127]. Theoretical investigations of fluorinated SWCNTs by Bettinger et al. [128] indicated that while the thermodynamic stability of fluorinated SWCNTs decreases with tube diameter, the mean bond dissociation energy for the C–F bonds increases as the diameter decreases: C–F bonds were found to be covalent (1.41 to 1.45 Å in length). It was further predicted that endohedral fluorination of (5,5) SWCNTs is thermodynamically possible.

4.3.2
Chlorination of Carbon Nanotubes

A promising theoretical suggestion for controlled functionalization of SWCNTs with atoms and molecules, among others, chlorine, was recently presented by S.B. Fagan et al. [129]. Their idea was to substitute carbon atoms from the hexagon network of nanotubes with Si atoms, thus gaining an sp^3-like stable defect center as a trapping site for the chemisorption of other atoms or molecules. These defects can be readily modified to Si-X derivatives by different routes, or the Si-X complex directly incorporated. The greatest experimental difficulty will certainly be the insertion of Si into the tubes' carbon network, given that it would prefer an sp^3 rather than an sp^2 hybridization. However, there is vast experimental evidence about the possibility of substitutional doping of fullerenes by Si atoms. As a consequence, it is conceivable to similarly dope CNTs with Si. This approach suggests the functionalization of CNTs by F, Cl, H, CH_3, and SiH_3 [129].

The chemical functionalization of MWCNTs using electrolysis was described by Unger et al. of Infineon Technologies [88]. By electrolytical evolution of chlorine on an anode made from a foil of CNTs, the halogen atoms were coupled to the nanotube lattice. Furthermore, oxygen-bearing functional groups such as hydroxyl and carboxyl are formed at the same time enhancing the solvation of the tubes in water or alcohol. The functionalized CNTs can be converted with sodium amide or triphenylmethyl lithium to add the corresponding functional groups [88].

A more mechano-chemical functionalization of MWCNTs, in the chemical sense actually not belonging to this chapter, was described by Konya et al. in 2002 [130]. From ball-milling of purified MWCNTs in the presence of the reactant gas, Cl_2, cleavage of the tubes' C-C bonds and breaking of the tubes occured and chlorine-functionalized tubes were obtained. Similar functionalizations succeeded with ball-milling in NH_3, H_2S, CH_3SH, etc. reactive atmospheres. The efficiency of the reaction depended strongly on the reactant: uniform short CNTs containing different chemical functional groups such as amine, thiol, mercapto, etc. were formed [130].

4.3.3
Bromination of MWCNTs

The low susceptibility of CNTs to bromination was utilized as a means of purification for MWCNTs contaminated by other carbon particles [131, 132].

4.4
Hydrogenation of Carbon Nanotubes

Several methods for hydrogenation of SWCNTs have been described via dissolved metal reduction. Haddon et al. first reported the Birch reduction of SWCNTs using lithium in diaminoethane [66]. The hydrogenated derivatives were thermally stable up to 450 °C, whilst above 450 °C a characteristic decomposition took place. Transmission electron micrographs showed corrugation and disorder of the nanotube walls due to hydrogenation and the formation of C-H bonds was suggested. The average hydrogen contents determined by Pekker et al. from thermogravimetric analysis (TGA)/MS investigations of SWCNT hydrogenation experiments using lithium in ammonia corresponded to the composition $C_{11}H$ [133]. Hydrogenation occurred even on the inner tubes of MWCNTs, as shown by the chemical composition and the overall corrugation. It may be of interest that, despite the lower reactivity of graphite compared to CNTs (due to the curvature), turbostratic graphitic flakes and powder resulted in a C:H=4:1 ratio upon hydrogenation [133].

Owens and Iqbal [134] succeeded in an electrochemical hydrogenation of open ended SWCNTs synthesized by CVD. Sheets of SWCNT bucky paper were used as the negative electrode in an electrochemical cell containing an aqueous solution of KOH as electrolyte. With laser Raman IR evidence and the release of hydrogen by thermogravimetric measurements at 135 °C the authors claimed to have incorporated up to 6 wt% of hydrogen into the tubes [134]. However, the stability of exohydrogenated CNTs and the low temperature of hydrogen release (135 °C in [134]) is in contradiction with the 400–500 °C cited in references [66, 133].

Hydrogen bound to SWCNTs should not be released until ca. 500 °C, indicating robust attachment [133]. Theoretical first-principles total energy and electronic structure calculations of fully exohydrogenated zigzag and armchair SWCNTs (C_nH_n) points to crucial differences in the electronic and atomic structures with respect to hydrogen storage and device applications. C_nH_ns were estimated to be stable up to the radius of an (8,8) nanotube, with binding energies proportional to $1/R$. Attaching a single hydrogen to any nanotube is always exothermic. By calculation, zigzag nanotubes were found more likely to be hydrogenated than armchair tubes with equal radius [135].

4.5
Addition of Radicals

The addition of perfluorinated alkyl radicals, obtained by photoinduction from heptadecafluorooctyl iodide, to SWCNTs yielded perfluorooctyl-derivatized CNTs (Scheme 9). No remarkable difference in the solubility of the fluoroalkyl-substituted nanotubes and the starting materials was observed [136]. A pathway to the radical-functionalization of CNTs' sidewalls was predicted by classical molecular dynamics simulations of a bombardment of a bundle of SWCNTs by CH_3 radicals [137].

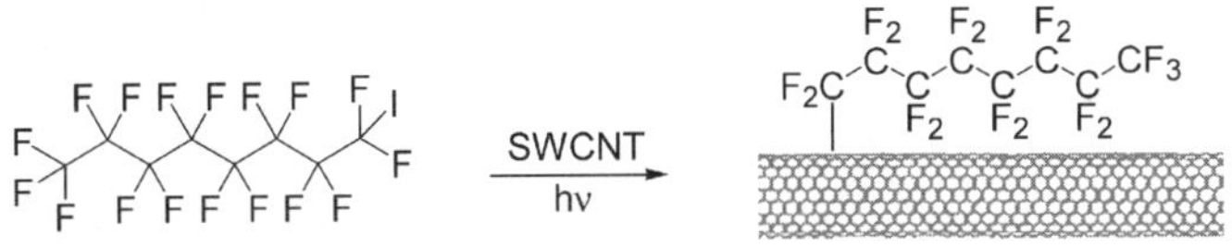

Scheme 9 Addition of heptadecafluorooctyl radicals obtained from photoinduction of heptafluorooctyl iodide

Pristine SWCNTs and their fluorinated derivatives, F-SWCNTs, were reacted with organic peroxides to functionalize them by covalent sidewall attachment of free radicals (Scheme 10), and to compare the reactivity of their corresponding polyaromatic and conjugated polyene π-systems toward radical addition [138]. The characterization of the functionalized SWCNTs and F-SWCNTs was performed by Raman, FT-IR, and UV/Vis-NIR spectroscopy as well as TGA/MS, TGA/FT-IR, and TEM data. In the Raman spectra the decrease in the typical purified HiPCO-SWCNTs' breathing and tangential mode peaks at 200–263 and 1,591 cm^{-1}, respectively, along with the substantial increase in the sp^3 carbon peak at 1,291 cm^{-1} provided diagnostic indication of disruption of

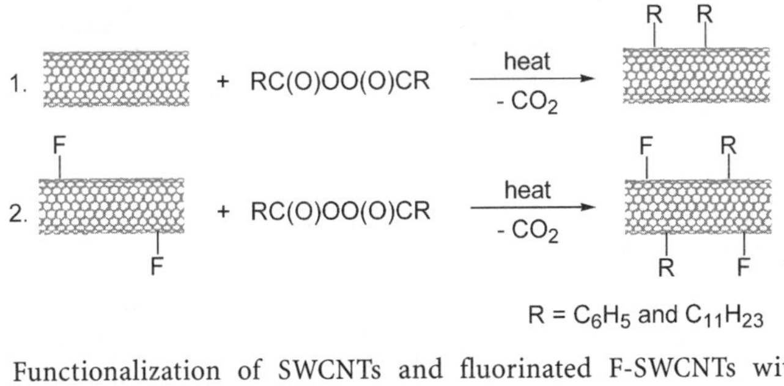

Scheme 10 Functionalization of SWCNTs and fluorinated F-SWCNTs with benzoyl (R=C_6H_5) and lauroyl (R=$C_{11}H_{23}$) peroxides

the graphene π-bonded electronic structure of the sidewalls, suggesting their covalent functionalization. The solution-phase UV/Vis-NIR spectra showed complete loss of the van Hove absorption band structures, typical for functionalized SWCNTs [138].

4.6
Addition of Nucleophilic Carbenes

The reaction of a nucleophilic dipyridyl imidazolidene with the electrophilic SWCNT π-system to give zwitterionic polyadducts was reported in 2001 by Hirsch et al. [136]. Each covalently bound imidazolidene addend bears a positive charge; one negative charge/addend is transferred to the delocalized tube surface (Scheme 11) [136].

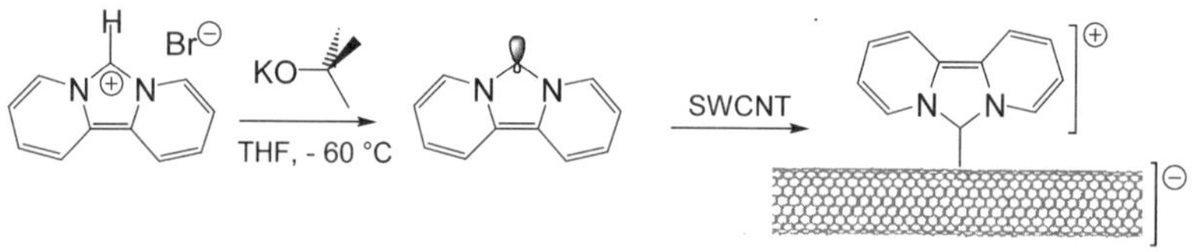

Scheme 11 Sidewall functionalization by the addition of nucleophilic dipyridyl imidazolidene to the electrophilic SWCNT π-system

Sufficiently derivatized nanotubes were quite soluble in DMSO, permitting the separation of DMSO-insoluble, unreacted and insufficiently functionalized SWCNTs [136]. Although the gain in solubility of these carbene derivatives could also be achieved using other chemical decorations of nanotubes, the n-doping of the tubes' surface offers a new way to modify tube properties. This functionalization represents a controlled intervention into the CNT's electronic properties.

4.7
Sidewall Functionalization Through Electrophilic Addition

In 2002, Prato et al. [139] reported the derivatization of SWCNTs through electrophilic addition of $CHCl_3$ in the presence of $AlCl_3$. In addition, from hydrolysis of the so-produced labile chlorinated intermediate species, hydroxy-functionalized SWCNTs were obtained, the coupling of which with propionyl chloride led to the corresponding ester SWCNT derivatives (Scheme 12). The functionalities allowed not only the identification of the structure of the modified tubes, but also provided better solubility in organic solvents [139].

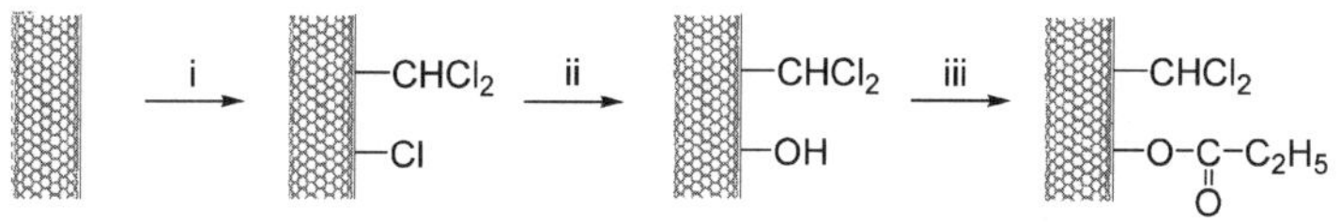

Scheme 12 Electrophilic addition of CHCl₃ to SWCNTs (*i*), followed by chlorine substitution (*ii*) and esterification (*iii*), using *i* CHCl₃, AlCl₃; *ii* OH⁻, MeOH; *iii* C₂H₅CO-Cl

4.8 Cycloadditions

4.8.1 Carbene Addition

In the course of a study on organic functionalization of CNTs, Haddon et al. discovered in 1998 that dichlorocarbene was covalently bound to soluble SW-CNTs (Scheme 13) [76]. The carbene was first generated from chloroform with potassium hydroxide [66], and later from phenyl(bromodichloromethyl)mercury [76]. However, the degree of functionalization was low: a chlorine atom amount of only 1.6 wt% was determined by XPS [140]. Because of impure starting material and a large amount of amorphous carbon, the site of reaction could not be ascertained [66].

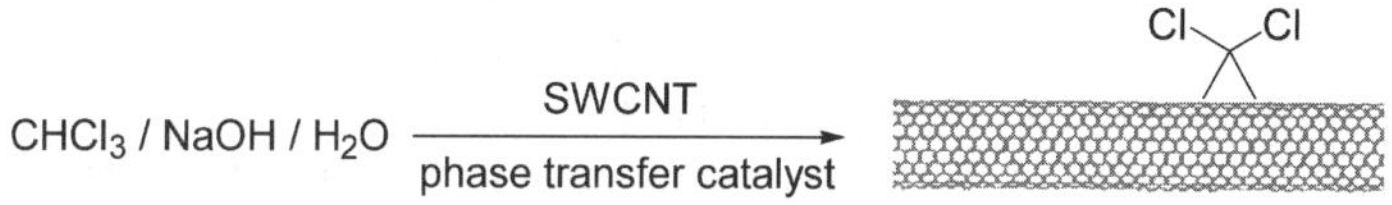

Scheme 13 Dichlorocarbene addition to SWCNTs

4.8.2 Addition of Nitrenes

Sidewall functionalization of SWCNTs was achieved via the addition of reactive alkyloxycarbonyl nitrenes obtained from alkoxycarbonyl azides. The driving force for this reaction is the thermally induced N_2-extrusion (Scheme 14). Such nitrenes attack nanotube sidewalls in a [2+1] cycloaddition forming an aziridine ring at the tubes' sidewalls [136, 141, 142].

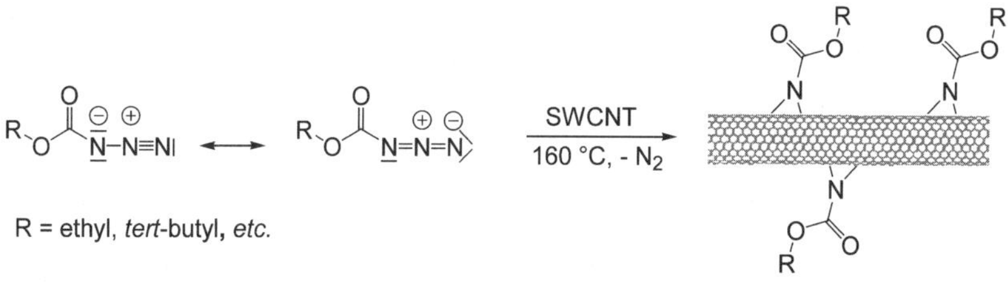

R = ethyl, *tert*-butyl, etc.

Scheme 14 Sidewall functionalization of SWCNTs by the [2+1] cycloaddition of alkoxycarbonyl nitrenes obtained from azides

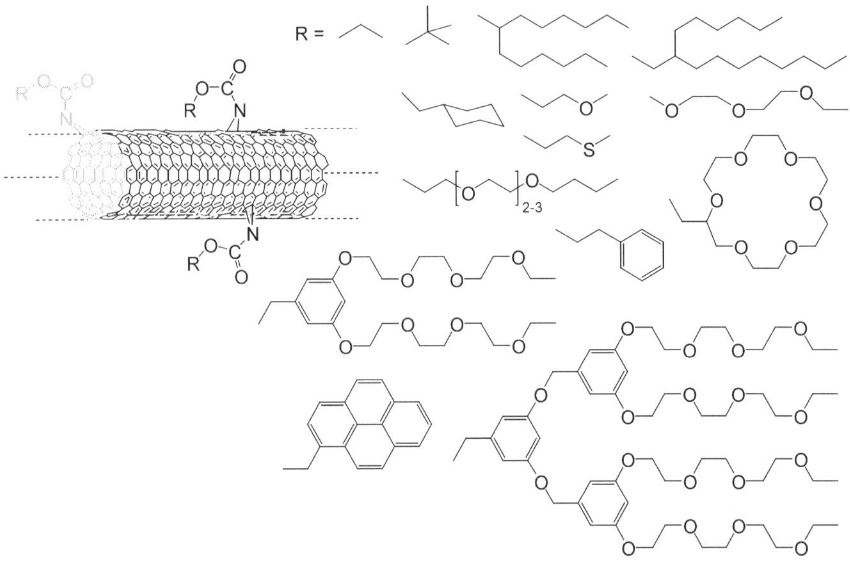

Fig. 10 Sidewall functionalization of SWCNTs via addition of (R)-oxycarbonyl nitrenes

A broad range of addends was obtained by a nanotube chemists group from Erlangen, including such addends like alkyl chains, aromatic groups, crown ethers, and oligoethylene glycol units (Fig. 10) [136, 141, 142].

Nitrene additions led to a considerable increase in the solubility in organic solvents such as 1,1,2,2-tetrachloroethane, DMSO, and 1,2-dichlorobenzene. The highest solubilities of 1.2 mg/mL were found for SWCNT adducts with nitrenes containing crown ethers of oligoethylene glycol moieties in DMSO and TCE, respectively. AFM and TEM revealed that the functionalized tubes formed thin bundles with typical diameters of 10 nm. The presence of thin bundles in solution is supported by ^{1}H NMR spectroscopy. The elemental composition of the functionalized SWCNT was determined by XPS [142, 143]. The use of Raman and electron absorption spectroscopy (UV/Vis-NIR) showed that the electronic properties of the SWCNTs are mostly retained after functionalization, indicating less then 2 wt% addend per C-atom of the tube sidewalls within this series of SWCNT derivatives [141]. Nitrene addition to nanotubes of different origins and production methods was compared with unfunctionalized pristine tubes by investigating their XPS, Raman and UV/Vis/NIR spectra and TEM images [144].

Interconnections between individual SWCNTs using α,ω-dinitrenes as the molecular linker, were first reported in 2003 and the covalent linkage characterized by preliminary microscopy studies [145].

4.8.3
Nucleophilic Cyclopropanation

Fullerenes are known to react easily with bromomalonates to form cyclopropanated methanofullerenes [146, 147]. A similar reaction was performed by Coleman et al. [148] using purified SWCNTs and diethyl bromomalonate as addend (Scheme 15). The authors have developed a chemical tagging technique which allows the functional groups to be visualized by AFM. Immobilized SW-CNT derivatives were transesterified with 2-(methylthio)ethanol, and by exploiting the gold–sulfur binding interaction the cyclopropane group was "tagged" using preformed 5 nm gold colloids [148].

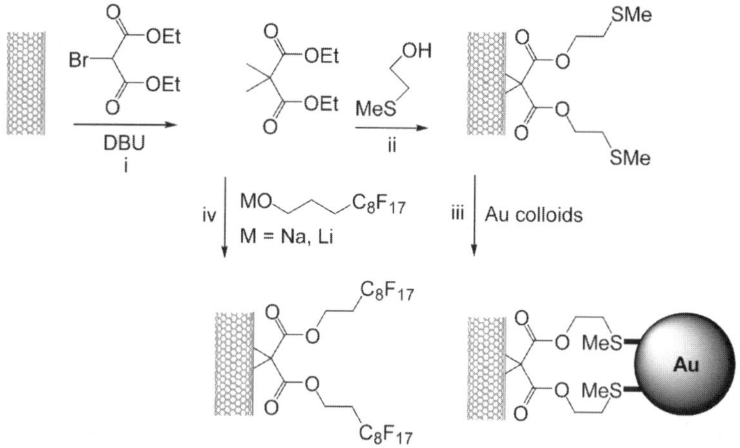

Scheme 15 Schematic representation of the cyclopropanation of SWCNTs and the introduction of chemical markers for AFM visualization, and ^{19}F NMR and XPS spectroscopy. *i* diethyl bromomalonate, DBU, RT; *ii* 2-(methylthio)ethanol, diethyl ether; *iii* preformed 5 nm gold colloids; *iv* sodium or lithium salt of 1H,1H,2H,2H-perfluorodecan-1-ol

Gold colloids were observed both on the sides and at the ends of the nanotubes, indicating sidewall and end modification. To further confirm the derivatization of the SWCNTs, a perfluorinated marker was introduced by transesterification to allow the nanotubes to be probed by ^{19}F NMR and XPS spectroscopy [148].

4.8.4
Azomethine Ylides

Prato et al. recently discovered a method of functionalizing SWCNTs using the 1,3-dipolar addition of azomethine ylides [149, 150], originally developed for modification of C_{60}. Treatment of SWCNTs in DMF with an aldehyde and an N-substituted glycine derivative at elevated temperature resulted in the formation of substituted pyrrolidine moieties on the SWCNT surface (Scheme 16), and in the solubilization of the functionalized tubes in most common organic solvents. The approach was reported to work for all types of CNTs tested, including native and oxidatively etched SWCNTs, with SWCNTs prepared by several different methods (and different diameters), and MWCNTs. The functionalized CNTs were investigated by analytical techniques as well as TEM [149, 150].

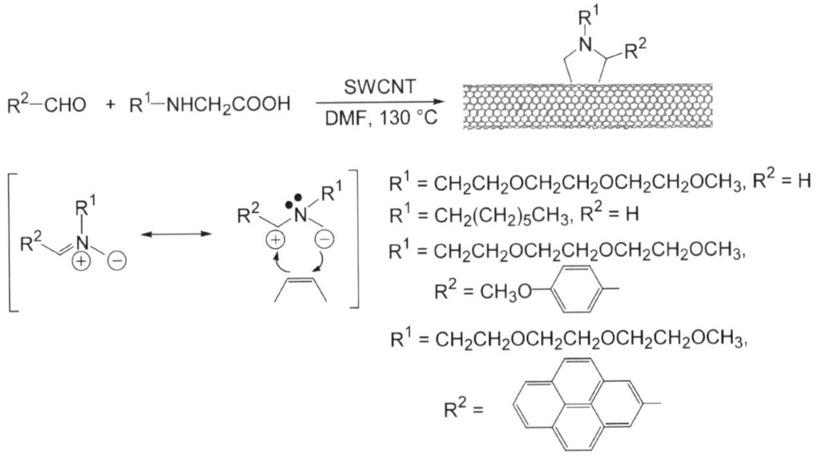

Scheme 16 1,3-Dipolar addition of azomethine ylides to SWCNTs and MWCNTs

The azomethine ylide functionalization technique (above) was also used for the purification of HiPCO-SWCNTs [151].

The viability of 1,3-dipolar cycloadditions of a series of 1,3-dipolar molecules, such as azomethine ylide, ozone, nitrile imine, nitrile ylide and others, onto the sidewalls of CNTs has been assessed theoretically by means of a two-layered ONIOM approach. The calculations confirmed that both azomethine ylide and ozone can be readily attached on the nanotube sidewall. It was predicted that next to azomethine ylides and ozone, nitrile imines and nitrile ylides would be the best candidates for researchers to try [152, 153].

4.8.5
Diels-Alder Reaction

The viability of Diels-Alder [4+2] cycloadditions of conjugated dienes onto the sidewalls of SWCNTs was assessed by means of a computational [two-layered ONIOM(B3LYP/6–31G*:AM1)] approach [154]. Whereas the Diels-Alder reaction of 1,3-butadiene on the sidewall of an armchair (5,5) nanotube is found to be unfavorable, the cycloaddition of quinodimethane is predicted to be viable due to the aromaticity stabilization at the corresponding transition states and products [154].

4.8.6
Sidewall Osmylation of Individual SWCNTs

Sidewall osmylation of individual metallic SWCNTs has been achieved by exposing the tubes to OsO_4 vapor under UV photoirradiation (Scheme 17a) [155]. The covalent attachment of osmium oxide was found to increase the electrical resistance of the tubes by up to several orders of magnitude. Cleavage of the cycloaddition product was accomplished with the aid of UV light or under an oxygen atmosphere, whereby the original resistance is restored. The reversible cycloaddition of OsO_4 represents the first example of a covalent modification electrically monitored at the single-tube level (Scheme 17b) [155].

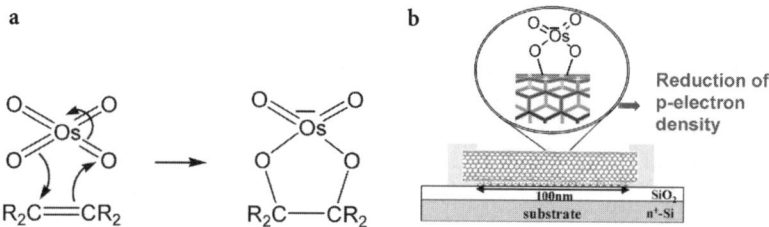

Scheme 17 a Reversible sidewall osmylation of SWCNT. b Electrical monitoring of the resistance at the single tube level

4.9
Aryl Diazonium Chemistry – Electrochemical Functionalization – Cathodic and Anodic Coupling

In 2001 Tour et al. [156] reacted a series of aryl diazonium salts with purified SWCNTs in an electrochemical reaction, using a bucky paper as working electrode, to prepare a variety of functionalized SWCNTs [156]. The corresponding reactive aryl radicals are generated from the diazonium salts via one-electron reduction (Scheme 18, a). Soon after, the same authors described in similar

Scheme 18 Reaction of SWCNTs with aryl diazonium compounds: electrochemical reaction with preformed diazonium salts (**a**), and thermally promoted reaction with in-situ generated diazonium compounds (**b**)

reactions the in situ generation of the diazonium ions from the corresponding anilines (Scheme 18, b) [157].

Based on thermal gravimetric and elemental analysis, the authors were able to achieve functionalization of 1-in-40 up to 1-in-20 nanotube carbons. Some of the materials prepared did not exhibit dramatic changes in solubility, and AFM analysis did not indicate significant exfoliation of SWCNT ropes [156]. For the preparation of polystyrene-CNT nanocomposites the same authors functionalized SWCNTs by in situ generation of a diazonium compound from 4-(10-hydroxydecyl)aminobenzoate [158].

These basic studies of diazonium chemistry to functionalize CNTs led to the two most efficient techniques of derivatization of SWCNTs, namely the "solvent-free functionalization" [159] and the "functionalization of individual (unbundled) nanotubes" [160]. The solvent-free functionalization methodology (Scheme 19) gives heavily functionalized and soluble material, and the nanotubes disperse in polymer more efficiently than pristine SWCNTs [159]. With the second method, the authors show that aryldiazonium salts react efficiently with individual HiPco-produced and SDS-coated SWCNTs in water to deliver individual functionalized SWCNTs. The resulting tubes remained unbundled throughout their entire lengths and were incapable of re-roping. This protocol yielded a very high degree of functionalization, up to one in nine carbons on the nanotube having an organic addend [160].

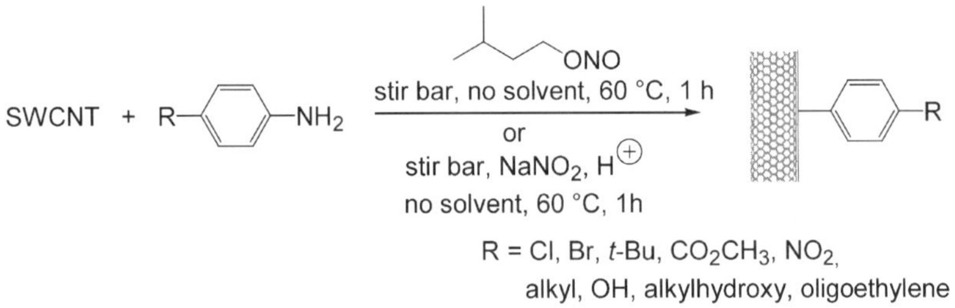

$R = Cl, Br, t\text{-Bu}, CO_2CH_3, NO_2,$
alkyl, OH, alkylhydroxy, oligoethylene

Scheme 19 Solvent-free functionalization of carbon nanotubes performed with various substituted anilines and isoamyl nitrite or $NaNO_2$/acid. R can be Cl, Br, NO_2, CO_2CH_3, alkyl, OH, alkylhydroxy, oligoethylene

In 2002, Burghard et al. described an elegant method for the electrochemical modification of individual SWCNTs [161, 162]. To address individual SWCNTs and small bundles electrically, the purified tubes were first deposited on surface-modified Si/SiO_2 substrates, and then placed in contact with electrodes which had been shaped by electron-beam lithography. Electrochemical modification was carried out in a miniaturized electrochemical cell. The cathodic coupling was achieved by the reduction of, e.g., $4\text{-}NO_2C_6H_4N_2^+BF_4^-$ in DMF with $NBu_4^+BF_4^-$ as the electrolyte (Scheme 20a). Anodic coupling to SWCNTs was accomplished with two aromatic amines in dry ethanol with $LiClO_4$ as the electrolyte salt (Scheme 20b) [161, 162]. The functional groups of the amines allow

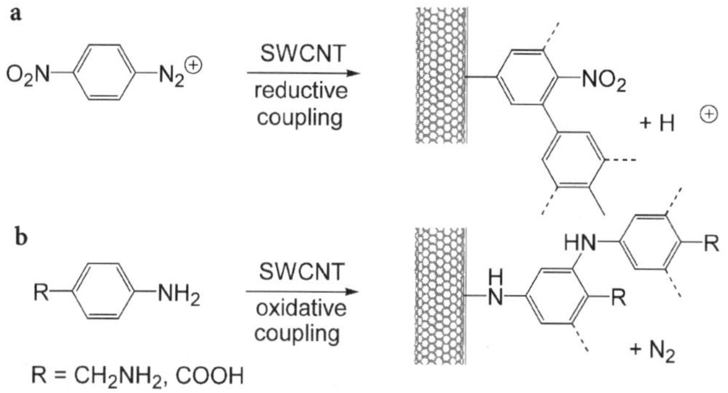

Functionalization of Carbon Nanotubes

Scheme 20a, b Electrochemical modification of SWCNTs. **a** Reductive coupling of *p*-nitrophenyl diazonium salts to SWCNTs. **b** Oxidative coupling of anilins to SWCNTs. The *dotted lines* mark positions of further linkages that could be formed during the growth of a polymerized layer of phenyl units on the SWCNTs

for the linking of further molecules, which may facilitate tailoring of the surface properties of the nanotubes.

5
Noncovalent Exohedral Functionalization

The prospect of functionalizing the outer surface of CNTs in a noncovalent way in order to preserve the extended π-network of the tubes is very attractive. In the search for nondestructive purification methods, it has been shown that nanotubes can be transferred to the aqueous phase through noncovalent functionalization of surface-active molecules such as SDS or benzylalkonium chloride [163–165].

With the use of the surfactant Triton X-100 [166], the surfaces of the CNTs were changed from hydrophobic to hydrophilic. Consequently, the hydrophilic surface of the CNT-Triton conjugate interacted with the hydrophilic surface of biliverdin reductase, thus creating a water-soluble complex of the immobilized enzyme. Ultracentrifugation, gel electrophoresis and Raman spectroscopy revealed the presence of the enzyme and interactions with the CNT-Triton conjugates [166].

Images of the assembly of surfactants on the surface of CNTs were obtained by a French research group using TEM [167] Above the critical micellar concentration, SDS forms supramolecular structures consisting of rolled-up half-cylinders on the nanotube's surface (Fig. 11). Depending on the symmetry and the diameter of the CNT, rings, helices, or double helices were observed [167].

Similar self-assemblies have been obtained with synthetic single-chain lipids designed for the immobilization of histidine-tagged proteins [167]. At the nanotube-water interface, permanent assemblies were produced from mixed

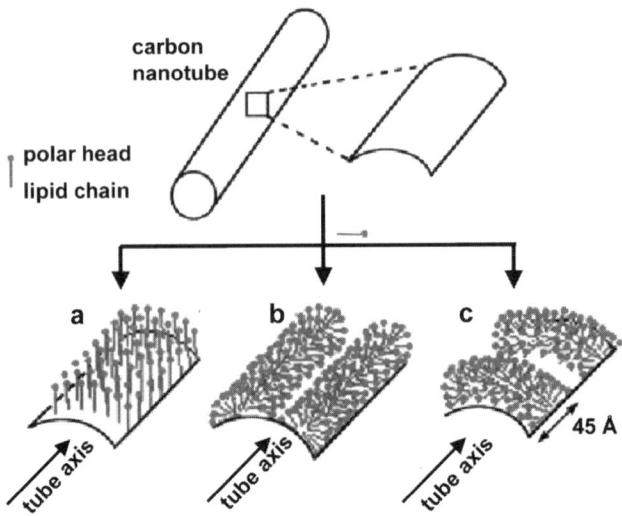

Fig. 11 Different organizations of the SDS molecules on the surface of a CNT. **a** Adsorbed perpendicular to the surface, forming a monolayer, **b** organized into half-cylinders oriented parallel to the tube axis, and **c** half-cylinders oriented perpendicular to the tube axis oriented perpendicular to the tube axis. Copyright with permission from Science International, UK [167]

micelles of SDS and different water-insoluble double-chain lipids after dialysis of the surfactant (Fig. 12) [167].

Immobilization of biomolecules on CNTs motivated the use of nanotubes as potentially new types of biosensor materials [168–171]. The electronic properties of nanotubes coupled with the specific recognition properties of the immobilized biosystems would indeed make for an ideal miniaturized sensor. A prerequisite for research in this area is the development of chemical methods to immobilize biological molecules onto CNTs in a reliable manner. Thus far, only limited work has been carried out with MWCNTs [168–171]. Streptavidin was found to adsorb on MWCNTs, presumably via hydrophobic interactions between the nanotubes and hydrophobic domains of the proteins (Fig. 13) [171]. DNA molecules adsorbed on MWCNTs via nonspecific interactions were also observed [169, 170].

The interactions between CNTs and six-membered ring molecules including benzene, cyclohexane, and 2,3-dichloro-5,6-dicyano-1,4-benzoquinone were theoretically studied using first principles calculations [172]. The equilibrium tube-molecule distance, adsorption energy, and charge transfer were obtained. The hybridization between the benzoquinone molecular level and nanotube valence bands transforms the semiconducting tube into a metallic one. Coupling of π-electrons between tubes and aromatic molecules were observed. The results showed that noncovalent functionalization of CNTs by aromatic molecules may be an efficient way to control the electronic properties of CNTs [172].

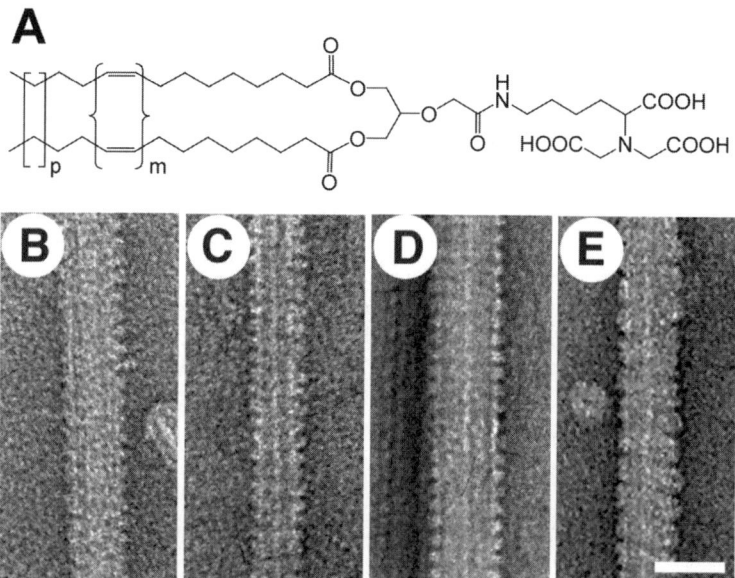

Fig. 12A–E Lipid molecules to functionalize the surface of CNTs. **A** Representation of the molecules: (a) $m=0, p=1$; (b) $m=1, p=1$; (c) $m=1, p=3$; (d) $m=1$ and $p=5$. **B** Organization of (a) on a MWNT after dialysis of SDS and negatively stained with uranyl acetate. **C** Supramolecular organization of (b) on a MWNT. **D** Organization of (c) on the nanotube. **E** Supramolecular organization of (d) on a MWNT. *Scale bar*, 20 nm

Fig. 13a, b Electron micrographs showing MWCNTs non-covalently coated with the protein streptavidin and negatively stained with uranyl acetate. **a** Stochastic binding of streptavidin molecules on a MWCNT with a diameter smaller than 15 nm. **b** Helical organization of streptavidin molecules on a CNT with a suitable diameter of 16 nm (*bar*=50 nm). Reprinted from Balavoine F. et al., Angew Chem Int Ed 1999, 38:1912 [171]. Copyright with permission from Wiley-VCH, Weilheim, Germany

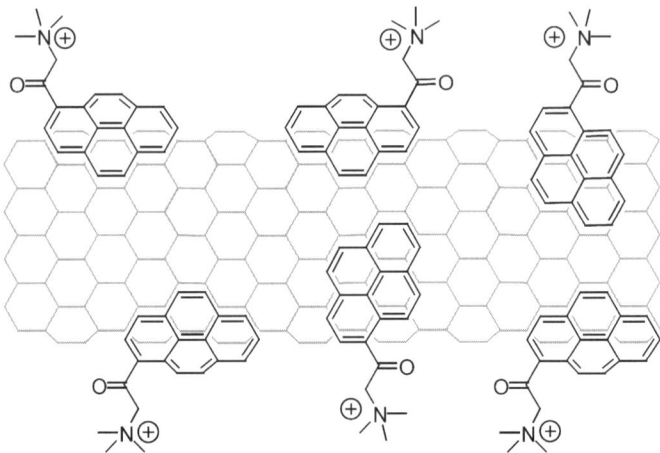

Fig. 14 Water-soluble SWCNTs via noncovalent sidewall functionalization with a pyrene-carrying ammonium ion

Sonication of solid SWCNTs in an aqueous solution of a pyrene-carrying ammonium ion (Fig. 14) gave a transparent dispersion/solution of the nanotubes, which was characterized by TEM, UV/Vis absorption, fluorescence and ^{1}H NMR spectroscopies. There was evidence for the interaction of the nanotube sidewall with the pyrene moiety in the aqueous dispersion/solution [173].

Simultanous non-covalent and covalent functionalizations occur in the "inter-tubulary" dimers and excimers already depicted in Scheme 4. While the pyrene tether is covalently bound through an ester linkage with one single SWCNT, a SWCNT dimer is formed by π-π-interactions of the polyaromatic pyrene system with a neighbouring CNT [112].

The nanodimensions, graphitic surface chemistry and electronic properties of SWCNTs make such a material an ideal candidate for chemical or biochemical sensing. Proteins adsorb individually, strongly, and noncovalently along the nanotube lengths. The resulting nanotube-protein conjugates are readily characterized at the molecular level by AFM. Several metalloproteins and enzymes have been bound on both the sidewalls and termini of SWCNTs. Though coupling can be controlled, to a degree, through variation of tube oxidative pre-activation chemistry, careful control experiments and observations made by AFM suggest that immobilization is strong, physical, and does not require covalent bonding [174]. Importantly, in terms of possible device applications, protein attachment appears to occur with retention of native biological structure. Nanotube electrodes exhibit useful voltammetric properties, with direct electrical communication possible between a redox-active biomolecule and the delocalized π-system of its CNT support [174].

Because of very effective π-stacking interactions of aromatic molecules strong interactions with the graphitic sidewalls of SWCNTs may result. This ef-

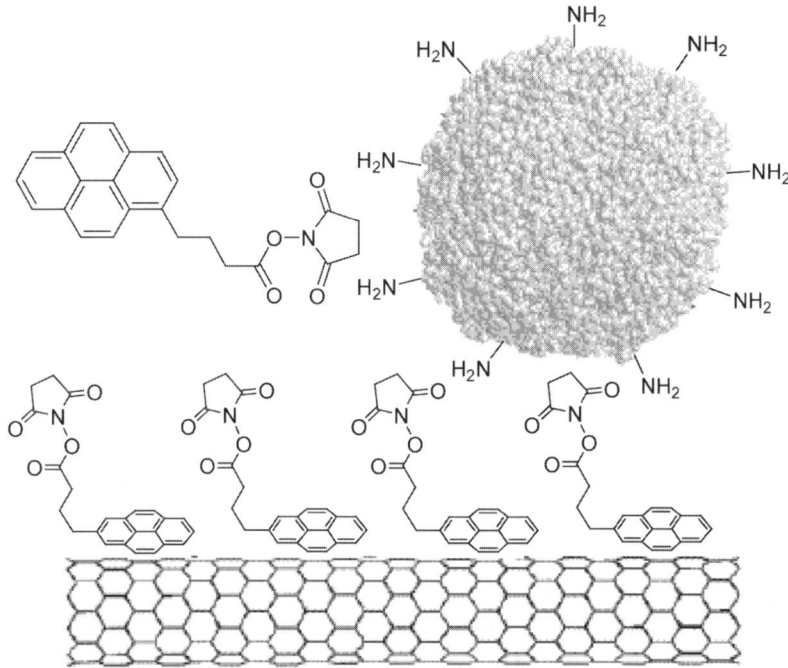

Fig. 15 1-Pyrenebutanoic acid succinimidyl ester irreversibly adsorbed onto the sidewall of a SWCNT via π-stacking. Amino groups of a protein react with the anchored succinimidyl ester to form amide bonds for protein immobilization. Printed with permission from R.J. Chen et al. [175]. Copyright American Chemical Society, USA

fect was demonstrated in the aggregation with the bifunctional N-succinimidyl-1-pyrenebutanoate [175], irreversibly adsorbed onto the inherently hydrophobic surfaces of the SWCNT. In these conjugates, the succinimidyl group could nucleophilically be substituted with primary or secondary amino groups from proteins such as ferritin or streptavidin (Fig. 15) and caused immobilization of the biopolymers on the tubes [175].

The same π-stacking of SWCNTs with polyaromates served Liu et al. [176] in the self-assembling of gold nanoparticles to soluble SWCNTs. 17-(1-pyrenyl)-13-oxa-heptadecanethiol (PHT) was non-covalently bound onto the nanotubes' surface by π-π-interactions and subsequently treated with a colloid gold solution (Fig. 16). A dense coverage of the nanotubes' surface was obtained by self-assembling of additional gold nanoparticles [176].

Direct, nonsurfactant-mediated immobilization of metallothionein proteins [168–170] and streptavidin [171] has also been carried out on MWCNTs. The hydrophobic regions of the proteins are probably important for adsorption. A monoclonal antibody, specific for C_{60} fullerenes, could also be bound directly to SWCNTs [177]. It was shown that the binding site of this IgG antibody is a domain of hydrophobic amino acids.

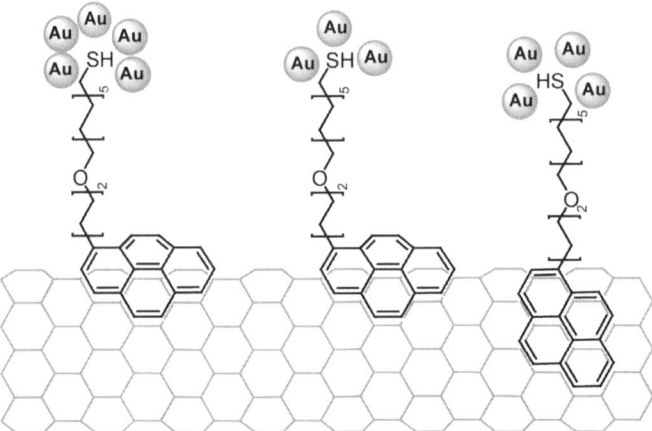

Fig. 16 π-stacking of 17-(1-pyrenyl)-13-oxa-heptadecanethiol (PHT) with SWCNTs and self-assembling of gold nanoparticles from a colloid gold solution

In a recent paper, an amphiphilic alpha-helical peptide specifically designed to coat and solubilize CNTs and to control the assembly of the peptide-coated nanotubes into macromolecular structures through peptide-peptide interactions between adjacent peptide-wrapped nanotubes was described [178]. The authors claim that the peptide folds into an amphiphilic alpha-helix in the presence of CNTs and disperses them in aqueous solution by noncovalent interactions with the nanotube surface. Electron microscopy and polarized Raman studies reveal that the peptide-coated nanotubes assemble into fibers with the nanotubes aligned along the fiber axis. The size and morphology of the fibers could be controlled by manipulating solution conditions that affect peptide-peptide interactions [178].

Pronounced noncovalent interactions were also found between SWCNTs and anilines [179] and several types of alkyl amines [180]. This interaction was detected, for example, by the alteration of the electrical conductivity of SWCNTs upon adsorption of primary amines, and through their very high solubility (up to 8 mg/mL) in anilines. Presumably, as in the case of C_{60} fullerenes [181], donor-acceptor complexes are formed, as the curvature present in both classes of materials lends acceptor character to the corresponding molecular or macromolecular carbon networks [182]. The stable aniline solutions of SWCNTs can be diluted with other organic solvents such as acetone, DMF or THF, without causing precipitation of the tubes.

Polymers have also been used in the formation of supramolecular complexes of CNTs. Thus, the suspension of purified MWCNTs and SWCNTs in the presence of conjugated polymers such as poly(m-phenylene-*co*-2,5-dioctoxy-p-phenylenevinylene) (PmPV), in organic solvents led to hybrid systems, the polymer wrapping around the tubes [183–185]. The properties of these supramolecular compounds were markedly different from those of the individual components. For example, the SWCNT/PmPV complex exhibited a conductivity eight times higher than that of the pure polymer, without any restriction of its luminescence properties. AFM showed that the polymer coated the tubes uniformly. The small average diameter of the complexes (about 7.1 nm) suggested that the bundles were mostly broken upon complex formation. The promising optoelectronic properties of the SWCNT/PmPV complexes have been used in the manufacture of photovoltaic devices [184]. Mono- to triple-layer devices of this CNT composite have been tested as electron transport layers in organic light-emitting diodes [185].

By coating with a conjugate polymer, the authors cited above also succeeded in developing a new experimental technique for a high-yield, nondestructive purification and quantification method for MWCNTs [186]. The polymer host selectively suspends nanotubes relative to impurities, and after removal by filtration gave a 91% pure nanotube material [186]. The wrapping of SWCNTs with polymers carrying polar side-chains such as polyvinylpyrrolidone (PVP) or polystyrenesulfonate (PSS) led to stable solutions of the corresponding SWCNT/polymer complexes in water [187]. The thermodynamic driving force for this complex formation is the avoidance of unfavorable interactions between the apolar tube walls and the solvent water.

For MWCNTs it was demonstrated that the wrapping of the polymer ropes around the tube lattice occurs in a well-ordered periodic fashion [188]. The authors suggested that at low loading fractions the polymer intercalated between the nanotubes, which leads to unravelling of ropes and causes a decrease in the interactions between the individual SWCNTs. The functionalized SWCNTs were characterized by optical absorption spectroscopy, electron microscopy, and Raman spectroscopy [189]. Moreover, Raman and absorption studies suggested that the polymer interacts preferentially with nanotubes of specific diameters or a range of diameters [189].

A range of other, mostly ionic polymers, such as bovine serum albumin, are also capable of coating SWCNTs. Polymers such as poly(ethylene glycol) (PEG) and poly(vinyl alcohol) (PVA) are ineffective in this respect. It is thought that multi-helical wrapping of the tubes with the polymers is most favorable for reasons of strain. Changing to a less polar solvent such as THF caused the polymer complexes to dissociate once more. The SWCNT/PVP complexes exhibited liquid-crystalline properties [187]. Grafting shortened SWCNTs with PEG by microwave-assisted heating produced a soluble derivative of SWCNTs. Compared to conventional heating, remarkable enhanced reaction rates were observed [190].

Using diamine-terminated oligomeric PEG in different functionalization reactions under various conditions, American researchers from Clemson and Oak Ridge succeeded in the solubilization of as-prepared and purified SWCNTs [191]. The soluble tubes were characterized by spectroscopic, microscopic, and gravimetric techniques.

A new nonwrapping approach to noncovalent engineering of CNT surfaces by short, rigid, functional conjugated polymers, poly(aryleneethynylene)s, is reported [192]. The technique not only enables the dissolution of various types of CNTs in organic solvents, which represents the first example of solubilization of CNTs via π-stacking without polymer wrapping, but could also introduce numerous neutral and ionic functional groups onto the CNT surfaces [192].

A family of poly[(m-phenylenevinylene)-co-(p-phenylenevinylene)]s, functionalized in the synthetically accessible C-5 position of the meta-disubstituted phenylene rings have been designed and synthesized [193]. They have been prepared both (1) by the polymerization of O-substituted 5-hydroxyisophthaldehydes, and (2) by chemical modifications carried out on polymers bearing reactive groups at the C-5 positions. PAmPV polymers solubilize SWCNT bundles in organic solvents by wrapping themselves around the nanotube bundles. PAmPV derivatives which bear tethers or rings form pseudorotaxanes with rings and threads, respectively. The formation of the polypseudorotaxanes has been investigated in solution by NMR and UV/vis spectroscopies, as well as on silicon oxide wafers in the presence of SWCNTs by AFM and surface potential microscopy [193]. Wrapping of these functionalized PAmPV polymers around SWCNTs results in the grafting of pseudorotaxanes along the walls of the nanotubes in a periodic fashion. The results hold out the prospect of being able to construct arrays of molecular switches and actuators [193].

The solubilization of oxidized CNTs has been achieved through derivatization using a functionalized organic crown ether. The resultant synthesized adduct yielded concentrations of dissolved nanotubes on the order of approximately 1 g/L in water as well as in methanol, according to optical measurements [194]. The nanotube-crown ether adduct was readily redissolved in ten different organic solvents at substantially high concentrations. Characterization of these solubilized adducts was performed with ^{1}H NMR spectroscopy; ^{7}Li NMR was also used to examine the ability of the crown ether's macrocyclic ring to bind Li$^+$ ions. The solutions were further analyzed using UV/Vis, photoluminescence, and FT-IR spectroscopies and were structurally characterized using AFM and TEM. Adduct formation likely results from a noncovalent chemical interaction between carboxylic groups on the oxidized tubes and amine moieties attached to the side chain of the crown ether derivative [194].

No extensive theoretical investigations on the interactions between polymer and CNTs were reported until 2003. A molecular dynamics computer simulation in conjunction with experimental evidence was used to elucidate the nature of the interaction [195]. Computational time was reduced by representing CNTs as a force field. The calculations indicated an extremely strong noncovalent binding energy. Furthermore, the correlation between the chirality of the nanotubes and mapping of the polymer onto the lattice was discussed [195].

SWCNTs were exploited as a platform for investigating surface–protein and protein–protein binding and developing highly specific electronic biomolecular detectors [196]. The nonspecific binding on nanotubes (Fig. 17), a phenomenon found with a wide range of proteins, was overcome by immobilization of poly(ethylene oxide) chains. A general approach is then advanced to enable the selective recognition and binding of target proteins by conjugation

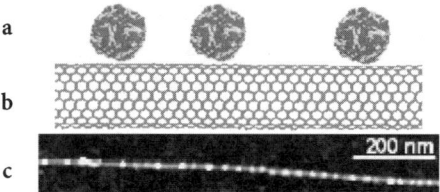

Fig. 17 a Schematic illustration of globular protein. b Adsorption onto a nanotube. c An AFM image showing protein (bright dot-like structures decorating the line-like nanotube) nonspecifically adsorbed onto a nanotube. Reprinted with the permission of the National Academy of Science of the USA [196]

of their specific receptors to polyethylene oxide-functionalized nanotubes. This scheme, combined with the sensitivity of nanotube electronic devices, enables the production of highly specific electronic sensors for detecting clinically important biomolecules such as antibodies associated with human autoimmune diseases [196].

Prakash et al. [197] demonstrated that individual CNTs can be visualized with fluorescence microscopy through noncovalent labeling with conventional fluorophores. Reversal of contrast in fluorescence imaging of the CNTs was observed when the labeling procedure was performed in a nonpolar solvent. The results were consistent with a CNT-fluorophore affinity mediated by hydrophobic interaction; the reverse-contrast images also provided clear indication of nanotube location [197].

Norbornene polymerization was initiated selectively on the surface of SW-CNTs via a specifically adsorbed pyrene-linked ring-opening metathesis polymerization initiator (Scheme 21). The adsorption of the organic precursor was

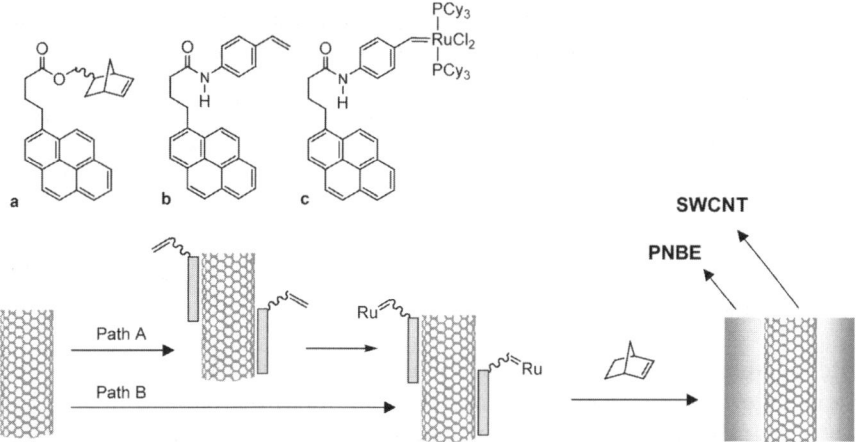

Scheme 21 Functionalization strategy for a polynorbornene (*PNBE*) coating of SWCNTs. *Path A* Adsorption of organic precursors (**a** or **b**) followed by cross-metathesis with a ruthenium alkylidene. *Path B* Adsorption of a pyrene-substituted ruthenium alkylidene (**c**)

followed by cross-metathesis with a ruthenium alkylidene, resulting in a homogeneous noncovalent poly(norbornene) coating [198].

6
Endohedral Functionalization

The inner cavity of SWCNTs offers space for the storage of guest molecules [199–204]. For example, gold and platinum nanothreads have been created in the capillaries of the tubes, by treating the SWCNTs with the corresponding perchlorometallic acids at high temperatures. In some cases, the resulting nanothreads are single crystals. The incorporation of fullerenes such as C_{60} [205, 206] or metallofullerenes such as $Sm@C_{82}$ [207] are especially impressive examples of the endohedral chemistry of SWCNTs, and may enable the creation of new materials for molecular scale devices. This incorporation is executed at defect sites localized at the ends or in the sidewalls. The encapsulated fullerenes tend to form chains that are coupled by van der Waals forces. Such arrays are sometimes called bucky peapods [206].

Upon annealing, the encapsulated fullerenes coalesce in the interior of the SWCNTs, which results in new, concentric, endohedral tubes with a diameter of 0.7 nm. The progress of such reactions inside the tubes could be monitored in real time by use of high-resolution electron microscopy [207].

The structures and energetics of several types of all-carbon peapods made by C_{60} encapsulated inside SWCNTs were calculated, and the interactions between the tubes and the C_{60} molecules revealed a minimal tube diameter for exothermic encapsulation of 11.74 Å and a radial deformation of the C_{60} cage as a linear function of the tube diameter [208] The activation energy for the translation of isolated C_{60} molecules inside the tubes is very low. An isolated fullerene C_{60} can diffuse freely inside the pod; the major factor controlling the dynamics of molecular transport in the peapods is therefore the C_{60}-C_{60} interaction [208].

Small proteins have been entrapped in the inner hollow channel of opened nanotubes by simple adsorption, thus forming natural nano test tubes [168, 169, 209].

7
Conclusions

The examples described of the functionalization of SWCNTs demonstrate that the chemistry of this new class of molecules represents a promising field within nanochemistry. The basis for the exploitation of macroscopic quantities of this carbon allotrope has been laid. The unique properties of SWCNTs can now be coupled with those of other classes of materials. For the successful future development of this research area, however, it is important to improve the quality of the nanotube raw material, with particular emphasis on the uniformity of the samples. Furthermore, comparative studies on individual nanotubes before and after functionalization are necessary, so that both the dependence of reactivity on electronic structure and the effect of chemical modification on the electrical and mechanical properties can be determined.

References

1. Iijima S (1991) Nature 354:56
2. Iijima S, Ichihashi T (1993) Nature 363:603
3. Behune DS (1993) Nature 363:605
4. Dresselhaus MS, Dresselhaus G, Avouris P (eds) (2001) Carbon nanotubes: synthesis, structure properties and applications. Springer, Berlin Heidelberg New York
5. Collins PG, Hersam M, Arnold M, Martel R, Avouris P (2001) Phys Rev Lett 86:3128
6. Yao Z, Kane CL, Dekker C (2000) Phys Rev Lett 84:2941
7. Radosavljevi M, Lefebvre J, Johnson AT (2001) Phys Rev B 64:241307
8. Berger C, Poncharal P, Yi Y, de Heer W (2003) J Nanosci Nanotechnol 3:171
9. Collins PG, Zettl A, Bando H, Thess A, Smalley RE (1997) Science 278:100
10. de Heer WA, Chatelain A, Ugarte DA (1995) Science 270:1179
11. Dai HJ, Hafner JH, Rinzler AG, Colbert DT, Smalley RE, Nature 384:147
12. Tans SJ, Verschueren A, Dekker C (1998) Nature 393:49
13. Hafner J, Cheung C, Lieber C (1999) Nature 398:761
14. Martel R, Schmidt T, Shea HR, Hertel T, Avouris P (1998) Appl Phys Lett 73:2447
15. Soh H, Morpurgo A, Kong J, Marcus C, Quate C, Dai H (1999) Appl Phys Lett 75:627
16. Kong J, Franklin NR, Zhou C, Chapline MG, Peng S, Cho K, Dai H (2000) Science 287:622
17. Postma HWCh, Teepen T, Yao Z, Grifoni M, Dekker C (2001) Science 293:76
18. Maurin G, Bousquet Ch, Henn F, Bernier P, Almeirac R, Simon B (1999) Chem Phys Lett 312:14
19. Chen P, Wu X, Lin J, Tan KL (1999) Science 285:91
20. Dillon AC, Jones KM, Bekkedahl TA, Kong CH, Bethune DS, Heben MJ (1997) Nature 386:377
21. Liu C, Fan YY, Liu M, Cong HT, Cheng HM, Dresselhaus MS (1999) Science 286:1127
22. Hone J, Batlogg B, Benes Z, Johnson AT, Fischer JE (2000) Science 289:1730
23. Wong EW, Sheehan PE, Lieber CM (1997) Science 277:1971
24. Popov VN, Van Doren VE, Balkanski M (2000) Phys Rev B 61:3078
25. Qian D, Dickey EC, Andrews R, Rantell T (2000) Appl Phys Lett 76:2868
26. Salvetat JP, Kulik AJ, Bonard JM, Briggs GAD, Stoeckli T, Metenier K, Bonnamy S, Beguin F, Burnham NA, Forro L (1999) Adv Mater 11:161
27. Yu MF, Files BS, Arepalli S, Ruoff RS (2000) Phys Rev Lett 84:5552
28. Saito R, Dresselhaus G, Dresselhaus MS (eds) (1998) Physical properties of carbon nanotubes. Imperial College Press, London
29. Ajayan PM (1999) Chem Rev 99:1787
30. Xie SS, Chang BH, Li WZ, Pan ZW, Sun LF, Mao JM, Chen XH, Qian LX, Zhou WY (1999) Adv Mater 11:1135
31. Dai L, Mau AWH (2001) Adv Mater 13:899
32. Rao CNR, Satishkumar BC, Govindaraj A, Nath M (2001) Chem Phys Chem 2:78
33. Yao Z, Postma HWC, Balents L, Dekker C (1999) Nature 402:273
34. Fuhrer MS, Nygard J, Shih L, Forero M, Yoon YG, Mazzoni MSC, Choi HJ, Ihm J, Loui SG, Zettl A, McEuen PL (2000) Science 288:494

35. Bahr JL, Mickelson ET, Bronikowski MJ, Smalley RE, Tour JM (2001) Chem Commun 193
36. Niyogi S, Hamon MA, Hu H, Zhao B, Bhowmik P, Sen R, Itkis ME, Haddon RC (2002) Acc Chem Res 35:1105
37. Bahr JL, Tour JM (2002) J Mater Chem 12:1952
38. Hirsch A (2002) Angew Chem Int Ed 41:1853
39. Sinnott SB (2002) J Nanosci Nanotechnol 2:113
40. Tasis D, Tagmatarchis N, Georgakilas V, Prato M (2003) Chemistry 9:4000
41. Chen Z, Thiel W, Hirsch A (2003) Chem Phys Chem 4:93
42. Rinzler AG, Liu J, Dai H, Nikolaev P, Huffman CB, Rodriguez-Macias FJ, Boul PJ, Lu AH, Heymann D, Colbert DT, Lee RS, Fischer JE, Rao AM, Eklund PC, Smalley RE (1998) Appl Phys A 67:29
43. Dujardin E, Ebbesen TW, Krishnan A, Treacy MMJ (1998) Adv Mater 10:611
44. Vaccarini L, Goze C, Aznar R, Micholet V, Journet C, Bernier P (1999) Synth Met 103:2492
45. Vaccarini L, Goze C, Aznar R, Micholet V, Journet C, Bernier P, Metenier K, Beguin F, Gavillet J, Loiseau A (1999) C R Acad Sci Ser IIb 327:925
46. Nagasawa S, Yudasaka M, Hirahara K, Ichihashi T, Iijima S (2000) Chem Phys Lett 328:374
47. Holzinger M, Hirsch A, Bernier P, Duesberg GS, Burghard M (2000) Appl Phys A 70:599
48. Sumanasekera GU, Allen JL, Fang SL, Loper AL, Rao AM, Eklund PC (1999) J Phys Chem B 103:4292
49. Liu J, Rinzler AG, Dai H, Hafner JH, Bradley RK, Boul PJ, Lu A, Iverson T, Shelimov K, Huffman CB, Rodriguez-Macias F, Shon YS, Lee TR, Colbert DT, Smalley RE (1998) Science 280:1253
50. Morishita K, Takarada T (1994) J Mater Sci 34:1169
51. Tohji K, Goto T, Takehashi H, Shinoda Y, Shimizu N, Jeyadevan B, Matsuoka I, Saito Y, Kasuya A, Ohsuna T, Hiraga K, Nishina Y (1996) Nature 383:679
52. Mawhinney DB, Naumenko V, Kuznetsova A, Yates JT, Liu J, Smalley RE (2000) J Am Chem Soc 122:2383
53. Mawhinney DB, Naumenko V, Kuznetsova A, Yates JT, Liu J, Smalley RE (2000) Chem Phys Lett 324:213
54. Kuznetsova A, Popova I, Yates JT, Bronikowski MJ, Huffman CB, Liu J, Smalley RE, Hwu H, Chen JG (2001) J Am Chem Soc 123:10699
55. Deng JP, Mou CY, Han CC (1997) Full Sci Technol 5:1033
56. Ajayan PM, Iijima S (1993) Nature 361:333
57. Ugarte D, Chatelain A, deHeer WA (1996) Science 274:1897
58. Colomer JF, Piedigrosso P, Willems I, Journet C, Bernier P, VanTendeloo G, Fonseca A, Nagy JB (1988) J Chem Soc Faraday Trans 94:3753
59. Kinoshita K (1988) Carbon electrochemical and physicochemical properties. Wiley, New York
60. McKay SF (1992) J Appl Phys 35:1992
61. Monthioux M, Smith BW, Burteaux B, Claye A, Fischer JE, Luzzi DE (2001) Carbon 39:1251
62. Itkis ME, Niyogi S, Meng M, Hamon M, Hu H, Haddon RC (2002) Nano Lett 2:155
63. Tohji K, Takahashi H, Shinoda Y, Shimizu N, Jeyadevan B, Matsuoka I, Saito Y, Kasuya A, Ito S, Nishina Y (1997) J Phys Chem B 101:1974
64. Bandow S, Asaka S, Zhao X, Ando Y (1998) Appl Phys A 67:23
65. Dillon AC, Gennet T, Jones KM, Alleman JL, Parilla PA, Heben MJ (1999) Adv Mater 11:1354

66. Chen Y, Haddon RC, Fang S, Rao AM, Eklund PC, Lee WH, Dickey EC, Grulke EC, Pendergrass JC, Chavan A, Haley BE, Smalley RE (1998) J Mater Res 13:2423
67. Zhou W, Ooi YH, Russo R, Papanek P, Luzzi DE, Fischer JE, Bronikowski MJ, Willis PA, Smalley RE (2001) Chem Phys Lett 350:6
68. Yang Y, Zou H, Wu B, Li Q, Zhang J, Liu Z, Guo X, Du Z (2002) J Phys Chem B 106:7160
69. Hamon MA, Hu H, Bhowmik P, Niyogi S, Zhao B, Itkis ME, Haddon RC (2001) In: Carbon'01, internernational conference on carbon, Lexington, US, July 14–19, 2001, p 872
70. Zhang N, Xie J, Kvaradan V (2002) Smart Mater Struct 11:962
71. Wu B, Zhang J, Wie Z, Cai S, Liu Z (2001) J Phys Chem B 105:5075
72. Yanagi H, Sawada E, Manivannan A, Nagahara LA (2001) Appl Phys Lett 78:1355
73. Chattopadhyay D, Galeska I, Papadimitrakopoulos F (2001) J Am Chem Soc 123:9451
74. Mamedov AA, Kotov NA, Prato M, Guldi DM, Wicksted JP, Hirsch A (2002) Nature Mater 1:190
75. Sotiropoulou S, Chaniotakis NA (2003) Anal Bioanal Chem 375:103
76. Chen J, Hamon MA, Hu H, Chen Y, Rao AM, Eklund PC, Haddon RC (1998) Science 282:95
77. Holzinger M (2002) Functionalization of single-walled carbon nanotubes. PhD thesis, University Erlangen-Nürnberg
78. Li B, Lian YF, Shi ZJ, Gu ZN (2000) Gaodeng Xuexiao Huaxue Xuebao 21:1633
79. Hennrich F, Lebedkin S, Malik S, Tracy J, Barczewski M, Rösner A, Kappes M (2002) Phys Chem Chem Phys 4:273
80. Jung A (2003) Defektgruppen-Funktionalisierung von Kohlenstoffnanoröhren mit Newkome-Derimeren. Diploma thesis, University Erlangen-Nürnberg
81. Hamon MA, Chen J, Hu H, Chen Y, Itkis ME, Rao AM, Eklund PC, Haddon RC (1999) Adv Mater 11:834
82. Hamon MA, Hu H, Bhowmik P, Niyogi S, Zhao B, Itkis ME, Haddon RC (2001) Chem Phys Lett 347:8
83. Pompeo F, Resasco DE (2002) Nano Lett 2:369
84. Liu Z, Shen Z, Zhu T, Hou S, Ying L, Shi Z, Gu Z (2000) Langmuir 16:3569
85. Basiuk VA, Basiuk EV, Saniger-Blesa JM (2001) Nano Lett 1:657
86. (a) Chiu PW, Duesberg GS, Dettlaff-Weglikowska U, Roth S (2002) Appl Phys Lett 80:2811; (b) Dettlaff-Weglikowska U, Benoit JM, Chiu PW, Graupner R, Lebedkin S, Roth S (2002) Curr Appl Phys 2:497
87. Kiricsi I, Konya Z, Niesz K, Koos AA, Biro LP (2003) Proc SPIE Int Soc Opt Eng 5118 (Nanotechnology), p 280
88. Unger E, Graham A, Kreupl F, Liebau M, Hoenlein M (2002) Curr Appl Phys 2:107
89. Pokrovs'kyi VO, Gromovyi TY, Chuiko OO, Basyuk OV, Basyuk VO, Saniger-Blesa JM (2002) Dop Nat Akad Nauk Ukr 172
90. Chen J, Rao AM, Lyuksyutov S, Itkis ME, Hamon MA, Hu H, Cohn RW, Eklund PC, Colbert DT, Smalley RE, Haddon RC (2001) J Phys Chem B 105:2525
91. Hamon MA, Chen J, Hu H, Itkis ME, Bhowmik P, Rozenzhak SM, Rao AM, Haddon RC (2000) Dissolution of single-walled carbon nanotubes. In: Abstr Pap – Am Chem Soc 2000, 220th IEC-130
92. Chattopadhyay D, Lastella S, Kim S, Papadimitrakopoulos F (2002) J Am Chem Soc 124:728
93. Cai L, Bahr JL, Yao Y, Tour JM (2002) Chem Mater 14:4235
94. Wong SS, Woolley AT, Joselevich E, Cheung CL, Lieber CM (1998) J Am Chem Soc 120:8557

95. Sun YP, Fu K, Lin Y, Huang W (2002) Acc Chem Res 35:1096
96. Sun YP, Huang W, Lin Y, Fu K, Kitaygorodskiy A, Riddle LA, Yu YJ, Carroll DL (2001) Chem Mater 13:2864
97. Sano M, Kamino A, Okamura J, Shinkai S (2001) Langmuir 17:5125
98. Ham HT, Koo CM; Kim SO; Choi YS, Chung IJ (2002) Hwahak Konghak 40:618
99. Lin Y, Rao AM, Sadanadan B, Kenik EA, Sun YP (2002) J Phys Chem B 106:1294
100. Holzinger M, Hirsch A, Duesberg GS, Burghard M (2000) Novel purification procedure and derivatization method of single-walled carbon nanotubes (SWNTs). In: AIP Conf Proc 544 (Electronic properties of novel materials-molecular nanostructures) 246
101. Huang W, Taylor S, Fu K, Lin Y, Zhang D, Hanks TW, Rao AM, Sun YP (2002) Nano Lett 2:311
102. Ravindran S, Chaudhary S, Ozkan M, Ozkan C (2003) Carbon nanotube bioconjugates for in vivo applications. In: Abstr Pap 225th ACS National Meeting, New Orleans, US, March 23–27, 225th IEC-161
103. Dwyer C, Guthold M, Falvo M, Washburn S, Superfine R, Erie D (2002) Nanotechnology 13:601
104. Hazani M, Naaman R, Hennrich F, Kappes MM (2003) Nano Lett 3:153
105. Baker SE, Cai W, Lasseter TL, Weidkamp KP, Hamers RJ, Nano Lett 2:1413
106. Williams KA, Veenhuizen PTM, de la Torre BG, Eritjia R, Dekker C (2002) Nature 420:761
107. Cai H, Cao X, Jiang Y, He P, Fang Y (2003) Anal Bioanal Chem 375:287
108. Liu L, Qin Y, Guo ZX, Zhu D (2003) Carbon 41:331
109. Hamon MA, Hui H, Bhowmik P, Itkis HME, Haddon RC (2002) Appl Phys A 74:333
110. Riggs JE, Guo Z, Carroll DL, Sun YP (2000) J Am Chem Soc 122:5879
111. Fu K, Huang W, Lin Y, Riddle LA, Carroll DL, Sun YP (2001) Nano Lett 1:439
112. (a) Qu L, Martin RB, Huang W, Fu K, Zweifel D, Lin Y, Sun YP, Bunker CE, Harruff BA, Gord JR, Allard LF (2002) J Chem Phys 117:8089; (b) Qu L, Martin RB, Huang W, Fu K, Zweifel D, Lin Y, Bunker CE, Harruff BA, Gord JR, Allard LF, Sun YP (2002) Proc Electrochem Soc 12 (Fullerenes – vol. 12: the exciting world of nanocages and nanotubes), p 563
113. Lim JK, Yun WS, Yoon M, Lee SK, Kim CH, Kim K, Kim SK (2003) Synth Met 139:521
114. Nakajima T, Kasamatsu S, Matsuo Y (1996) Eur J Solid State Inorg Chem 33:831
115. Hamwi A, Alvergnat H, Bonnamy S, Beguin F (1997) Carbon 35:723
116. Mickelson ET, Huffman CB, Rinzler AG, Smalley RE, Hauge RH, Margrave JL (1998) Chem Phys Lett 296:188
117. Mickelson ET, Chiang IW, Zimmerman JL, Boul PJ, Lozano J, Liu J, Smalley RE, Hauge RH, Margrave JL (1999) J Phys Chem B 103:4318
118. Peng H, Gu Z, Yang J, Zimmerman JL, Willis PA, Bronikowski MJ, Smalley RE, Hauge RH, Margrave JL (2001) Nano Lett 1:625
119. Marcoux PR, Schreiber J, Batail P, Lefrant S, Renouard J, Jacob G, Albertini D, Mevellec JY (2002) Phys Chem Chem Phys 4:2278
120. Touhara H, Inahara J, Mizuno T, Yokoyama Y, Okanao S, Yanagiuch K, Mukopadhyay I, Kawasaki S, Okino F, Shirai H, Xu WH, Kyotani T, Tomita A (2002) J Fluor Chem 114:181
121. Taylor R, Holloway JH, Hope EG, Avent AG, Langley GJ, Dennis TJ, Hare JP, Kroto HW, Walton DRM (1992) Chem Commun 9:665
122. Kudin KN, Bettinger HF, Scuseria GE (2001) Phys Rev B: Condensed Matter 63:045413
123. An KH, Heo JG, Jeon KG, Bae DJ, Jo C, Yang CW, Park CY, Lee YH, Lee YS, Chung YS (2002) Appl Phys Lett 80:4235
124. Chiang IW, Mickelson ET, Boul PJ, Hauge RH, Smalley RE, Margrave JL (2000) Abstr Pap Am Chem Soc 2000, 220th IEC-153

125. Boul PJ, Liu J, Mickelson ET, Huffman CB, Ericson LM, Chiang IW, Smith KA, Colbert DT, Hauge RH, Margrave JL, Smalley RE (1999) Chem Phys Lett 310:367
126. Gu Z, Peng H, Zimmerman JL, Chiang IW, Khabashesku VN, Hauge RH, Margrave JL (2002) Fluorination of polymeric C60 and carbon nanotubes: a starting point for various chemical modifications of nanostructured materials. In: Abstr Pap 223rd ACS National Meeting, Orlando, FLUO-012
127. Kelly KF, Chiang IW, Mickelson ET et al. (1999) Chem Phys Lett 313:445
128. Bettinger HF, Kudin KN, Scuseria GE (2001) J Am Chem Soc 123:12849
129. Fagan SB, da Silva AJR, Mota R, Baierle RJ, Fazzio A (2003) Phys Rev B 67:033405
130. Konya Z, Vesselenyi I, Niesz K, Kukovecz A, Demortir A, Fonseca A, Delhalle J, Mekhalif Z, Nagy JB, Koos AA, Osvath Z, Kocsonya A, Biro LP, Kiricsi I (2002) Chem Phys Lett 360:429
131. Chen YK, Green MLH, Griffin JL, Hammer J, Lago RM, Tsang SC (1996) Adv Mater 8:1012
132. Hou PX, Bai S, Yand QH, Liu C, Cheng HM (2002) Carbon 40:81
133. Pekker S, Salvetat JP, Jakab E, Bonard JM, Forro L (2001) J Phys Chem B 105:7938
134. Owens FJ, Iqbal Z (2002) Electrochemical functionalization of carbon nanotubes with hydrogen. In: 23rd Army Science Conference, Session L/LP-11, 2–5 Dec. 2002, Orlando (http://www.asc2002.com/summaries/l/LP-11.pdf)
135. Yildirim T, Gülseren O, Ciraci S (2001) Phys Rev B 64:075404
136. Holzinger M, Vostrowsky O, Hirsch A, Hennrich F, Kappes M, Weiss R, Jellen F (2001) Angew Chem Int Ed 40:4002
137. Ni B, Sinnott SB (2000) Phys Rev B 61:16343
138. Peng H, Reverdy P, Khabashesku VN, Margrave JL (2003) Chem Commun 362
139. Tagmatarchis N, Georgakilas V, Prato M, Shinohara H (2002) Chem Commun 2010
140. Lee WH, Kim SJ, Lee JG, Haddon RC, Reucroft PJ (2001) Appl Surf Sci 181:121
141. Holzinger M, Abraham J, Whelan P, Graupner R, Ley L, Hennrich F, Kappes M, Hirsch A (2003) J Am Chem Soc 125:8566
142. Vencelova A, Graupner R, Ley L, Abraham J, Holzinger M, Whelan P, Hirsch A, Hennrich F (2003) AIP Conf Proc 685:112
143. Graupner R, Vencelova A, Ley L, Abraham J, Hirsch A, Hennrich F, Kappes MM (2003) AIP Conf Proc 685:120
144. Abraham J, Whelan P, Hirsch A, Hennrich F, Kappes M, Samaille D, Bernier P, Vencelova A, Graupner R, Ley L, (2003) AIP Conf Proc 685:291
145. Holzinger M, Steinmetz J, Samaille D, Bernier P (2003) Eur Network Meeting, Functionalization of Single-Walled Carbon Nanotubes, Montpellier, France 20.3.–21.3
146. Camps X, Hirsch A (1997) J Chem Soc Perkin Trans 1:1595
147. Hirsch A, Vostrowsky O (2001) Eur J Org Chem 829
148. Coleman KS, Bailey SR, Fogden S, Green MLH (2003) J Am Chem Soc 2125:8722
149. Georgakilas V, Kordatos K, Prato M, Guldi DM, Holzinger M, Hirsch A (2002) J Am Chem Soc 124:760
150. (a) Tasis D, Tagmatarchis N, Georgakilas V, Gamboz C, Soranzo MR, Prato M (2003) C R Chimie 6:597; (b) Georgakilas V, Tagmatarchis N, Voulgaris D, Prato M, Kukovecz A, Kuzmany H, Hirsch A, Zerbetto F, Melle-Franco M (2002) AIP Conf Proc 633:73
151. Georgakilas V, Voulgaris D, Vázquez E, Prato M, Guldi DM, Kukovecz A, Kuzmany H (2002) J Am Chem Soc 124:14318
152. Lu X, Tian F, Xu X, Wang N, Zhang Q (2003) J Am Chem Soc 125:10459
153. Lu X, Tian F, Xu X, Wang N, Zhang Q (2002) Which 1,3-dipoles can be attached onto the sidewalls of single-wall carbon nanotubes by 1,3-dipolar cycloaddition? In: Abstr Pap, 224th ACS National Meeting, Boston, US, August 18–22, 2002, PHYS-236
154. Lu X, Tian F, Wang N, Zhang Q (2002) Org Lett 4:4313

155. Cui J, Burghard M, Kern K (2993) Nano Lett 3:613
156. Bahr JL, Yang J, Kosynkin DV, Bronikowski MJ, Smalley RE, Tour JM (2001) J Am Chem Soc 123:6536
157. Bahr JL, Tour JM (2001) J Mater Chem 12:3823
158. Mitchell CA, Bahr JL, Arepalli S, Tour JM, Krishnamoorti R (2002) Macromolecules 35:8825
159. Dyke CA, Tour JM (2003) J Am Chem Soc 125:1156
160. Dyke CA, Tour JM (2003) Nano Lett 3:1215
161. Kooi SE, Schlecht U, Burghard M, Kern K (2002) Angew Chem Int Ed 41:1353
162. Knez M, Sumser M, Bittner AM, Wege C, Jeske H, Kooi S, Burghard M, Kern K (2002) J Electroanal Chem 522:70
163. Bandow S, Rao AM, Williams KA, Thess A, Smalley RE, Eklund PC (1997) J Phys Chem B 101:8839
164. Duesberg GS, Burghard M, Muster J, Philipp G, Roth S (1998) Chem Commun 435
165. Krstic V, Duesberg GS, Muster J, Burghard M, Roth S (1998) Chem Mater 10:2338
166. Panhuis M, Salvador-Morales C, Franklin E, Chambers G, Fonseca A, Nagy JB, Blau WJ, Minett AI (2003) J Nanosci Nanotechnol 3:209
167. Richard C, Balavoine F, Schultz P, Ebbesen TW, Mioskowski C (2003) Science 300:775
168. Tsang SC, Davis JJ, Green MLH, Hill HAO, Leung YC, Sadler PJ (1995) Chem Commun 2579
169. Tsang SC, Guo Z, Chen YK, Green MLH, Hill HAO, Hambley TW, Sadler PJ (1997) Angew Chem 109:2292; Angew Chem Int Ed 36:2198
170. Guo Z, Sadler PJ, Tsang SC (1998) Adv Mater 10:701
171. Balavoine F, Schultz P, Richard C, Mallouh V, Ebbesen TW, Mioskowski C (1999) Angew Chem 111:2036; Angew Chem Int Ed 38:1912
172. Zhao J, Lu JP, Han J, Yang CK (2003) Appl Phys Lett 82:3746
173. Nakashima N, Tomonari Y, Murakami H (2002) Jap Chem Lett 6:638
174. Davis JJ, Coleman KS, Azamian BR, Bagshaw CB, Green MLH (2003) Chemistry 9:3732
175. Chen RJ, Zhang Y, Wang D, Dai H (2001) J Am Chem Soc 123:3838
176. Liu L, Wang T, Li J, Guo ZX, Dai L, Zhang D, Zhu D (2003) Chem Phys Lett 367:747
177. Erlanger BF, Chen BX, Zhu M, Brus L (2001) Nano Lett 1:465
178. Dieckmann GR, Dalton AB, Johnson PA, Razal J, Chen J, Giordano GM, Munoz E, Musselman IH, Baughman RH, Draper RK (2003) J Am Chem Soc 125:1770
179. Sun Y, Wilson SR, Schuster DI (2001) J Am Chem Soc 123:5348
180. Kong J, Dai H (2001) J Phys Chem B 105:2890
181. Hirsch A (1994) Chemistry of the fullerenes. Thieme, New York
182. Hamon MA, Itkis ME, Niyogi S, Alvaraez T, Kuper C, Menon M, Haddon RC (2001) J Am Chem Soc 123:11292
183. Curran SA, Ajayan PM, Blau WJ, Carroll DL, Coleman JN, Dalton AB, Davey AP, Drury A, McCarthy B, Maier S, Strevens A (1998) Adv Mater 10:1091
184. Star A, Stoddart JF, Steuerman D, Diehl M, Boukai A, Wong EW, Yang X, Chung SW, Choi H, Heath JR (2001) Angew Chem 113:1771; Angew Chem Int Ed 40:1721
185. (a) Coleman JN, Dalton AB, Curran S, Rubio A, Davey AP, Drury A, McCarthy B, Lahr B, Ajayan PM, Roth S, Barklie RC, Blau WJ (2000) Adv Mater 12:213; (b) Fournet P, Coleman JN, Diarmuid F, Lahr B, Drury A, Hoerhold HH, Blau WJ (2002) Proc SPIE-Intern Soc Opt Eng 4464
186. Murphy R, Coleman JN, Cadek M, McCarthy B, Bent M, Drury A, Barklie RC, Blau WJ (2002) J Phys Chem B 106:3087
187. O'Connell MJ, Boul P, Ericson LM, Huffman C, Wang Y, Haroz E, Kuper C, Tour J, Ausman KD, Smalley RE (2001) Chem Phys Lett 342:265

188. McCarthy B, Coleman JN, Curran SA, Dalton AB, Davey AP, Konya Z, Fonseca A, Nagy JB, Blau WJ (2000) J Mater Sci Lett 19:2239
189. Dalton AB, Stephan C, Coleman JN, McCarthy B, Ajayan PM, Lefrant S, Bernier P, Blau WJ, Byrne HJ (2000) J Phys Chem B 104:10012
190. Della Negra F, Meneghetti M, Menna E (2003) Fullerenes Nanotubes Carbon Nanostruct 11:25
191. Huang W, Fernando S, Allard LF, Sun YP (2003) Nano Lett 3:565
192. Chen J, Liu H, Weimer WA, Halls MD, Waldeck DH, Walker GC (2002) J Am Chem Soc 124:9034
193. Star A, Liu Y, Grant K, Ridvan L, Stoddart JF, Steuerman DW, Diehl MR, Boukai A, Heath JR (2003) Macromolecules 36:553
194. Kahn MGC, Banerjee S, Wong SS (2002) Nano Lett 2:1215
195. Panhuis M, Maiti A, Dalton AB, van den Noort A, Coleman JN, McCarthy B, Blau WJ (2003) J Phys Chem B 107:478
196. Chen RJ, Bangsaruntip S, Drouvalakis KA, Kam NWS, Shim M, Li Y, Kim W, Utz PJ, Dai H (2003) Proc Natl Acad Sci USA 100:4984
197. Prakash R, Washburn S, Superfine R, Cheney RE, Falvo MR (2003) Appl Phys Lett 83:1219
198. Gomez FJ, Chen RJ, Wang D, Waymouth RM, Dai H (2003) Chem Commun 190
199. Han W, Fan S, Li Q, Hu Y (1997) Science 277:1287
200. Sloan J, Hammer J, Zwiefka-Sibley M, Green MLH (1998) Chem Commun 347
201. Dujardin E, Ebbesen TW, Krishnan A, Treacy MMJ (1998) Adv Mater 10:1472
202. Matsui K, Pradhan BK, Kyotani T, Tomita A (2001) J Phys Chem B 105:5682
203. Govindaraj A, Satishkumar BC, Nath M, Rao CNR (2000) Chem Mater 12:202
204. Wilson M, Madden PA (2001) J Am Chem Soc 123:2101
205. Smith BW, Monthioux M, Luzzi DE (1999) Chem Phys Lett 315:31
206. Smith BW, Luzzi DE (2000) Chem Phys Lett 321:169
207. Okazaki T, Suenaga K, Hirahara K, Bandow S, Iijima S, Shinohara H (2001) J Am Chem Soc 123:9673
208. Melle-Franco M, Kuzmany H, Zerbetto F (2003) J Phys Chem B 107:6986
209. Davis JJ, Green MLH, Hill OAH, Leung YC, Sadler PJ, Sloan J, Xavier AV, Tsang SC (1998) Inorg Chim Acta 272:261

Equilibrium Structure of Dendrimers – Results and Open Questions

Christos N. Likos[1] · Matthias Ballauff[2] (✉)

[1] Institut für Theoretische Physik II, Heinrich-Heine-Universität Düsseldorf, Universitätsstrasse 1, 40225 Düsseldorf, Germany
likos@thphy.uni-duesseldorf.de
[2] Physikalische Chemie I, Universität Bayreuth, Universitätsstrasse 30, 95440 Bayreuth, Germany
Matthias.Ballauff@uni-bayreuth.de

1	Introduction	239
2	Scattering methods	242
3	Theory of Dendrimers	247
4	Open Questions and Directions of Further Research	250
	References	251

Abstract We review the problem of the structure of dendrimers in solution. Special emphasis is placed on recent theoretical work and computer simulations of the equilibrium structure of dissolved dendrimers. These investigations have led to the irrefutable conclusion that flexible dendrimers exhibit a dense-core structure in which the terminal groups are partially folded back. Hence, flexible dendrimers do not have a well-defined surface or interior. These conclusions are in total accord with recent experimental studies employing small-angle neutron scattering.

Keywords Dendrimers · Small-angle neutron scattering · Small-angle X-ray scattering · Monte-Carlo simulation · Molecular Dynamics · Effective interaction

1
Introduction

Dendrimers and dendritic polymers present one of the most active fields in modern research of supramolecular chemistry. Since the first pioneering studies by Vögtle et al. [1] and by Tomalia et al. [2] in the late 1970s and early 1980s there has been an exponential growth of the number of papers devoted to dendrimers and related systems [3, 4]. In particular, the regular tree-like molecular dendrimers have been a major challenge for modern synthesis. Hence, a tremendous amount of activity has been devoted to purely synthetic aspects in this field in the last 20 years. Surveys of work done on the synthesis of dendrimers may be found in recent reviews [3, 4].

Much less attention has been paid to the physical chemistry and physics of dendrimers. The physical properties of dendritic molecules dissolved in suitable solvents, however, are decisive for most of the applications now envisioned for dendrimers. Indeed, the further development of the field will now depend mainly on the number of applications that may materialize in the course of research over the next few years.

A question that is central to this point is the equilibrium structure of dendrimers in solution. Figure 1 illustrates the main point by showing the chemical structure of a flexible fourth-generation dendrimer with well-defined endgroups. This picture suggests that the density of segments grows from the center to the periphery of the molecule. This "dense-shell" picture of dendrimers has dominated the field for nearly two decades. It obtained strong support owing to a much-cited paper of Hervet and de Gennes from 1983 [5], which gave the first theoretical treatment of this dense-shell model of dendrimers. Recently, Zook and Pickett [6] have re-examined this approach. These authors have delineated the weak points of the approach used by Hervet and de Gennes [5] and demonstrated that the most probably conformation of a dendrimer has its maximum in the center of the molecule. This is in general accord with a great number of theoretical studies that have appeared since 1990, when Lescanec and Muthukumar [7] demonstrated for the first time the validity of the "dense-core" picture of dendrimers. Owing to the flexible dendritic scaffold (see Fig. 1), the overall structure is an average of a large number of possible conformers. In particular, in many conformers the endgroups will fold back into the interior of the molecule. Hence, theory demonstrated unambiguously that flexible dendrimers would have no well-defined structure in solution that could be characterized by a "surface" and internal holes. A survey of the large number of theoretical studies has been given recently [8]. It is fair to say that the theory of dendritic structure is now well-developed and has come to irrefutable conclusions.

Scattering methods such as small-angle neutron scattering (SANS) and small-angle X-ray scattering (SAXS) are the only experimental tools available to analyze the spatial structure of dissolved dendrimers in detail [9]. The analysis of dendrimers by SANS and SAXS is less straightforward than is anticipated in many recent studies, however. As discussed at length recently [8, 10], the information furnished by the scattering intensity measured for dilute solutions of dendrimers is rather limited and subject to a number of experimental uncertainties. Under no circumstances does small-angle scattering give the sort of complete information obtained from, e.g., the analysis of a crystal structure by wide-angle scattering. As demonstrated recently, studies of small-angle scattering must be supplemented by computer simulations in order to avoid faulty conclusions [11].

Despite these problems, studies using small-angle scattering have provided strong support for the dense-core picture of dendrimers. In particular, a recent investigation by SANS demonstrated directly that endgroups fold back into the interior of the molecule as has long been predicted by theory. For a sur-

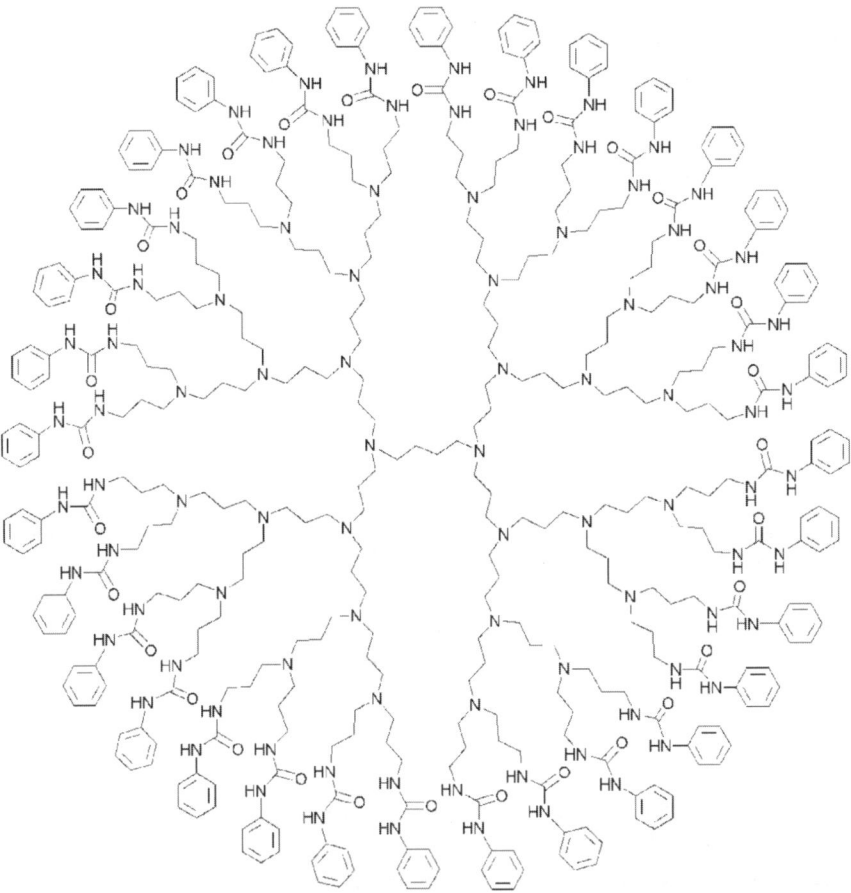

Fig. 1 The chemical structure of the urea-functionalized poly(propyleneamine) fourth-generation dendrimer G4-H

vey of recent studies employing small-angle scattering the reader is referred to [8].

In this review we wish to give a brief overview of the problem of the equilibrium structure of dissolved dendrimers. The main emphasis is placed on the combination of scattering methods and simulations. Rather than presenting a complete survey of all studies done in this field (see [8]), we shall give the main results of this analysis and delineate the most important conclusions reached so far in the realm of dilute solutions.

A further question to be discussed in this context is the interaction of dendrimers at higher concentrations. Recent work by Likos and coworkers clearly demonstrated that dendrimers exhibit an effective interaction potential that may be described by a Gaussian function [12, 13]. Hence, dendrimers provide an experimental system of particles interacting through a bounded potential.

In this way, dendrimers have become a new model system in colloid physics. This new and exciting development has passed almost unnoticed by the community of chemists working on the synthesis of dendrimers.

The review is organized as follows: section 2 outlines the main results by recent studies of dissolved dendrimers employing SANS and SAXS. It also surveys the problems for this type of analysis and the need for simulations to circumvent these problems. Section 3 in turn reviews recent theoretical developments and how these results may be used to interpret the SANS and SAXS data. Here, we shall mainly discuss recent results that are in direct relation to quantitative experimental findings. A final section will be devoted to open questions and directions for further research.

2
Scattering methods

The analysis of the equilibrium of dissolved dendrimers proceeds by the determination of the small-angle intensity $I(q)$ as a function of the magnitude of the scattering vector q [$q=(4\pi/\lambda)\sin(\theta/2)$; λ is the wavelength of radiation and θ the scattering angle] [9, 10]. Hence, $I(q)$ is determined by SAXS or SANS at a given number density N. In general, the intensity thus obtained may be rendered as the product of the intensity $I_0(q)$ related to the isolated molecule and the structure factor $S(q)$:

$$I(q) = NI_0(q) S(q) \tag{1}$$

Both $I_0(q)$ as well as $S(q)$ can be obtained from theory and simulations and thus directly compared to the experimental result [8]. In particular, the region of small angle, i.e., of small q leads to the radius of gyration R_g [9] that may be taken as a measure of the overall size of the dissolved dendrimer. With full generality this quantity follows from $I_0(q)$ by Guinier's law [9]

$$I_0(q) \cong N \cdot V_p^2 \cdot (\bar{\rho} - \rho_m)^2 \cdot \exp\left(-\frac{R_g^2}{3} \cdot q^2\right) \tag{2}$$

where V_p is the volume occupied by the dendrimer in the particular solvent and $\bar{\rho} - \rho_m$ is the contrast of the dissolved dendrimer (cf. [10]). The latter quantity follows from the difference of the average scattering length density $\bar{\rho}$ of the dendrimer and the scattering length density ρ_m of the solvent. In principle, Eq. 2 gives directly the overall size of the dissolved molecule in terms of V_p and R_g. V_p in turn leads to the molecular weight M through $M = V_p/\bar{v}_2$ where $\bar{v}_2$ is the partial specific volume. Moreover, the structural details of the dissolved molecule follow from $I_0(q)$ by comparison with model calculations or inversion into the real space through appropriate techniques. As mentioned in the Introduction, however, this comparison is more difficult than in the well-known case of linear polymers [8, 10]:

1. Dendrimers are small structures with diameters of a few nanometers only. Their radius of gyration R_g is hence of same the order of magnitude. Reliable structural information can only be obtained from $I(q)$ if $q \cdot R_g$ is considerably larger than unity [9, 10]. The measured scattering data must therefore extend far beyond the Guinier region. Because of the small V_p of typical dendrimers, $I_0(q=0)$ is much lower than the scattering intensities measured from, e.g., high polymers in solution. Moreover, $I_0(q)$ rapidly decays with higher $q \cdot R_g$. This is followed by poor statistical treatment of the data in the q-range where most of the information is to be gained. As a remedy to this problem, data are often taken at higher concentrations. This in turn requires a careful extrapolation to vanishing concentration [13, 14].
2. As a consequence of this, the experimental determination of the scattering intensity at high $q \cdot R_g$ requires special care. Here the influence of instrumental problems and of possible *incoherent* contributions must be taken into account and carefully subtracted prior to further interpretation of the data. SANS data have a strong incoherent part of the measured scattering intensity and its removal must be done by special procedures [14].
3. Small-angle scattering intensities $I_0(q)$ of fluctuating objects often exhibit only a few structural features such as, e.g., side maxima. For small dendrimers, $I_0(q)$ of dissolved dendrimers is a monotonically decaying function of q. Hence, the information embodied in such a curve is rather small and can under no circumstances be compared to the results of a crystallographic analysis. The inversion of $I_0(q)$ into the real space therefore requires special care. It can be done only in conjunction with simulations, as discussed in section 3.
4. Dendritic molecules are often envisioned as an assembly of units that have the same scattering power. Hence, it is assumed that inversion of $I_0(q)$ would give the structural information sought. In general, this is not the case and an analysis along these lines may lead to erroneous results [10].

Problems 1 and 2 are related to experimental difficulties and may be solved by appropriate treatment of the data. Problem 3 can be solved by resort to computer simulation and theory. This will be discussed further below. Problem 4 is the main difficulty with small-angle scattering and requires special attention. Contrast variation provides a means to solve this problem [9, 10, 15]. The intensity $I_0(q)$ is determined at different contrasts $\bar{\rho} - \rho_m$ and subsequently split into three partial intensities [10]:

$$I_0(q) = [\bar{\rho} - \rho_m]^2 I_S(q) + 2 \cdot [\bar{\rho} - \rho_m] I_{SI}(q) + I_I(q) \qquad (3)$$

The first term ("shape term") $I_S(q)$ is related to the scattering intensity extrapolated to infinite contrast. Here the difference between the scattering power vanishes and $I_S(q)$ gives the intensity of an assembly of scattering units that have equal magnitude. It is the Fourier-transform of the shape function $T(r)$ that describes the statistical average over all possible conformations of the dendrimers. $I_S(q)$ and $T(r)$ can therefore be compared directly to the result of simulations to be discussed further below [8]. The term $I_I(q)$, on the other hand, is solely related to internal differences in the scattering length density. If a dendrimer is set up of groups containing deuterons in place of hydrogen atoms, there is a strong difference in scattering power in the case of SANS. This will lead to a finite scattering intensity even at the match point, i.e., in a solvent where the contrast $\bar{\rho} - \rho_m$ is zero. On the other hand, $I_I(q)$ furnishes further useful information about the internal structure and may be use to localize the terminal groups ([15]; see below). The term $I_{SI}(q)$ presents the cross term between the former contributions. These three terms furnish all the information that can be obtained by a scattering experiment. Obviously, the three independent functions $I_S(q)$, $I_{SI}(q)$ and $I_I(q)$ furnish a lot more information than just a single intensity $I_0(q)$ taken at a given contrast.

All terms, as well as the dependence of R_g on contrast, have been discussed at length recently [9, 10]. Here we only demonstrate the power of contrast variation for the structural analysis of a fourth-generation dendrimer [15]. Its chemical structure is given in Fig. 1. Two types of molecules have been studied: in one case the terminal groups were bearing hydrogen atoms, whilst in the other the terminal groups were fully deuterated [15]. As mentioned before, the SANS intensity originating from these two types of endgroups will be markedly different. As a consequence of this, $I_I(q)$ will be different from zero and contrast variation should give all terms enumerated in Eq. 3.

As shown recently, a SANS analysis of both dendrimers can indeed furnish this information [15, 16]. The full information to be obtained from a scattering experiment, namely the three partial intensities $I_S(q)$, $I_{SI}(q)$ and $I_I(q)$ can be obtained and compared to data obtained from simulations. Contrast variation can be done by measurements of solutions in a mixture of protonated and deuterated dimethylacetamide. These mixtures differ widely with regard to ρ_m. Thus, the contrast $\bar{\rho} - \rho_m$ can be varied widely and the partial intensities can be securely obtained by a fit of Eq. 3 to the experimental data. Details of this procedure are given in [15, 16].

Figure 2 shows the shape term $I_S(q)$ obtained from the deuterated and the protonated dendrimer. As anticipated from the previous discussion, $I_S(q)$ must be independent of the partial deuteration of the molecule. As demonstrated in Fig. 2, this is indeed found from the experimental data. This term contains all structural information on the spatial arrangement of the dendritic units, irrespective of their contrast. The solid line represents a Gaussian function, which upon inversion leads the shape function $T(r)$, also a Gaussian function [11]. The problems of this inversion will be discussed in the subsequent section. Here, it suffices to state that $T(r)$ has its maximum at $r=0$, i.e., at the center of the molecule. This result hence demonstrates that flexible dendrimers have a dense-core structure. It is supported by many other investigations employing scattering methods [8].

If $T(r)$ has its maximum right at the center, some of the terminal groups must fold back. As shown in [15, 16], the localization of the endgroups can be done through an analysis of the terms $I_{SI}(q)$ and $I_I(q)$. Figure 3 gives these terms

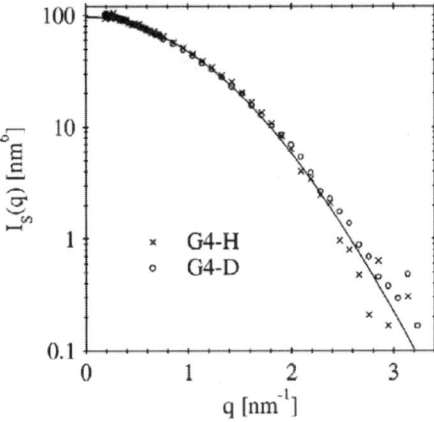

Fig. 2 The partial intensity $I_S(q)$ (cf. Eq. 3) determined by contrast variation for dendrimer G4-H (*crosses*; see also [16]) and for G4-D (*circles*). The *solid line* shows the fit of the data by a Gaussian function defined in Eq. 6. Taken from [15]

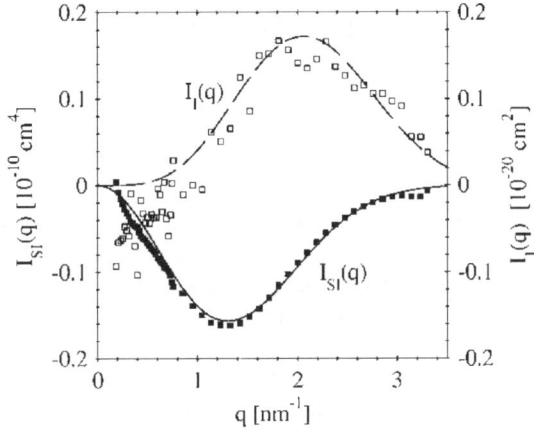

Fig. 3 The partial scattering functions $I_{SI}(q)$ (*black squares*) and $I_I(q)$ (*white squares*) as function of q (see Eq. 3) [15]

as function of q. Evidently, these partial intensities are more difficult to obtain, and there is some uncertainty of both functions in the vicinity of $q=0$. Theory states clearly, however, that $I_{SI}(q)$ and $I_I(q)$ must vanish in this limit. This renders a fit of the distribution of endgroups a do-able task. The solid line in Fig. 3 shows that both distributions can be fitted by a single endgroup distribution that will be discussed in further detail below (see also references [8, 15, 16]). Figure 4 displays the distribution of endgroups in terms of the function $\rho_p(r)$ taken from this analysis [15]. This function is proportional to the true distribution of endgroups (see the discussion in [15]) and demonstrates that the end-

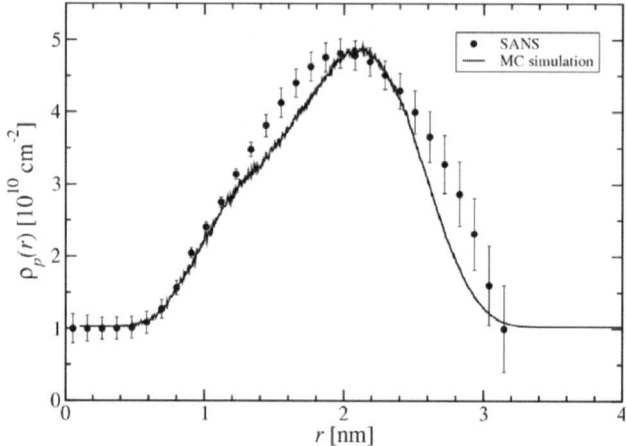

Fig. 4 Distribution $\rho_p(r)$ of endgroups in a fourth-generation dendrimer (see Fig. 1). The *points* refer to the experimental result [15] whereas the *solid line* was obtained from the theoretical model [19]

groups are dispersed throughout the dendritic molecule. Clearly they are not localized at the "surface" of the dendrimer, as anticipated by the dense-shell model.

Having discussed the scattering intensity $I_0(q)$ related to a single molecule, it remains to determine the structure factor $S(q)$ defined in Eq. 1. By virtue of its definition, $S(q)$ describes all alterations of the measured scattering intensity $I(q)$ by the interaction of the dissolved dendrimers that result from finite concentrations. Often this term is omitted by assuming that $S(q)=1$ for dilute solutions. Hence, concentrations of the order of 1wt% are defined as being small and the subsequent analysis is based on $S(q)=1$. The assumption that some arbitrary concentration is "small", however, is not correct (see the discussion in [14]). The effect of finite concentrations makes itself felt at small q and leads to radii of gyration that may be considerably lower. Working at low concentrations, on the other hand, leads to poor statistical resultsat high q.

As a matter of fact, no assumption of this sort is necessary but measurements should be done at several finite concentrations. Extrapolation to vanishing concentration according to well-established procedures provides no difficulty at all, thus leading to both $I_0(q)$ and $S(q)$ [14]. No excuses for incomplete experimental data should be accepted anymore in the course of a meaningful analysis of dissolved dendrimers or other structures. Hence, measurements at finite concentrations must become an integral part of the experimental protocol.

The interpretation of $S(q)$ can be done again with resort to theory and simulations [12, 13]. This will be the subject of the next section. There it will be demonstrated that low-generation dendrimers present a new kind of soft particle.

All previous discussion has been related solely to the case of *flexible* dendrimers, i.e., to dendritic structures that can assume a vast number of different shapes. If, on the other hand, the dendritic scaffold is set up of stiff units, no backfolding should occur and a dense-shell dendrimer must indeed result. This has been shown recently by a SANS study of a fully aromatic dendrimer synthesized by Müllen and coworkers [17]. Applying the method devised in [15, 16] to a fully aromatic fourth-generation dendrimer, it could be demonstrated that there is a crowding of endgroups in the periphery. Stiff units are hence suitable building blocks for true dense-shell dendrimers whereas flexible units inevitably lead to backfolding and a dense-core structure.

3
Theory of Dendrimers

When dealing with macromolecules that feature the complicated architecture of dendrimers, it is customary and useful to resort to simplified models that capture the salient characteristics of their architecture, while ignoring the specific details of the molecular bonding and interactions. This approach greatly simplifies the analysis by theory and simulation, and at the same time helps to shed light onto the basic physics that governs the conformations of single dendrimers as well as their mutual interactions in a concentrated solution. To this end, various so-called "coarse-grained" models have been introduced in the last few years; for an extensive review, see [8]. In what follows, we give only a brief overview of very recent work on this subject.

The conformations of isolated dendrimers have been examined by computer simulations employing a simplified model in which monomers are modelled as Lennard-Jones beads that are connected by inelastic springs [11]. This simple model turns out to reproduce the experimental results for the form factor of fourth-generation dendrimers. At the same time, other quantities of interest can also be resolved in a simulation, such as the distribution of monomers around the center of mass of the molecule (density profiles) as well as the distribution of endgroups. Figure 4 shows a characteristic example of the comparison of experimental results regarding the terminal group distribution with the prediction from simulations. The excellent agreement between the two demonstrates both the fact that endgroups are back-folded and the validity of the simplified simulation model.

An additional question that has been addressed in [11] is that of the importance of intramolecular fluctuations of the monomers. This is a central problem that has already been mentioned in the preceding section. It is related to the analysis of dissolved dendrimers by scattering methods: can the spatial structure be described by an average profile or do intramolecular fluctuations contribute significantly to the measured intensity? The experimental data provide no solution to this problem.

Here, simulations [11] have turned out to be an indispensable tool for a meaningful analysis of dendrimers by directly comparing the exact expression for the pair correlation function between the monomers

$$G(\vec{r}) = \frac{1}{N} \left\langle \sum_{i=1}^{N} \sum_{i=1}^{N} \delta(\vec{r} - \vec{r}_i - \vec{r}_j) \right\rangle; \tag{4}$$

where r denotes the magnitude of the separation vector $\vec{r}$ between any two monomers with coordinates $\vec{r}_i$ and $\vec{r}_j$; to the mean-field approximation for the same quantity

$$G(\vec{r}) = \frac{1}{N} \rho(\vec{r}) * \rho(\vec{r}); \tag{5}$$

where $\rho(r)$ is the density distribution around the center of mass and the asterisk denotes a convolution. In Eq. 4 above, the symbol <...> denotes an expectation value over all conformations of the dendrimer and N is the number of monomers. Equation 5 is the result that holds for rigid, colloidal particles without internal fluctuations. The excellent agreement between the exact result, Eq. 4, as was measured in the simulation, and the approximate expression, Eq. 5, demonstrates that the internal fluctuations of fourth-generation dendrimers are weak and correlated only at length scales of the order of a few Ångströms [11]. Thereby, the experimental procedure of employing an inverse Fourier transform of the form factor in order to obtain the shape function $T(r)$ that is subsequently identified with the density profile $\rho(r)$ has been explicitly confirmed [8].

The experimental findings point to a Gaussian-like density profile around the center of mass of a dendrimer. A Flory-type theory has been employed by Likos et al. [12, 13] that utilizes this information to derive a Gaussian effective interaction between the centers of the dendrimers. The generic form of this effective interaction, denoted $w(r)$, reads as:

$$w(r) = N^2 kTv \left(\frac{3}{4\pi R_g^2} \right)^{3/2} \exp\left(-\frac{3r^2}{4R_g^2} \right) \tag{6}$$

where v is the excluded volume parameter, R_g the gyration radius (see preceding section), k denotes Boltzmann's constant and T is the absolute temperature. Subsequently, standard tools from the theory of structure of classical fluids have been employed in order to calculate theoretically the static structure factors from concentrated dendrimer solutions and compare them with experimental results from SANS. Figure 5 shows a representative example of the agreement between theory and experiment, demonstrating that flexible dendrimers provide a physical realization of bounded, Gaussian interactions [13]. We emphasize that such types of interaction potentials are unknown in the realm of atomic or molecular physics, where the strong electrostatic or exclu-

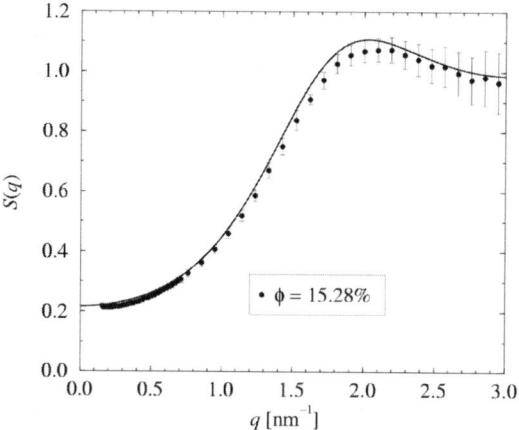

Fig. 5 Structure factor $S(q)$ (see Eq. 1) of a fourth-generation dendrimer [13]. The *points* mark the experimental data taken at a volume fraction of 15.28%. The *solid line* shows the result from the theory based on Eq. 6. See also [13] and [8]

sion-principle repulsions between the electrons bring about a concomitant interaction that strongly diverges at close approaches. Here, the soft character of the dendritic "colloids" reflects itself in a similarly soft interaction potential. The signature of the softness in the scattering profiles is a lack of pronounced correlation peaks in the structure factor.

An even simpler simulation model has been introduced by Götze and Likos [18]. Here, the monomers have been modelled as hard spheres connected by threads, with the ratio b between the maximal thread extension and the hard sphere diameter serving as a parameter that allows for tuning the flexibility of the dendrimer. Due to its simplicity and its purely entropic (excluded-volume) nature, the model is very well suited for fast Monte Carlo simulations. Thereby, dendrimers of generation numbers from 4 to 9 have been simulated and it has been demonstrated that with increasing generation number the dendrimers naturally bridge between polymeric entities and hard spheres. Indeed, with increasing generation numbers the form factors of the dendrimers, as measured in simulations, evolve towards shapes that characterize hard colloids, developing pronounced oscillations at high values of the wavevector. The latter denote an increasingly sharp boundary of the molecule and they follow Porod's law [9, 10].

A common feature of dendrimers of all generations, however, is the strong back-folding of the terminal groups in the molecule's interior. As a matter of fact, this tendency *grows* with increasing generation number, since the number of terminal monomers grows exponentially and the space available to the boundary of the molecule is not sufficient to accommodate them. Furthermore, a striking insensitivity of the density profiles was found, irrespective of the model used to describe the dendrimers: both the model of [18] and the aforementioned model of [11] yield form factors that are in excellent agreement with SANS.

Very recently, the hard-sphere-thread model has also been employed to measure in a simulation the effective interaction between dendrimers of various flexibilities and generations [19]. This simulation study nicely confirms the results of the Flory theory of [12] and [13]. The resulting effective interactions turn out to have a Gaussian form. The strength of the Gaussian interaction, compared to the thermal energy kT, can be tuned by the value of the flexibility ratio b and the terminal generation number. The smaller b is, the stiffer the interaction for a given terminal generation. On the other hand, increasing the generation for a given b brings about the same effect of making the interaction stronger. Thereby, a change of the interaction strength by as much as one order of magnitude can be achieved, pointing to the fact that dendrimers are excellent physical systems for realizing bounded and tuneable effective interactions.

4
Open Questions and Directions of Further Research

Most simulational models applied to date employ a coarse-grained description of the monomer interactions and their bonding, with a few notable exceptions in which atomistic, force-field models have been employed [20–22]. With the development of faster and more efficient computers, detailed atomistic simulations are feasible and certainly desirable as a future research perspective. The important question to be addressed in this respect is whether specific monomer–monomer or monomer–solvent interactions can bring about deviations from the dense-core model that is unanimously predicted by all coarse-grained simulation models. There are several conceivable factors that have to be taken into special consideration in this respect, namely the following.

1. The rigidity of the bonds connecting the monomers. In most studies to date, the bonds have been modelled as fully flexible, i.e., torsional constraints in their conformations have been ignored (see, however, [21] for an important exception). Though there is no doubt that flexibility causes backfolding of the endgroups, stiff bonds may lead to conformational changes in the dendrimers, as demonstrated recently experimentally for a system of stiff dendrimers ([17]; see preceding section). It would thus be of great importance to model and tune this rigidity in a simulation, in order to be able to make meaningful predictions that could be used as a guide for intelligent design of new materials. In this way simulations would become a direct guideline for the synthesis of dendrimers.
2. Specific interactions between the terminal monomers and the solvent. In most cases, solvent molecules have not been simulated explicitly. The possibility, however, that there exist specific chemical interactions between ter-

minal groups and solvent, leading, e.g., to the formation of hydrogen bonds, may again lead to a localization of the endgroups at the periphery of the molecule. Some preliminary investigations in which attractive, directional interactions between the endgroups have been introduced in a phenomenological way [23] show that there is no spectacular change of the conformation of the dendrimer under these conditions. Nevertheless, this question is worthy of further investigation.

3. Associated with the above is also the possibility of replacing the endgroups with ones that have a different size, or even shape, than those in the interior of the dendrimer. If there is a strong thermodynamic tendency that the species of the interior component and the species of the endgroups demix (e.g., spheres and rods under appropriate conditions), this may be reflected by a separation between the endgroups and the rest of the molecule, i.e., into a localization of the former at the periphery.

Another very important topic that is unsettled, and which deserves close consideration in the future, is that of *charged* dendrimers. In their pioneering work, Welch and Muthukumar [24] predicted a dramatic change of the dendrimer's size upon charging the endgroups, as well as a nontrivial dependence of the radius of gyration of the macromolecule on the salt concentration in the solution. Recent simulations of Lee et al. [25] are rather inconclusive as to the veracity of this phenomenon, which has been also called into question in the experimental work of Nisato et al. [26]. From the theoretical point of view, the great challenge lies therein, that the typical size of the dendrimers is rather small compared, say, with typical colloidal particles. The same holds, naturally, for the typical separations between the charged monomers in the dendrimer's interior, so that it becomes questionable whether a linear screening theory with a spatially uniform dielectric constant for the solvent is applicable in this case. In other words, it seems that here one has to take the solvent's granular nature explicitly into account and resort to more sophisticated simulations in which the solvent molecules are considered explicitly. This is similar to the case of dissolved proteins, for which it has been amply demonstrated that taking into account the water molecules brings about nontrivial effects that cannot be taken into account in any theory in which the solvent is modelled as a dielectric continuum [27].

References

1. Buhleier E, Wehner F, Vögte F (1978) Synthesis 155
2. Tomalia DA, Baker H, Dewald JR, Hall M, Kallos G, Martin S, Roeck J, Ryder J, Smith P (1985) Polym J 17:117
3. see e.g. Newkome GR, Morefield CN, Vögtle F (1996) Dendritic molecules: concepts, synthesis, perspectives. Wiley-VCH, Weinheim
4. Vögtle F, Gestermann, S, Hesse R, Schwierz H, Windisch B (2000) Progr Polym Sci 25:987
5. de Gennes PG, Hervet H (1983) J Phys Lett (Paris) 44:L351

6. Zook TC, Pickett G, (2003) Phys Rev Lett 90:105502
7. Lescanec RL, Muthukumar M (1990) Macromolecules 23:2280
8. Ballauff M, Likos CN (2004) Angew Chem Intl Ed 43:2998
9. Higgins JS, Benoit HC (1994) Polymers and Neutron Scattering. Clarendon Press, Oxford
10. Pötschke D, Ballauff M (2002) Structure of dendrimers n solution as probed by scattering experiments. In: Borsali R, Pecora R (eds) Structure and dynamics of polymer and colloidal systems. Kluwer, Dordrecht
11. Harreis HM, Likos CN, Ballauff M (2003) J Chem Phys 118:1979
12. Likos CN, Schmidt M, Löwen H, Ballauff M, Pötschke D, Lindner P (2001) Macromolecules 34:2914
13. Likos CN, Rosenfeldt S, Dingenouts N, Ballauff M, Lindner P, Werner N, Vögtle F (2002) J Chem Phys 117:1869
14. Rosenfeldt S, Dingenouts N, Ballauff M, Lindner, P, Likos CN, Werner N, Vögtle F (2002) Macromol Chem Phys 203:1995
15. Rosenfeldt, S, Dingenouts N, Ballauff M, Werner N, Vögtle F, Lindner P (2002) Macromolecules 35:8098
16. Dingenouts N, Rosenfeldt, S, Werner N, Vögtle F, Lindner P, Roulamo A, Rissanen K, Ballauff M, (2003) J Appl Cryst 36:674
17. Rosenfeldt, S, Dingenouts N, Pötschke D, Ballauff M, Berresheim AJ, Müllen K, Linder P (2004) Angew Chem 116:111
18. Götze IO, Likos CN (2003) Macromolecules 36:8189
19. Götze IO, Harreis HM, Likos CN (2003), J Chem Phys 120:7761
20. Cavallo L, Fraternali F (1998) Chem Eur J 4:927
21. Zacharopoulos N, Economou IG (2002) Macromolecules 35:1814
22. Mallamace F, Canetta E, Lombardo D, Mazzaglia A, Romeo A, Monsu Scolaro L, Maino G (2002) Physica A 304:235
23. Götze IO, Likos CN, Ballauff M (2004) to be published
24. Welch P, Muthukumar M (1998) Macromolecules 31:5892
25. Lee I, Athey DB, Wetzel AW, Meixner W, Baker JR Jr (2002) Macromolecules 35:4510
26. Nisato G, Ivkov R, Amis EJ (2000) Macromolecules 33:4172
27. Dzubiella J, Hansen JP (2003) J Chem Phys 119:1204

Functionalised Polyphenylene Dendrimers and Their Applications

Roland E. Bauer · Andrew C. Grimsdale · Klaus Müllen (✉)

Max-Planck-Institute für Polymerforschung, Ackermannweg 10, 55128 Mainz, Germany
muellen@mpip-mainz.mpg.de

1 Introduction 254

2 Functionalisation in the Core – Encapsulating Functionality 257

3 Functionalisation in the Scaffold 263

4 Functionalisation at the Surface 266
4.1 Desymmetrisation – Towards Functional Nano-Building Blocks 267
4.2 Dye-Functionalised Dendrimers – Emissive Nanoparticles 269
4.3 Dendrimers with Other Electronically Active Substituents 273
4.4 Dendrimers Bearing Polar Substituents 276
4.5 Dendrimers with Bioactive Substituents 277
4.6 Core–Shell Dendrimers as Supports for Metallocenes 279
4.7 Functional Dendrimers as Precursors to Functionalised 2-Dimensional Discs . . 281

5 Conclusions and Outlook 283

References 284

Abstract In this chapter is described the development of polyphenylene-based shape-persistent dendrimers as functional materials. Functional groups can be incorporated into such dendrimers in the core, in the scaffolding, or on the periphery. The rigidity and shape-persistence of the structure means that the functional groups are in well-defined spatial relationships to each other, which is advantageous for both the control of and the study of energy and charge transfer processes between them. Electrophoric and chromophoric units, including conjugated polymers, have been used as functional cores and the dendronisation has been found to have effects upon their opto-electronic properties. Efficient energy transfer occurs between dyes attached at the surface, in the scaffold and at the core, offering possible applications in light-harvesting systems. Dendrimers bearing biologically active moieties such as peptides or biotin have been made for bioassay applications, and as agents for RNA transfection. Attachment of suitable substituents, e.g. poly(ethylene oxide) chains, on the surface enables core–shell systems to be made with possible applications as supports for catalysts. Desymmetrisation opens the way to bifunctional nanoparticles, for example a dendrimer bearing both a biotin for binding to streptavidin and dye chromophores for visualisation.

Keywords Dendrimers · Functional nanoparticles · Energy transfer

Abbreviations

AFM	Atomic force microscopy
EL	Electroluminescence
GPC	Gel permeation chromatography
HBC	Hexa-*peri*-hexabenzocoronene
LED	Light emitting diode
NMR	Nuclear magnetic resonance spectroscopy
PEO	Poly(ethylene oxide)
PL	Photoluminescence
SMS	Single molecule spectroscopy
TIPS	Triisopropylsilyl

1
Introduction

Dendrimers [1, 2], as well-defined 3-dimensional macromolecules (as distinct from hyperbranched polymers), represent a key stage in the development of macromolecular chemistry. They have become of great scientific interest in recent years, and one particular area of recent research effort has been into using functionalised dendrimers in a variety of applications [3]. There are two classes of dendrimers identifiable in the literature, which are defined by the chemical and physical properties of their components: flexible "soft" dendrimers [4, 5] and inherently rigid "shape-persistent" dendrimers [6–8]. Polyphenylene-based dendrimers [9] fall into the second class owing to the nature of the phenyl-phenyl bonds and the large number of twisted interlocked phenyl rings with reduced mobility. Their rigidity and shape-persistence has been demonstrated by a number of techniques. Atomic force microscopy (AFM) studies of dendrimers deposited from a dilute solution onto mica revealed them to be discrete nano-objects whose measured height matched that calculated by molecular modelling [10]. As AFM typically produces underestimates of the height of soft objects owing to deformation, this confirms the rigidity and shape-persistence of these materials. The stiffness of the dendrimers was also demonstrated by pulsed-force mode AFM experiments [10, 11]. Solid-state NMR experiments showed that the phenyl rings in the dendrimers, especially those in the core and scaffold, display very limited dynamics with heavily restricted movements [12, 13]. Whereas the majority of experimental results for flexible dendrimers support the applicability of the kinetic growth model (dense-core) [14] where back-folding of some of the branches gives rise to a density distribution with a maximum in the core of the molecule [15], these NMR experiments [12], together with recent neutron diffraction experiments [15], clearly show that the density in polyphenylene dendrimers increases towards the periphery of the molecules, in concordance with the earlier developed self-consistent field (dense-shell) model [16].

This rigidity means that the dendrimers have a well-defined nanoporosity, and that functional groups in the core, interior layers (scaffold) and/or on the surface of the dendrimer are essentially fixed in well-defined and predictable spatial relationships with each other, which affects their interactions and so offers the possibility of controlling the chemical or physical properties of the dendrimers. For example, as the rates and efficiencies of energy and/or charge transfer processes are inversely dependent on the distance (or exponents thereof) between the interacting units, this means that within polyphenylene dendrimers such processes can be controlled. This is important not only for studying these processes, but also for the potential applications of functionalised dendrimers, as their materials' properties can be predicted. This permits fine-tuning of their opto-electronic properties by varying the number and positions of such groups within the dendrimer, thus potentially enabling a synthetic chemist to design and make materials with specific optical or electronic properties.

The synthetic approaches to and general properties of polyphenylene dendrimers have been covered in [9]. In this chapter we review the recent work done within our group into making functionalised polyphenylene dendrimers. Among the functional groups that have been incorporated are dye chromophores, charge-accepting and -transporting units including conjugated oligomers, and biologically active units such as biotin and peptides. We also discuss how the properties of the functional groups are affected by incorporation into the dendrimer framework. Some potential uses of these materials as monodisperse functional nanoparticles are also described. Dendrimers containing dye chromophores have been used as model compounds for energy transfer studies. Other compounds bearing polar groups have been tested as units in chemical sensing devices, while dendrimers bearing biologically active substituents can be used for bioassays. Another important potential application for functional dendrimers is as supports for metallocene catalysts for olefin polymerisation. Polyphenylene dendrimers have been shown to self-assemble into long nanofibres [17], and we show how the substitution of their surface with suitable groups can be used to control the self-assembly of the dendrimers or their complexation to other materials such as polymers or gold nanoparticles.

First, it is necessary to consider how the functional groups can be incorporated into the dendrimer. There are two basic synthetic approaches to polyphenylene dendrimers [9] which are exemplified in Scheme 1 for a second-generation dendrimer 3. Both utilise Diels-Alder cycloadditions of cyclopentadienones to alkynes. In the convergent approach the ethynyl-substituted core 1 is reacted with a cyclopentadienone 2 substituted with dendrons to make 3 directly. In the divergent approach the dendrons are grown outwards from the core in an iterative manner by addition to 1 of a cyclopentadienone 4 with silyl-protected ethynyl functionality to make 5, followed by deprotection and addition of a second cyclopentadienone 7 to the resulting ethynyl-substituted first-generation dendrimer 6. The convergent approach can be used only for first- and second-generation dendrimers as the cyclopentadienones needed to make higher generations are not synthetically accessible, but this approach is important for the synthesis of asymmetric dendrimers with different functional groups on different parts of their surfaces (*vide infra*). In contrast, the divergent approach can be used to make materials up to the fourth generation. Steric crowding on the surface of the fourth-generation dendrimers means that further Diels-Alder cycloaddition steps towards a potential fifth-generation dendrimer occur incompletely, i.e. not all of the peripheral ethynyl groups can be converted to the subsequent generation, so that only a partial fifth layer is formed.

As can be seen from Scheme 1 functionality can be introduced into the dendrimer at three places – at the core, in the scaffold (rings A and B in **3**) and at the surface (rings C and D in **3**). We will now consider functionalisation at each of these areas in turn.

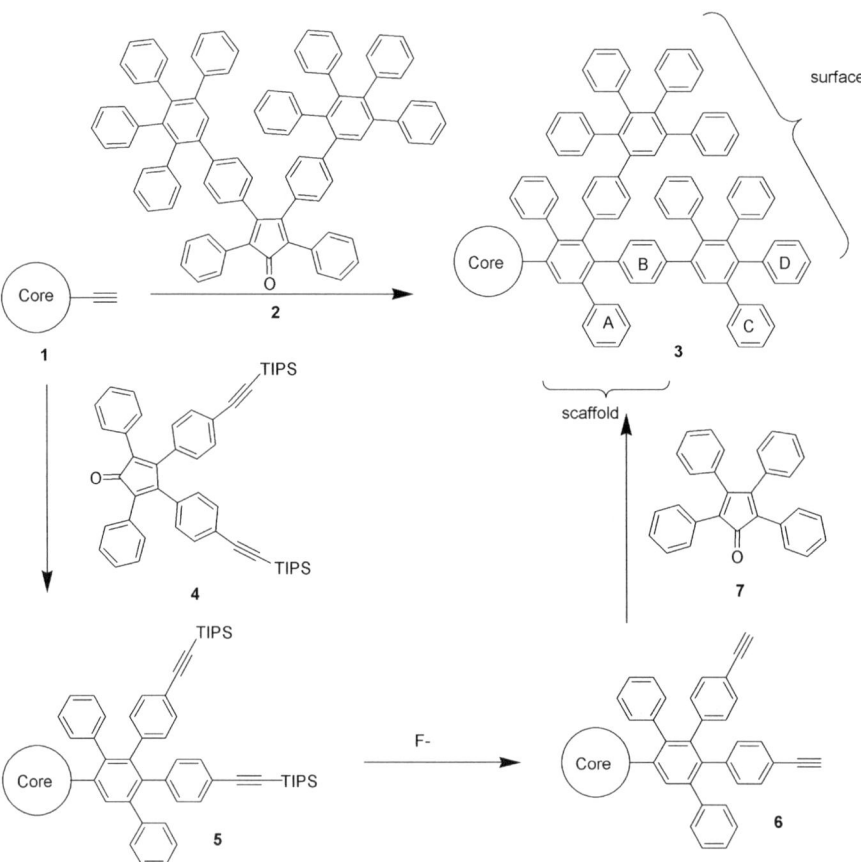

Scheme 1 Convergent (*top*) and divergent approaches to second-generation polyphenylene dendrimers

2
Functionalisation in the Core – Encapsulating Functionality

Essentially any group which can withstand the Diels-Alder cycloaddition conditions and can be substituted with ethyne units may be used as a core for a polyphenylene dendrimer. We have previously demonstrated that a wide variety of hydrocarbon cores e.g. tetraphenylmethane, hexaphenylbenzene and biphenyl can be used as cores for dendrimers, their molecular geometry influencing the overall shape of the resulting dendrimers [9]. To make a core-functional dendrimer one can thus start with an ethynyl-substituted functional molecule. It should be noted that whereas in our previous work a given core was selected merely to influence the overall shape of the dendrimer, here the core is being chosen in order to introduce a function, e.g. an optical or electronic property. In such a dendrimer the dendrons will act as a shield protecting the core from interacting with other cores or with other species such as solvents, which is particularly important for controlling the optical and electronic properties. One particularly important optical property we seek to control is the emission of light or luminescence – either light-induced [photoluminescence (PL)] or electrically induced [electroluminescence (EL)]. As interactions between chromophores are well known to lead to quenching of PL or to red-shifting in absorption or PL spectra owing to formation of aggregates or excimers [18], encasing a chromophore such as a dye in a dendrimer can be expected to lead to enhancement of its optical properties by suppression of these undesirable interactions. To prove this, dendrimers with perylenetetracarboxydiimide dyes as cores (Scheme 2) have been made. These dyes were chosen because of their excellent photo- and thermal stabilities, and high solid-state PL efficiencies. The cores used were **8** with ethynyl groups attached to the imide moieties [19], and **9** where they are in the bay positions [20].

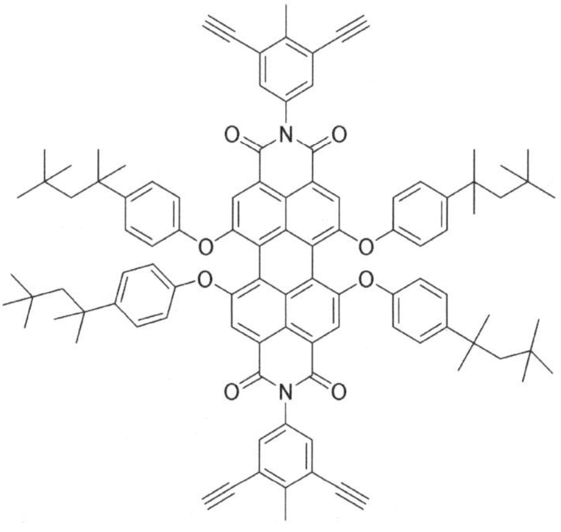

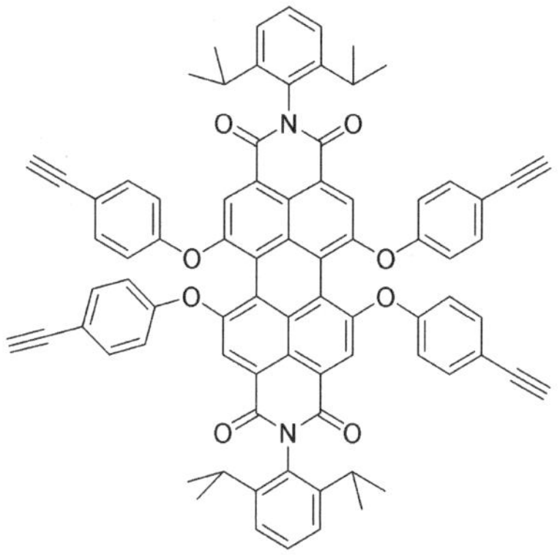

Scheme 2 Ethynyl-substituted perylenediimide dyes as functional cores for dendrimers

First- to third-generation dendrimers, e.g. **10**, have been prepared from **8** [19] and first- and second-generation dendrimers, e.g. **11**, from **9** [21] (Scheme 3). Molecular models clearly show that the dendrons in these materials shield the perylene core to hinder aggregation of the chromophores. The shielding is more effective for the material derived from **9** with the dendrons in the bay positions. The second-generation dendrimer derived from **8** has been shown to self-assemble into long fluorescent nanofibres [22].

In both cases the effect of the dendronisation is to produce a blue-shift in the solid-state absorption and emission spectra compared with an undendronised perylenediimide, thus showing that the chromophores have been prevented

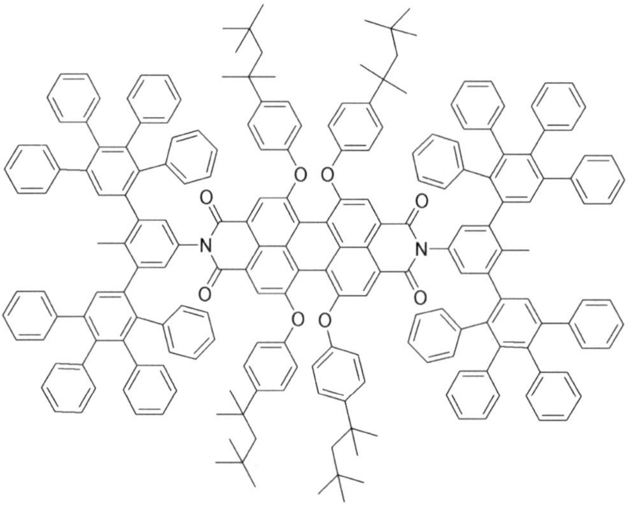

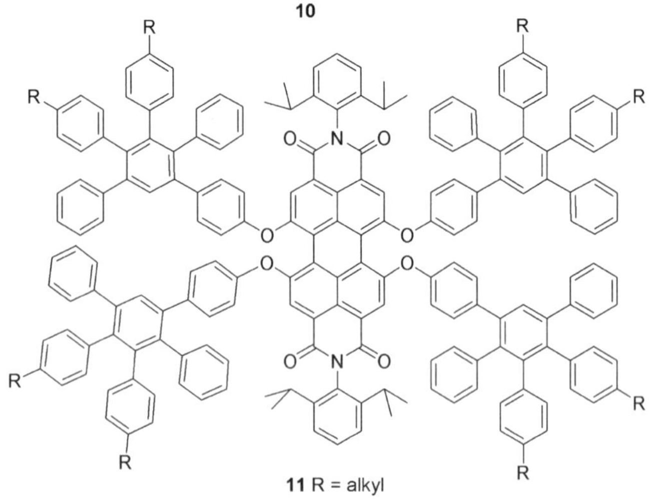

Scheme 3 Dendrimers with perylenetetracarboxydiimide dye cores

from aggregating [19, 21]. The PL efficiencies of the dendrimers derived from **8** decrease with increasing dendrimer generation, which is attributed to the opening up of new non-radiative deactivation pathways by the dendrons [19]. Light emitting diodes (LEDs) have been prepared using these dendrimers [21, 23], but show low efficiencies, especially for the higher generation dendrimers, which is attributed to poor charge transport through the dendrimer shell. Blending the imide-dendronised dyes such as **10** with polyfluorenes has been shown to improve the EL efficiency owing to better charge injection and transport [23]. In this case some of the emission also probably arises through energy transfer from the polymer to the perylene diimide.

Another type of photoresponsive material we have used as a core is azobenzene. This unit is photochromic and, as shown in Scheme 4 for a first-generation dendrimer, undergoes an isomerism from the *trans* **12** to the *cis* **13** form upon irradiation with ultraviolet light [24]. This produces a marked change in the shape of the dendrimer, which is detectable by gel permeation chromatography (GPC) as an increase in retention time on the column owing to a marked (24–38%) decrease in the hydrodynamic volume of the *cis* isomers. The contraction is greatest for the first-generation dendrimer. Photophysical measurements show that the rate of light-induced *trans–cis* isomerisation and the proportion of *cis* isomer in the photostationary state decreases with increasing generation or increasing branching of the dendrimers, while the rate of the thermally induced reverse isomerisation increases. Other azobenzene-cored dendrimers had been prepared previously [24], but because of the flexibility of their dendrons their photoisomerisation does not cause a marked, well-defined change in the dendrimer shape, thus illustrating the superiority of rigid polyphenylene dendrons.

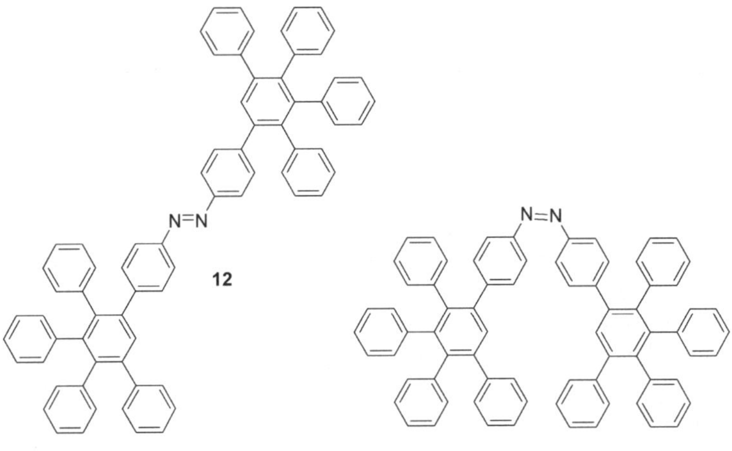

Scheme 4 First-generation azobenzene-cored dendrimers showing the marked change in volume upon *trans-cis* isomerisation

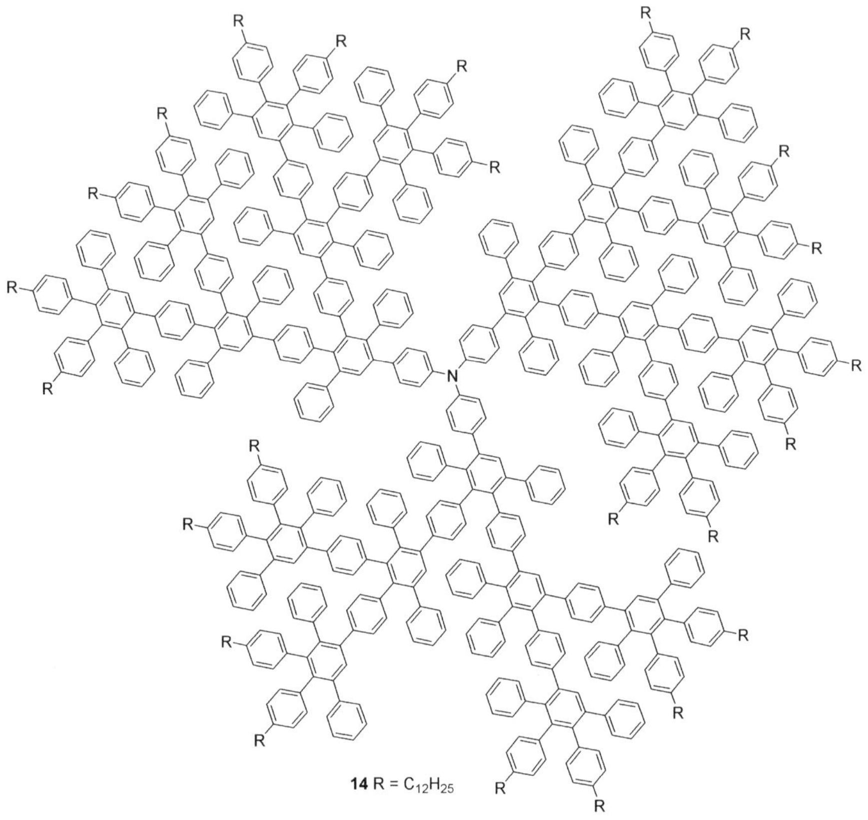

Scheme 5 Third-generation dendrimer with a triphenylamine core

Redox-active cores can be used. Thus dendrimers such as **14** have been synthesised with a hole-accepting triphenylamine core (Scheme 5) [25, 26].

Cyclic voltammetry experiments displayed an increase of 130 mV in the potential of the anodic peak in the oxidation cycle in going from the first to the third-generation dendrimer together with a two-thirds reduction in the peak current [25]. This shows that the effect of the dendrons is to hinder electron transfer from the hole-accepting triphenylamine core, which is consistent with the poor efficiencies seen for EL from the dendronised dyes discussed above.

Dendronisation can also be used to prevent aggregation of conjugated polymer chains. For a general discussion of dendronised polymers the reader is referred to the review by Schlüter and Rabe [27]. There are reports in the literature of flexible Fréchet-type dendrons being used to shield conjugated polymer backbones and so prevent red-shifting of the PL spectra in the solid state [28, 29]. However these have usually required at least second-generation dendrimers to achieve complete suppression of aggregation. We have shown that polyfluorenes **15** and **16** (Scheme 6) with rigid first-generation polyphenylene

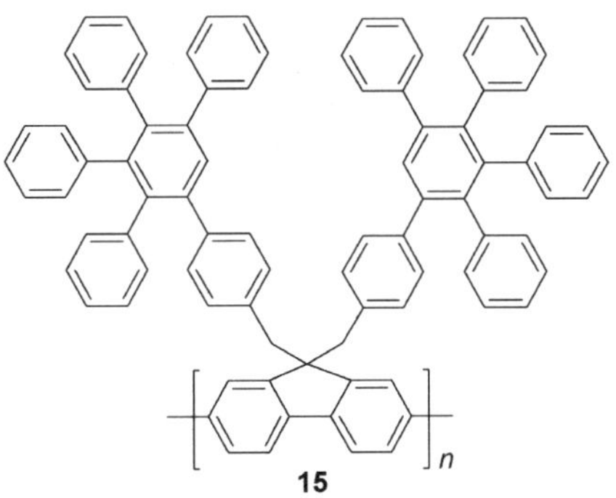

Scheme 6 Polyfluorenes with phenylene dendrimer side-chains

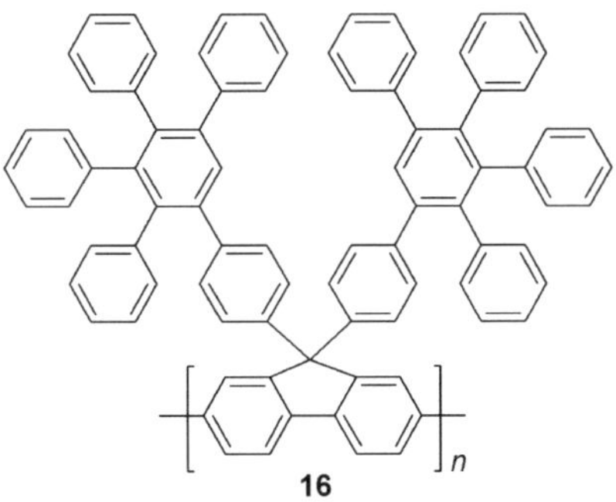

dendrons show complete suppression of aggregation [30], thus displaying the superior shielding ability of this type of dendron. The polymer **16** without the benzyl linkages displays more stable emission then does **15**, presumably because of its greater resistance to oxidation, and has been used to make LEDs which produce stable blue emission [31]. It has been demonstrated that the photophysics of these dendronised polyfluorenes differ from those of non-dendronised analogues with a change in the dynamics of singlet excitons in the solid state [32, 33], an increase in triplet lifetime [34] and a reduction in diffusion of triplet excitons [35].

In principle, dendronised polymers could be made by polymerisation of a dendronised monomer or by dendronisation of a suitable precursor polymer [27]. To date we have used only the former approach, as all attempts to make a suitable precursor polymer have been unsuccessful, since ethynyl sidegroups complex with the transition metal reagents used in the polymerisations and so deactivate them. The problem with this approach is that dendronised monomers tend to give low degrees of polymerisation, presumably due to steric congestion. To obtain high degrees of polymerisation, copolymerisation with non-dendronised monomers can be performed; this, however, dilutes the degree of screening of the polymer [28, 29].

Both convergent and divergent approaches to the dendronised fluorene monomers have been used. The monomer for **15** was made by addition of tetraphenylcyclopentadienone **7** to a diyne, while the monomer for **16** was made by addition of a pre-formed dendron **17** to a biphenyl ester **18** followed by ring closure (Scheme 7) [30]. This route is not suitable for attaching functionalised or higher generation dendrimers, as the steric congestion at the reaction site will hinder the reaction, and some functional groups are not compatible with the reaction conditions. More recently, we have developed a divergent synthesis of dendronised fluorenes in which we react **18** with an aryl-lithium bearing a protected ethynyl substituent. Deprotection then gives a dialkyne, which acts as the core for building up the dendrons, after which the fluorene forms by acid-, or Lewis acid-induced ring-closure (Scheme 8) [36]. This

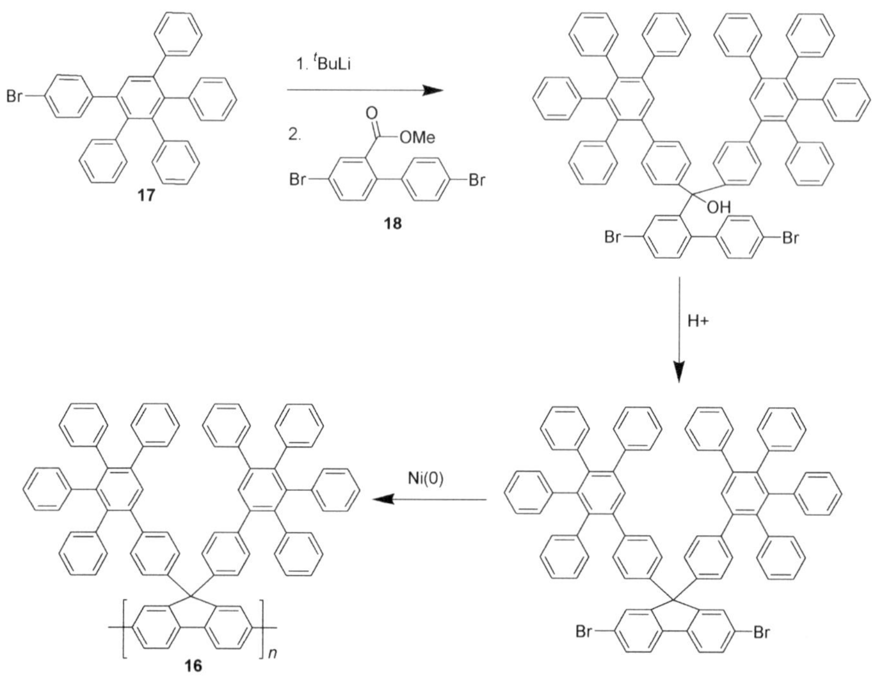

Scheme 7 Dendronised fluorene via a convergent synthesis

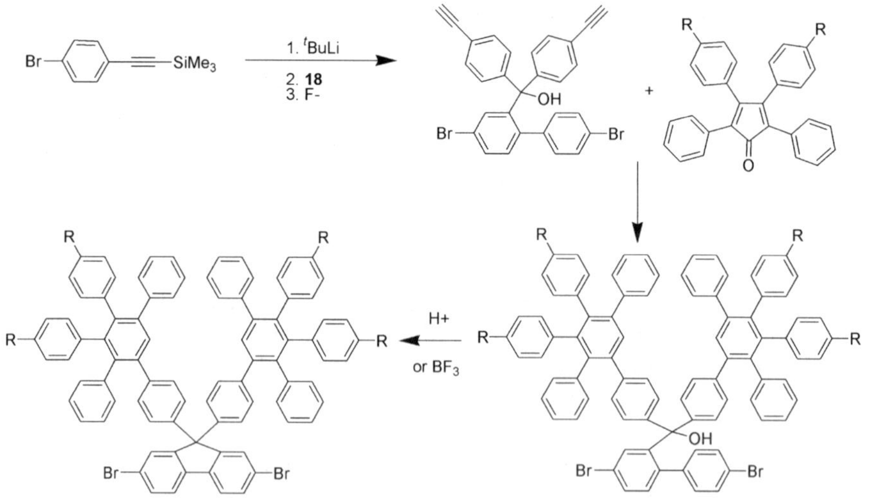

Scheme 8 Divergent route to fluorene with functional dendron substituents

route enables functionalised dendrons to be attached to a fluorene. By using the reagent **4** second or higher generation dendrimers have also been built up on a fluorene.

Dendrimers can also be used to suppress the aggregation of discotic 2-dimensional graphite molecules. We have previously shown that hexa-*peri*-hexabenzocoronones (HBCs) form extensive columns due to strong π–π interactions between the discs [37]. HBCs such as **19** (Scheme 9) with first-generation dendrimer substituents, however, cannot form such extended stacks due to the steric requirements of the bulky dendrons, and so in solution are present as dimers in equilibrium with isolated discs, as shown by UV-VIS absorption, NMR spectroscopy and light-scattering experiments [38]. The equilibrium can be controlled by varying the solvent, the temperature and the concentration of the dendronised discs.

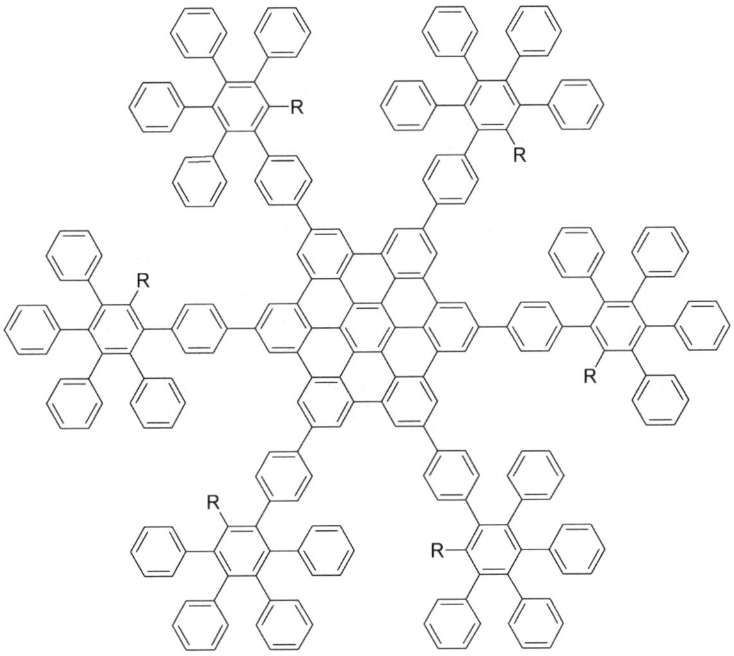

19 R = H, $C_{12}H_{25}$

Scheme 9 Hexa-*peri*-benzocoronenes (HBCs) with dendron substituents

3
Functionalisation in the Scaffold

Whereas functionalisation at the core is used primarily for preventing undesirable optical properties by encapsulation of the functional group within an inert shell, functionalisation of the scaffold can serve several purposes, e.g. fine-tuning of charge or energy transfer properties through placing a function at a well-defined distance from a functional core or surface, and the formation of functional cavities. While functionalisation at the core requires synthesis of a suitable functional core with ethynyl substituents, functionalisation at the scaffold or surface is generally achieved by use of suitably functionalised cyclopentadienones **20**. Such cyclopentadienones are prepared by Knoevenagel condensation of 1,3-disubstituted acetones **21** with functionalised benzyls **22** (Scheme 10) [9, 39].

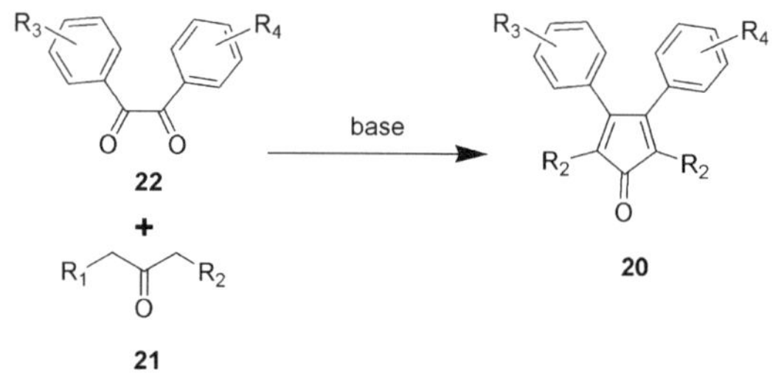

Scheme 10 Synthesis of functionalised cyclopentadienones

As the dibromocompound **22** is obtainable in high yield from commercially available precursors it is often used as the starting point for the synthesis of other disubstituted cyclopentadienones (Scheme 11). For example, **22** is convertible to the di-amine **23** by Buchwald coupling with benzophenone imine followed by hydrolysis [40], while palladium-catalysed trans-boronation produces the bis-boronate **24**, which can then be used to introduce many other functional groups via Suzuki coupling [41].

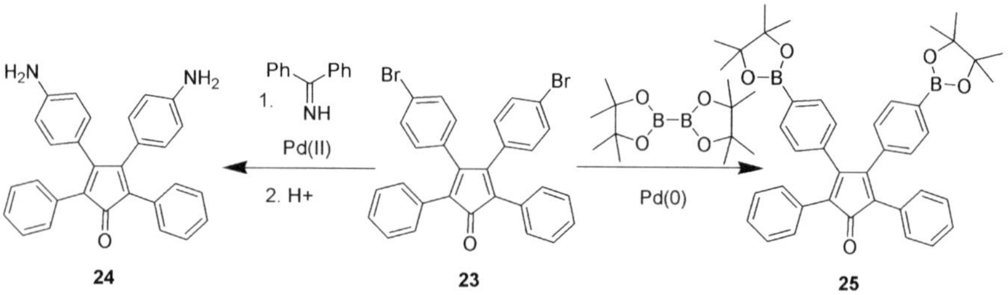

Scheme 11 Synthesis of disubstituted cyclopentadienones from dibromide **23**

Functionalisation of the scaffold requires modification of the ethynylcyclopentadienone **4** (Scheme 1) used to build the higher-generation dendrimers, which can be done in two ways (Scheme 12). Functional groups can be added into **4** either at the α-positions by replacement or substitution of the phenyl rings, e.g. **26** or **27** [42], or by inserting them between the β-phenyls and the ethyne bonds, e.g. **28** [43] or **29** [44] – both made via **25**. The second way is to

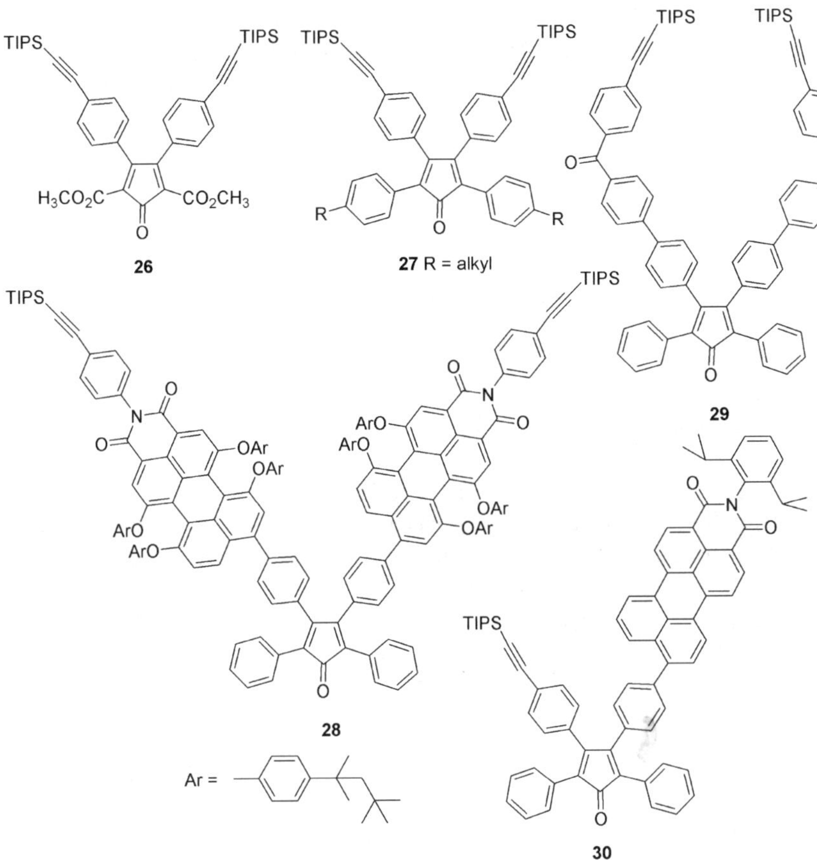

Scheme 12 Cyclopentadienones for introduction of functionality into scaffold. *TIPS* Triisopropylsilyl

desymmetrise **4** by replacing one of the ethynyl groups with a functional group, for example **30** with one perylene monoimide dye [45]. This is achieved by using a Knoevenagel condensation of diphenylacetone with an asymmetrically functionalised benzil [39]. The synthesis of the latter is synthetically demanding and bulky substituents on the benzil lower the yield of the condensation.

The effect of the substituents in **26** and **27** is to introduce functionality within the scaffold, which may affect the internal density of the dendrimer and the chemical properties of the cavities without affecting the possible number of branches or the number of potential functional groups at the surface. Thus by using the reagent **26** we have been able to incorporate up to 24 ester groups within a dendrimer, which can then be hydrolysed to carboxylates, thus producing charged cavities of defined sizes which could be used to form complexes with cationic species within the dendrimers [46]. In combination with suitable peripheral substitution these materials have potential applications as chemical sensors. The perylene monoimide groups in **28** not only introduce functionality, but also act to extend the space between the adjacent layers, which has the effect of reducing the peripheral density of the molecules, thus opening up new possibilities towards obtaining higher generations of polyphenylene dendrimers. It also allows the total number of chromophores in the dendrimers to be significantly increased, thus increasing the intensity of their overall absorption or emission. The effect of using desymmetrised reagents such as **30** is to introduce bulky functional groups into a layer of the scaffold without increasing the space between the layers. However this approach also means the number of branches in that layer is halved, which will obviously also halve the number of possible functional groups on the surface.

The advantage of incorporating functionality into the scaffolding by using reagents such as **26–28** is that it enables more functionality to be introduced than by using a functional core while avoiding any problem of isomer formation during the Diels-Alder reactions as can be seen for surface functionalisation (*vide infra*). Such functionalisation also provides a way of ensuring efficient energy or charge transfer between the core and the surface in larger dendrimers. We do not usually functionalise the scaffolding alone but in combination with functionalisation at the core and/or surface, precisely for this purpose, for example the triad **41** discussed below (Scheme 18).

Another concept we are exploring is to introduce functionality, which will allow subsequent chemical functionalisation of the interior of the dendrimer. Thus the ketone groups on **30** are sites for later chemistry, e.g. addition of lithium reagents [44].

4
Functionalisation at the Surface

Functionalisation at the surface is the simplest and most flexible method of functionalising dendrimers with the greatest possibilities for controlling the type, number and position of the substituents. A number of examples of this have been presented in our earlier review [9]. A potential complication with surface functionalisation not seen for functionalisation of the core and the scaffold is the formation of isomers. The surface can be functionalised either by aromatic substitution of a pre-formed dendrimer or by using a functionalised cyclopentadienone **20** (Scheme 10) to prepare the final layer of the dendrimer. The former approach presents obvious difficulties in controlling the degree and position of substitution. The latter approach allows a predictable number of functional groups to be attached, but isomers are still possible. If a cyclopentadienone with one functionalised β-phenyl ring is used to make the outermost layer of the dendrimer, then it can add in two ways to form equal amounts of "*para*"- and "*meta*"-isomers (Scheme 13). (it should be noted that the isomers

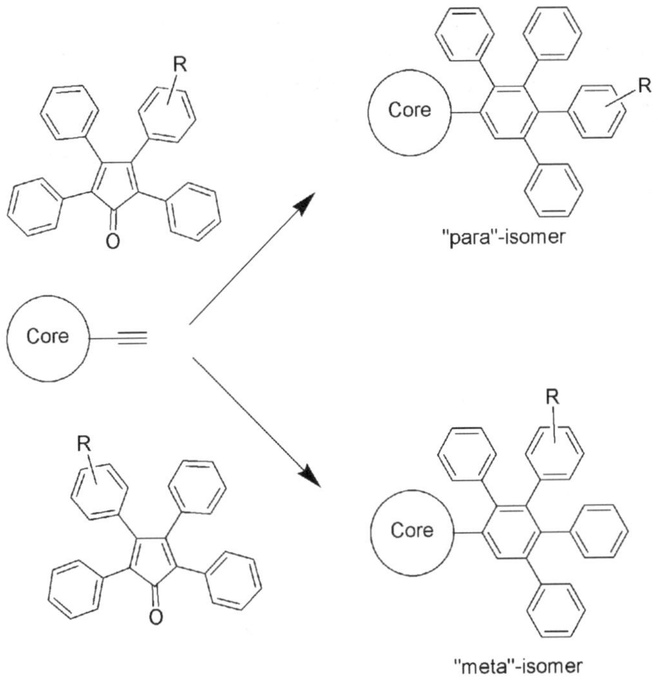

Scheme 13 Isomers from addition of mono-functionalised cyclopentadienones to ethynes

can be further defined in terms of the position of the substituent on the phenyl ring, so as to be, for example, *"para-para"*, *"para-meta"* or *"para-ortho"* isomers). In these isomers the substituents on different branches will be in slightly different spatial relationships with each other and with the scaffold and core. Similar problems would also arise from use of a reagent with one substituted α-phenyl ring.

4.1
Desymmetrisation – Towards Functional Nano-Building Blocks

An exciting possibility for tailoring the properties of functional dendrimers is desymmetrisation [39], that is, to have different functional groups on different regions of the surface. For example a dendrimer with part of its surface substituted with groups capable of binding to a biologically active material, while other parts bear chromophores might have potential applications in bioassays. To prepare such dendrimers the convergent approach must be used and so one first requires a desymmetrised core. As an example of such a method, we have demonstrated that treatment of the tetra(4-ethynylphenyl)methane **31** with less than 4 equivalents (eq.) of *n*-butyl lithium followed by an excess of triisopropyl chloride (TIPS-Cl) gives a statistical mixture of the mono- **32**, di- **33**, tri- **34**, and tetrasilylated **35** cores (Scheme 14), which can be separated by column chromatography [41]. The relative proportions of the mixture can be varied by controlling the amount of butyl lithium used, e.g. 2.2 eq. produces 36% of **32**, 47% of **33** and 6% of **34**. The unreactive tetrasilylated product **35** can be quantitatively deprotected back to **31** for recycling.

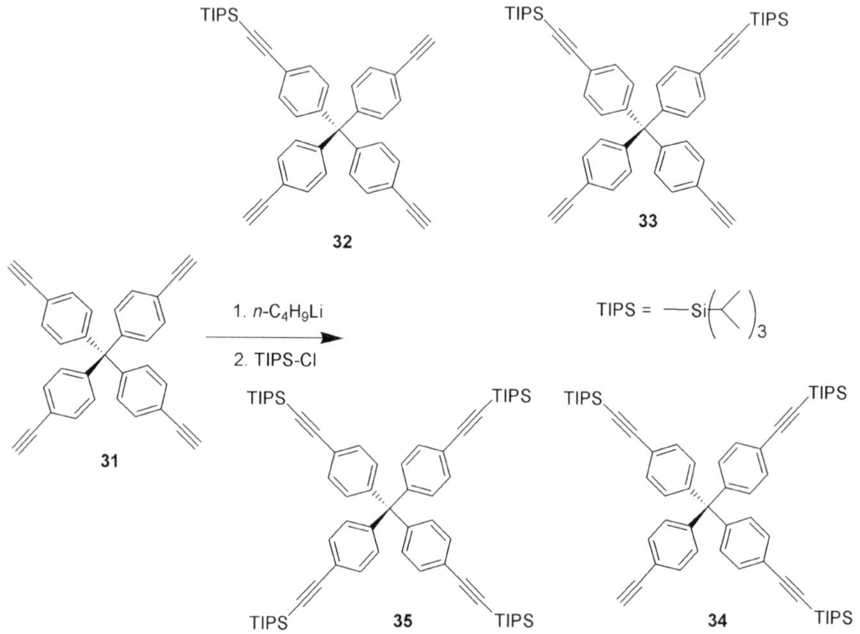

Scheme 14 Synthesis of desymmetrised dendrimer cores

These desymmetrised cores can then be used to make dendrimers bearing two different functionalities in either 1:3 (from **32** or **34**) or 1:1 (from **33**) ratios, by attaching dendrons with one functional group to the unsubstituted ethyne groups, followed by deprotection of the TIPS-protected alkynes and attachment of dendrons bearing a second functional group to them. This is illustrated in Scheme 15 for a first-generation dendrimer **36** bearing three perylene monoimide and one terrylene diimide dye substituent [41].

Having now shown the ways by which polyphenylene dendrimers can be functionalised at the surface, we now present some representative examples of the various types of functional groups we have attached to such dendrimers and of the potential uses of the resulting materials.

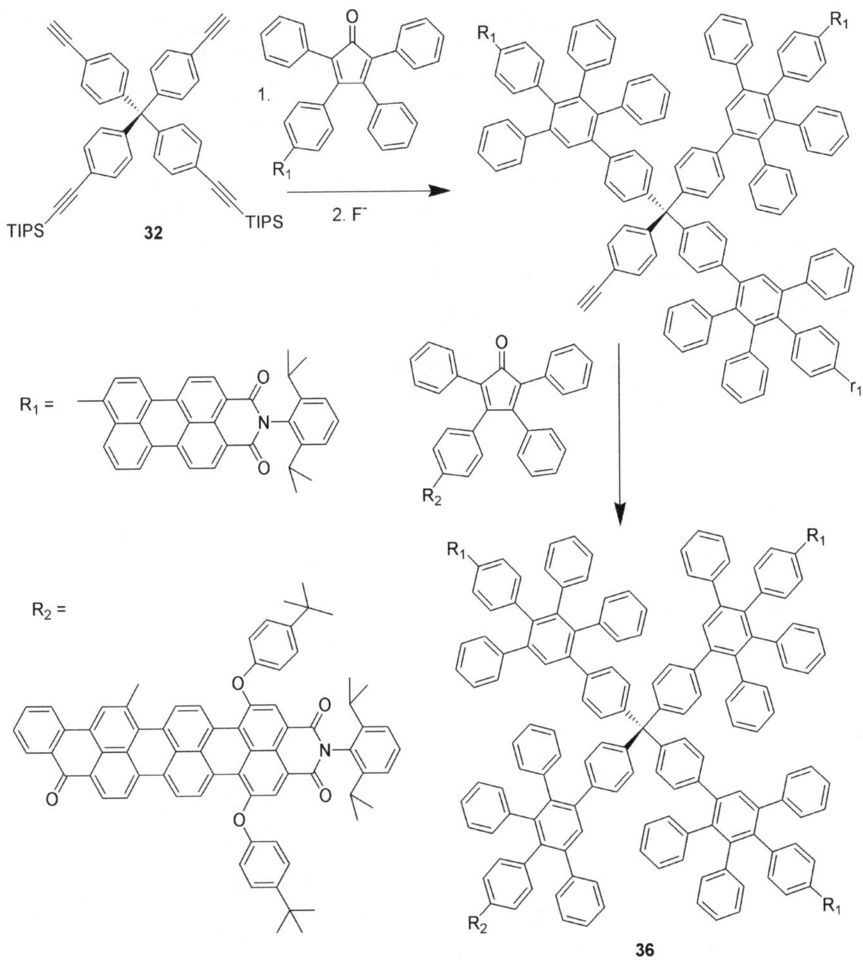

Scheme 15 Synthesis of desymmetrised first-generation dendrimer

4.2
Dye-Functionalised Dendrimers – Emissive Nanoparticles

Polyphenylene dendrimers with dyes attached to the surface not only have possible applications as emissive nanoparticles, but owing to their substituents being in well-defined spatial relationships to each other, are particularly suitable substrates for single molecule spectroscopy (SMS) studies on the interactions between chromophores. Studying such chromophore interactions is essential for an understanding of the photophysical processes that occur in solid-state optical systems and in multichromophore systems such as the light-harvesting systems involved in photosynthesis.

Dendrimers, e.g. **37**, based on a biphenyl core with up to 16 perylene monoimide chromophores attached on the surface have been prepared [47] and their properties studied by SMS techniques and compared with those of a model compound **38** with only one chromophore (Scheme 16) [47–52]. One unexpected result was that the dendrimer **37**, in which the chromophores are only weakly coupled, showed collective on/off fluorescence, which is usually seen only in strongly coupled systems. This has been attributed to energy transfer between the chromophores on **37**, with the collective off periods arising from a triplet state on one of the chromophores acting as a non-radiative trap [52]. This concept has been used as a model for understanding the energy transfer processes seen in natural light-harvesting systems [52]. A combination of the non-spherical shape of the dendrimer **37**, the fact that the dye is attached at a

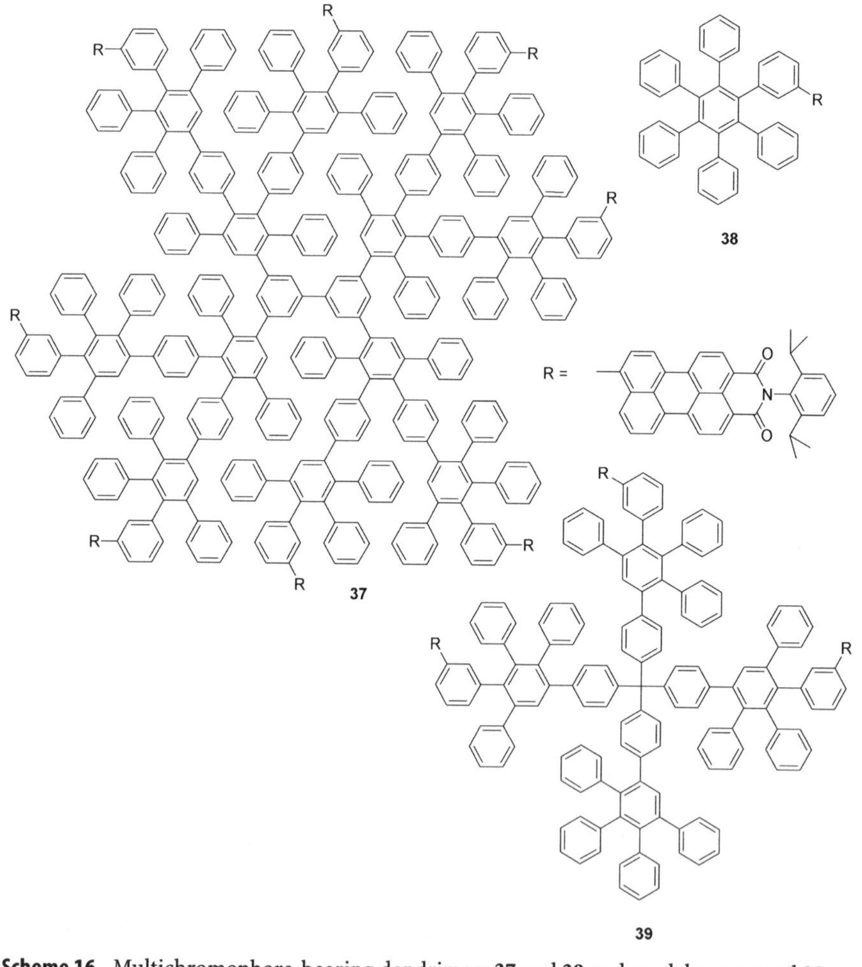

Scheme 16 Multichromophore-bearing dendrimers **37** and **39** and model compound **38**

meta-position of the phenyl ring and so has some rotational freedom, and the problems of isomerism shown in Scheme 9, means there are a range of interchromophore distances and orientations which lead to a loss of structure in the fluorescence spectra from **37**, and the appearance of extra components in time-resolved spectra. The spectra of two isomers of a first-generation dendrimer **39** bearing only three chromophores have been distinguished in SMS experiments [53], but this is not feasible for larger systems. Further studies are being performed on quasi-spherical dendrimers, with a tetrahedral core having the dyes attached at a *para* position. As these molecules have a more symmetrical shape and the dyes possess less rotational freedom, there is a smaller range of possible interchromophore distances, and so better resolved spectra can be achieved.

Better models for light-harvesting systems are dendrimers containing both donor and acceptor chromophores for energy transfer. Here one can attach both dyes to a desymmetrised surface or have each separately incorporated in different layers, i.e. as core, in the scaffold, or at the surface. Both approaches have been successfully used, as exemplified by the desymmetrised dendrimer **36** (Scheme 15) [41], and the dyad **40** (Scheme 17) [54]. The most advanced example of this approach incorporating three different dyes respectively at the core, in the scaffold and at the periphery is the triad **41** (Scheme 18) [45]. In **41**, the good overlap between the emission spectrum of the chromophore in each layer with the absorption spectrum of the chromophore on the next layer inwards is designed to promote efficient energy transfer from the periphery to

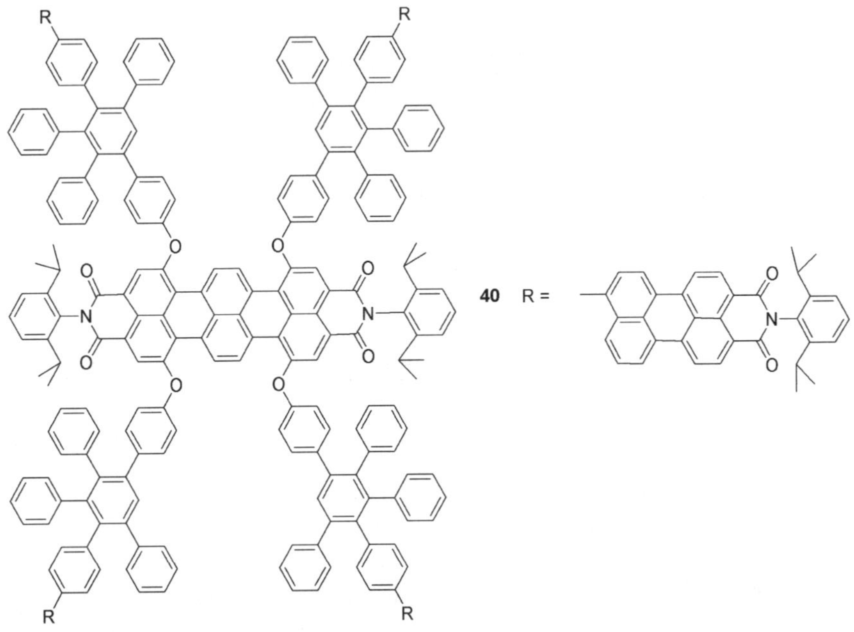

Scheme 17 Perylene-terrylene dyad

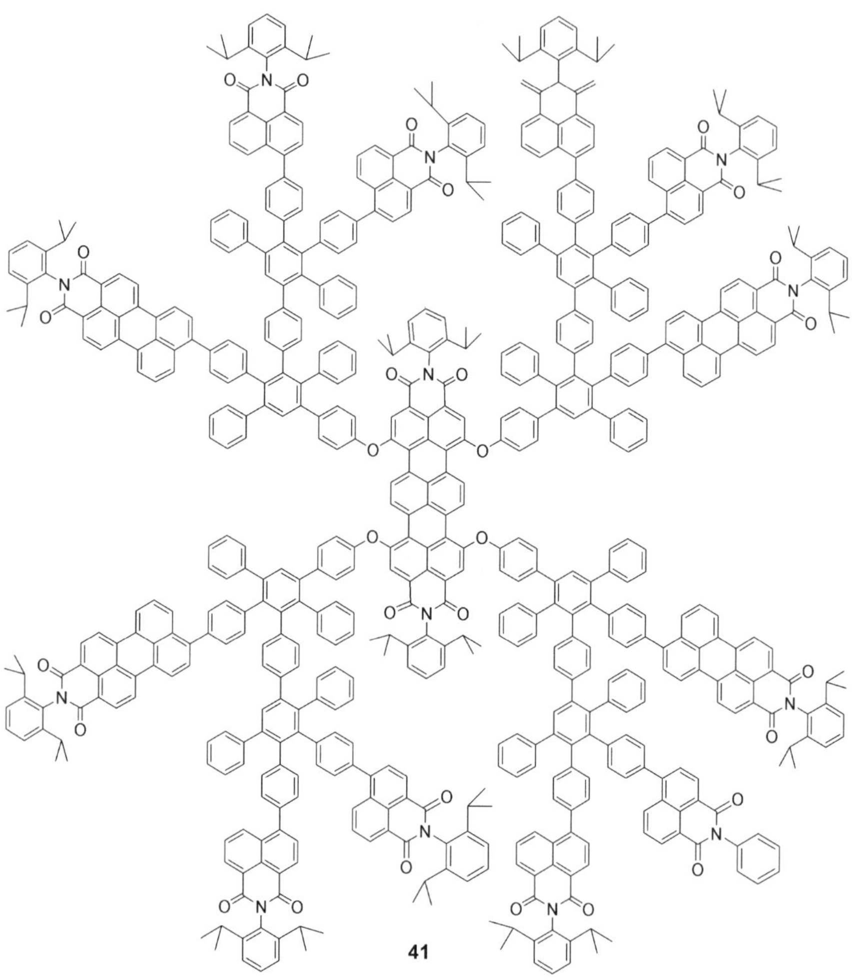

Scheme 18 Triad system for energy harvesting

the core. This triad is made by first adding the desymmetrised cyclopentadienone **30** bearing a perylene carboxydiimide dye unit (Scheme 12) to a terrylene carboxydiimide core, then deprotecting with fluoride, and finally reacting the deprotected alkynes with a cyclopentadienone bearing naphthalimide dye units. In all three molecules **36, 40** and **41**, efficient intramolecular Förster energy transfer is demonstrated, with the dominant PL emission coming from the terrylene diimide units. A blend of the triad (5 wt%) in charge-transporting violet-emitting poly(N-vinylcarbazole) produces red EL, showing that efficient energy transfer can also occur under electrical stimulation [55].

As perylene carboxydiimide chromophores are good electron acceptors, delayed fluorescence measurements on the dendrimer **42** (Scheme 19) with an

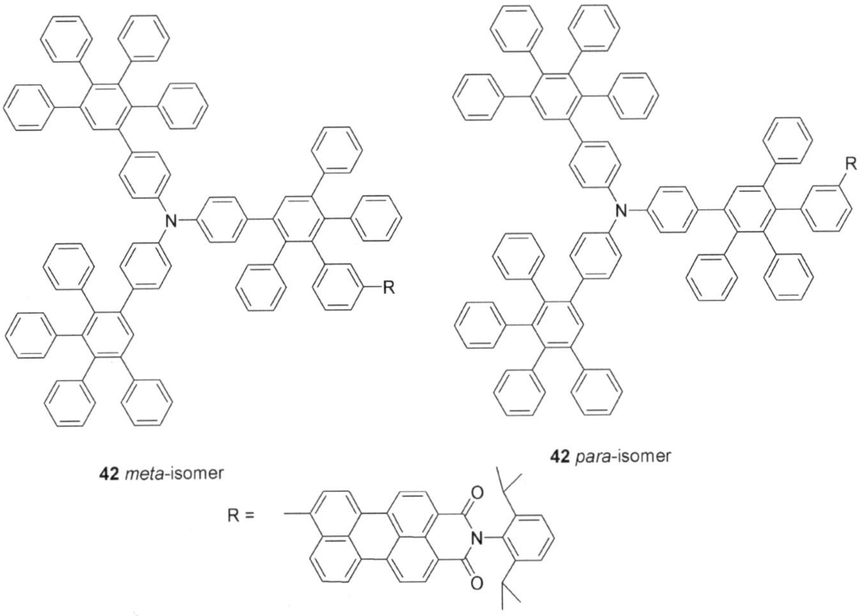

Scheme 19 Triarylamine-cored dendrimer with a perylene chromophore on periphery

electron-donating triarylamine core show reversible electron transfer occurs between the core and the dyes [56]. Here also the single molecule emission spectra of the two isomers from the Diels-Alder addition are resolvable, as the electron-transfer processes have different kinetics due to the different distances between donor and acceptor in the isomers (15 Å for the *meta*- vs. 17 Å for the *para*-isomer).

A similar system **43** (Scheme 20) with an electron-donating HBC core and six perylene dye acceptors on the periphery has been prepared and the energy and electron transfer between core and dyes is being investigated [57]. Such donor-acceptor systems have potential in organic solar cell applications.

4.3
Dendrimers with Other Electronically Active Substituents

Attaching other electronically active units to the periphery of a dendrimer offers the possibility of tuning its electronic properties. An obvious example is the use of charge-transporting units in order to improve the injection of electron or holes into an emissive dendrimer so as to improve its EL efficiency. Such an approach has been used by Thompson, Fréchet et al. [58] who obtained moderately efficient EL from a flexible dendrimer with a chromophore at the core and triarylamine hole-accepting units on the periphery. As the EL efficiencies

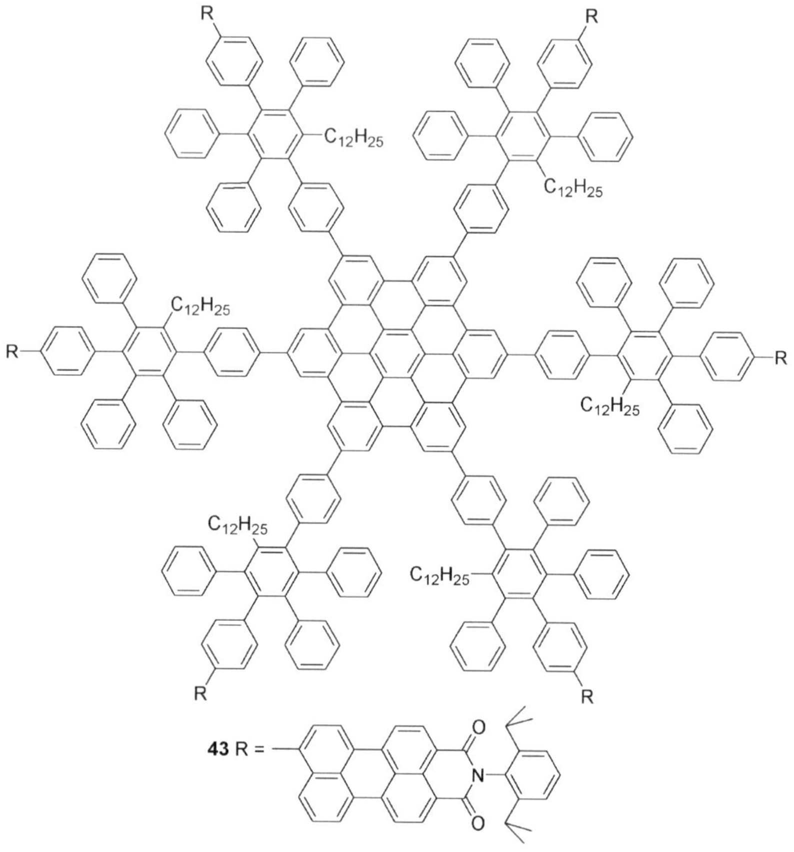

Scheme 20 HBC-dendrimer-dye system

of the dendronised dyes such as **10** and **11** discussed above are low, owing to poor charge transport through the dendrons, we prepared compounds, e.g. **44** (Scheme 21), bearing hole-transporting triphenylamine substituents [20]. However, we found that the substituents acted as electron donors to the electron-accepting core so that the fluorescence was quenched, particularly in polar solvents.

We are also investigating the attachment of conjugated rods to the surface of dendrimers. Such functionalised dendrimers could assemble to form 3-dimensional networks of conjugated rods. As an example of such an approach, dendrimers e.g. **45**, with terthienyl substituents (Scheme 22) have been prepared and their electronic properties are under investigation [59].

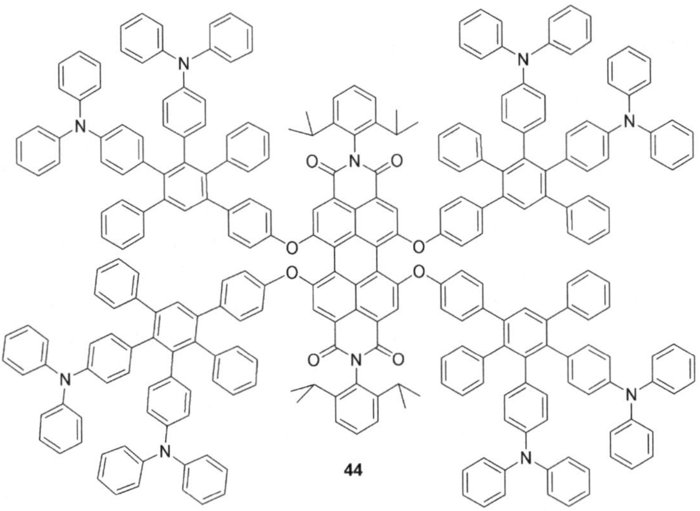

Functionalised Polyphenylene Dendrimers and Their Applications 275

Scheme 21 First-generation perylenecarboxydiimide cored dendrimer with triarylamine surface functionality

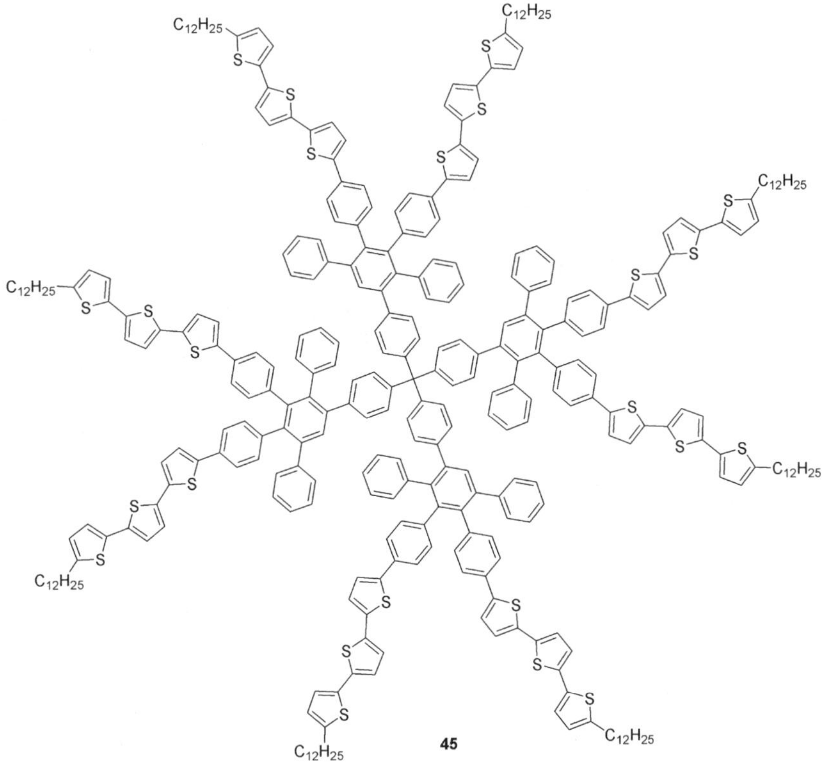

Scheme 22 Dendrimer with terthienyl substituents on periphery

4.4
Dendrimers Bearing Polar Substituents

Dendrimers have been prepared bearing a variety of polar groups, e.g. carboxylic acids **46**, nitriles **47**, amines **48**, and thiols **49, 50** (Scheme 23) [9]. Such substituents can be used to confer water solubility (**46**) or to assist in binding to substrates, e.g. the thiol-substituted dendrimers **49** and **50** bind to gold. Dendrimers **49** with thiomethyl substituents have been used as templates for formation of gold nanoparticles [60]. The particles self-assemble to form composites in which the gold nanoparticles are separated by dendrimer cross-linkers. Dendrimers **46–48** with carboxylic acid, nitrile and amine functionalities have been shown to selectively bind volatile organic molecules [40], as have composites made using dithiane-substituted dendrimers **50** [61, 62], and so these materials might be used as chemical vapour sensors.

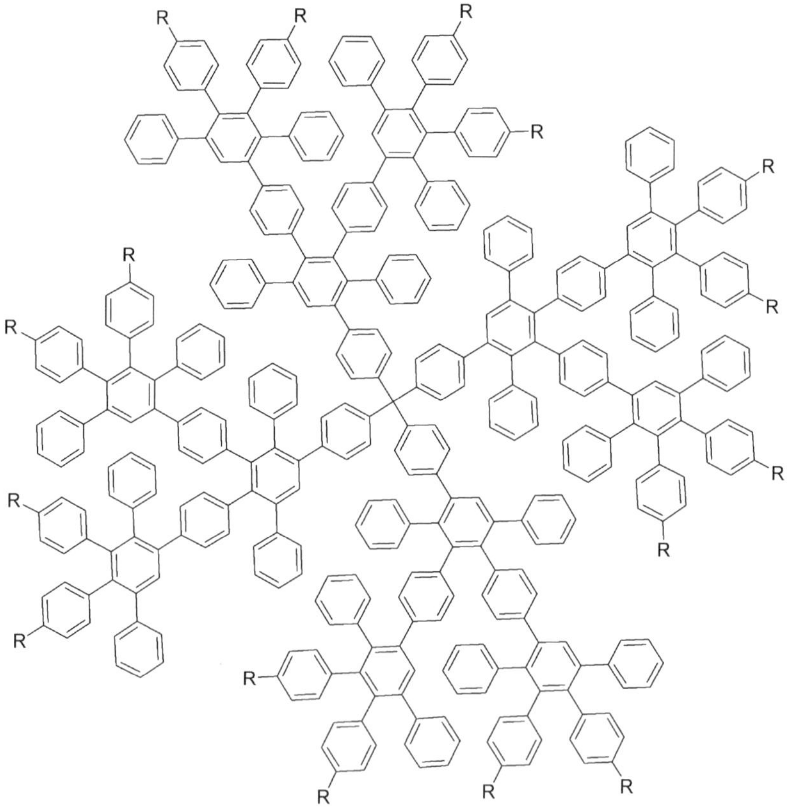

46 R = CO_2H, **47** R = CN **48** R = NH_2 **49** R = SCH_3 **50** R = —O–C(=O)–(CH$_2$)$_4$–(dithiolane)

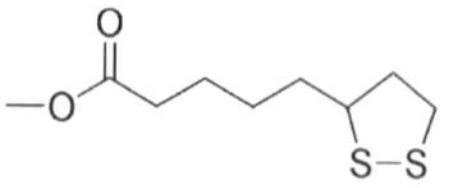

Scheme 23 Dendrimers with polar substituents

The carboxylate-substituted dendrimers 46 form aggregates as shown by light-scattering experiments, and AFM analysis of thin deposits from very dilute solutions reveal objects with heights corresponding to three to eight dendrimer layers [11]. These dendrimers also undergo Decher-style layer-by-layer self-assembly with poly(allylamine), which can be followed by surface plasmon spectroscopy [63]. A similar self-assembly process has been seen for these dendrimers with the lysine-substituted dendrimers discussed in the next section [46].

4.5
Dendrimers with Bioactive Substituents

Dendrimers have been the subject of much recent interest as possible delivery agents for drugs or DNA, e.g. [64]. Attachment of bioactive moieties (i.e. those capable of binding to biological materials such as proteins or DNA) to polyphenylene dendrimers offers prospects for preparing novel materials for this and other possible applications, e.g. biosensors. Polyphenylene dendrimers are particularly suitable substrates for biological applications as they are biologically inert and thus non-toxic, but can be readily and controllably functionalised. Dendrimers bearing oligopeptide chains (Scheme 24) have been prepared [65] either by using peptide-substituted cyclopentadienones, e.g. dendrimer 51, or by functionalising an amino-substituted dendrimer such as 48 with an N-alkyl maleimide followed by reaction with a thiol-endcapped peptide, e.g. dendrimers 52 and 53. Such dendrimers are being used as model compounds for studying DNA complexation, and as agents for the transfection of RNA [66]. As mentioned above, the lysine-substituted dendrimers 52 undergo layer-by-layer self-assembly with the carboxylic-acid-decorated dendrimers 46, while the glutamate-substituted dendrimers 53 form complexes with cationic polyionenes [67].

By utilising the desymmetrisation method discussed above (Scheme 14), we can make dendrimers bearing both a biologically active functionality and an optically or electrically active functional group, which may have bioassay applications in which the former group binds to a target biological moiety and the latter group is used for visualisation. Here the rigidity and shape-persistence of the dendrimers is important, as it maintains the biologically active moieties in fixed, well-defined relationships to each other and to the other active substituents. For example, the asymmetrically functionalised dendrimer 54 (Scheme 25) bearing one biotin unit which binds strongly with certain biologically active molecules, e.g. the protein streptavidin, and three dye units which act as fluorescent markers, has been prepared [68]. To obtain water solubility the dendrimer has to be mixed with a non-ionic detergent. The solubilised dendrimer has been shown to bind strongly to streptavidin in aqueous media to create a strongly fluorescent complex, thus demonstrating the potential of this approach for bioassay applications.

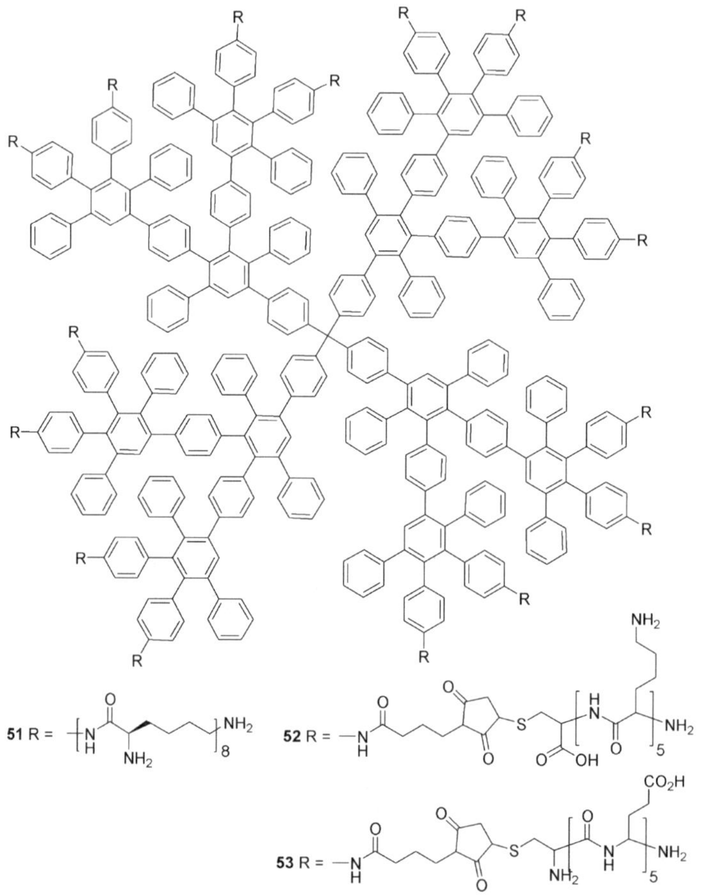

Scheme 24 Peptide-substituted dendrimers

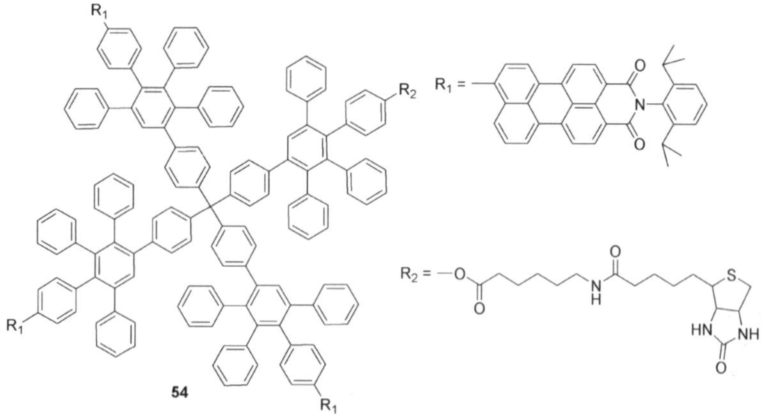

Scheme 25 Biotin-substituted dendrimer for bioassay applications

4.6
Core–Shell Dendrimers as Supports for Metallocenes

Solid-supported metallocenes have great industrial importance as catalysts for the polymerisation of olefins [69]. A key concern in obtaining high quality polymers from such processes is that the solid support must fragment down to small and preferably uniformly sized particles. As polyphenylene dendrimers are small discrete particles (less than 10 nm in diameter), they are attractive candidates as carriers for metallocene systems, and it has previously been shown that zirconocene catalysts for olefin polymerisation can be attached to

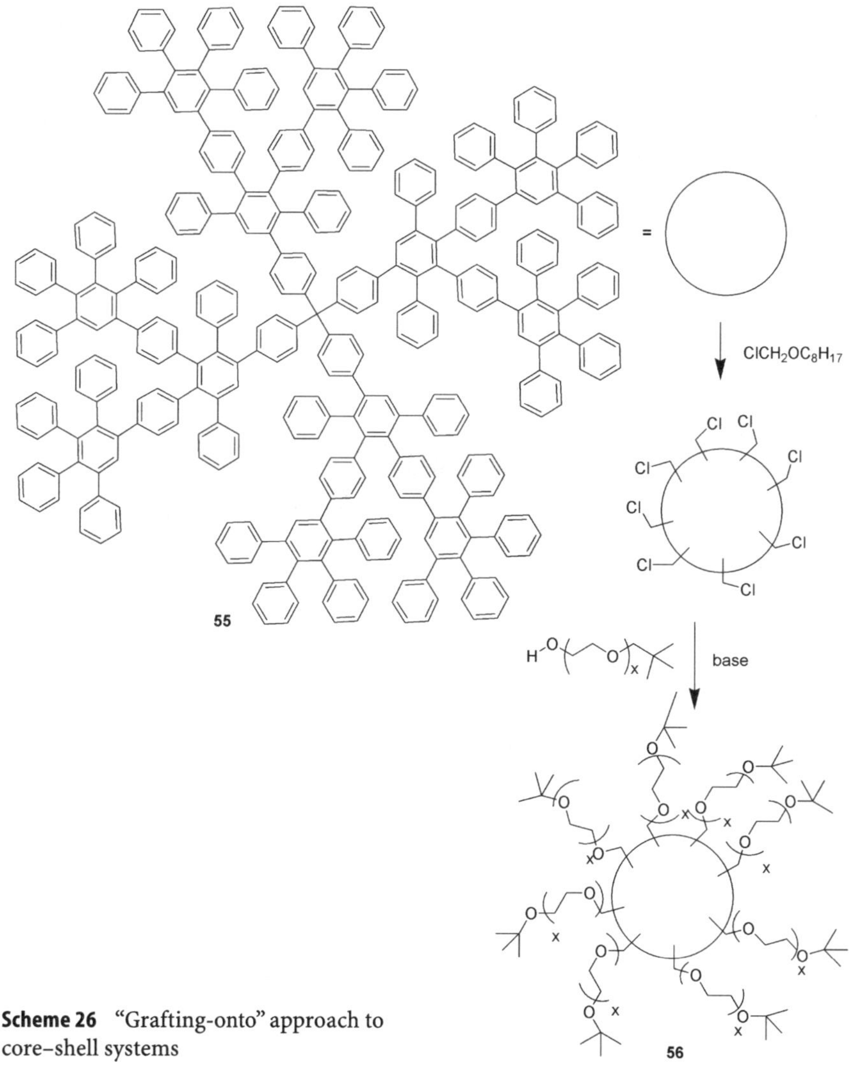

Scheme 26 "Grafting-onto" approach to core–shell systems

the surface of a polyphenylene dendrimer [9]. Immobilisation of the metal catalyst on the surface of a support by use of polar groups such as ethylene oxide has been found to improve the catalyst performance, for example latex particles with poly(ethylene oxide) (PEO) chains on their surface can be used as efficient supports for such catalyst systems [70]. Such particles, however, are much larger (typically 200–500 nm) than polyphenylene dendrimers. A dendrimer-based core–shell system with a rigid polyphenylene core surrounded by a shell of PEO is thus an attractive candidate to act as the support for such systems. The synthesis of such materials has accordingly been investigated by both a "grafting-onto" and a "grafting-from" approach [71]. In the former (Scheme 26), a dendrimer **55** was functionalised with chloromethyl groups to which pre-formed PEO units were attached by nucleophilic displacement of the halides (Williamson synthesis) to give the core–shell system **56**. The disadvantages of this approach are that there is no control over the position of the chloromethyl

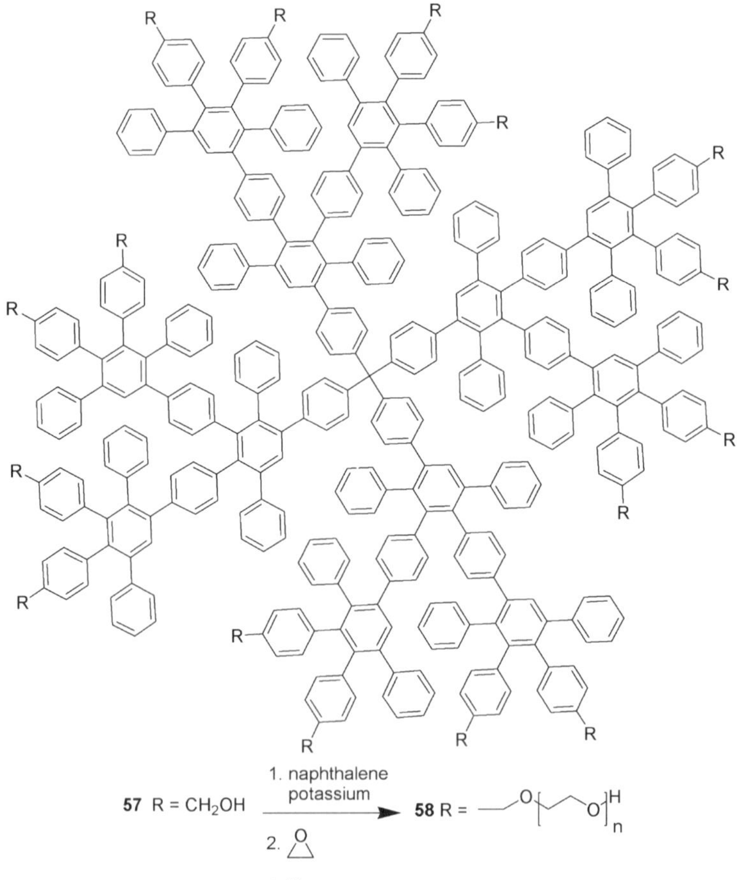

Scheme 27 "Grafting-from" approach to core–shell systems

substituents, and only partial control over their number by varying the ratio of reagent to substrate, and that steric crowding hinders the attack of the PEO chains for highly substituted systems so that only partial addition is achieved in these cases. The advantage is that the length of the PEO chains is known.

In the second approach (Scheme 27) a dendrimer 57 functionalised with hydroxymethyl units was converted into a macroinitiator for the anionic polymerisation of ethylene oxide by treatment with potassium naphthalide. Addition of ethylene oxide then gave the core–shell compound 58. The advantages of this method are that the sites of addition are well-defined and addition occurs at all of them. The disadvantage is that the length of each PEO chain is unknown, though the average can be determined from the estimated total amount of PEO addition. These materials are currently being tested as supports for metallocene-methylaluminoxane catalyst systems in olefin polymerisation.

4.7
Functional Dendrimers as Precursors to Functionalised 2-Dimensional Discs

Oxidative cyclodehydrogenation of polyphenylene dendrimers to give 2-dimensional graphite-like molecules has been developed in our group [37, 39]. For example the compound 59 with 222 carbon atoms has been made from the dendrimer 60 (Scheme 28) [72]. By changing the shape of the dendrimer the

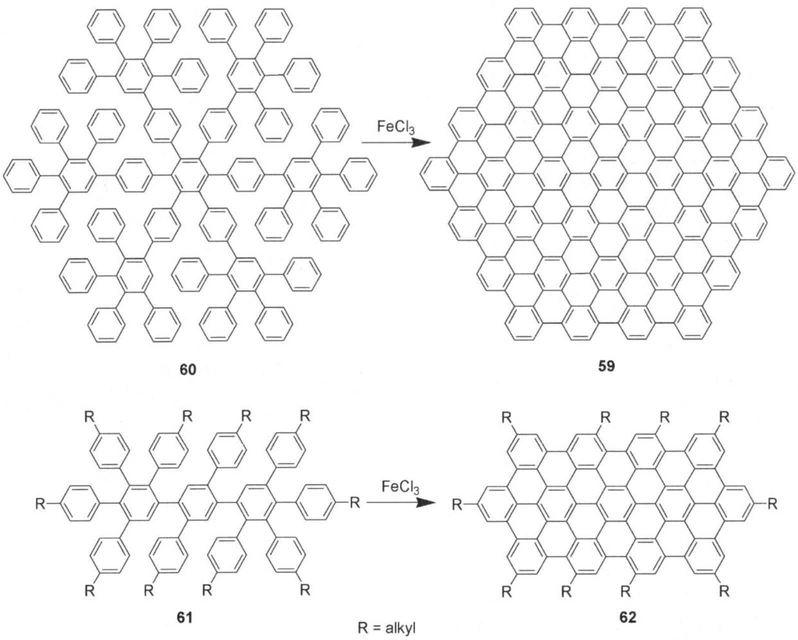

Scheme 28 Cyclodehydrogenation of dendrimers to graphite molecules

shape of the resulting graphite molecule can be controlled; for example, dehydrogenation of the alkyl-substituted dendrimer **61** gives **62** (Scheme 28) [73]. This opens the possibility of using functionalised dendrimers as precursors to functionalised graphite molecules, but the range of functional groups that do not inhibit or decompose during the oxidation is limited, and so we have found the best way to introduce a wide range of functionalities onto the disc molecules is to use a dendrimer with one or more bromine substituents which can be used to introduce other functionality later onto the resulting brominated graphite molecule [39].

The oxidation can be controlled to obtain only partial cyclodehydrogenation. For example oxidation of the dendrimer **63** produces mainly the two propeller-like structures **64** and **65** and only traces of the fully dehydrogenated product **66** (Scheme 29) [74, 75]. Molecules like **64** and **65** containing

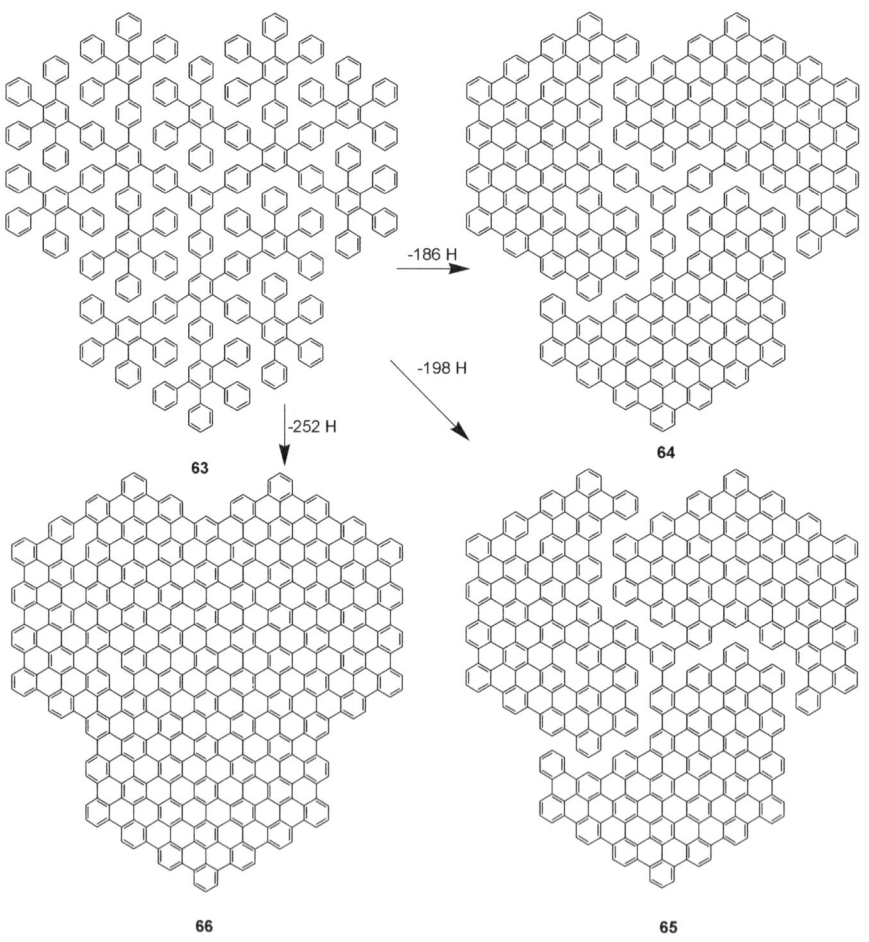

Scheme 29 Partial cyclodehydrogenation of dendrimers to "propeller" molecules

3-dimensional arrangements of graphene sheets are of interest for hydrogen storage for fuel-cell or similar applications and for lithium storage for battery applications.

5
Conclusions and Outlook

A large variety of rigid, shape-persistent polyphenylene dendrimers functionalised at the core, in the scaffold, at the surface, or some combination of these, with a wide range of functional groups including dyes, charge-transporting groups, electrolytes and biological moieties such as peptides, have been prepared by using an iterative Diels-Alder-deprotection sequence. These materials are of considerable interest both as model compounds for studying various processes e.g. energy transfer, and as active materials in applications as varied as chemical vapour sensors and supports for metallocene catalysts.

Dendrimers with functional cores such as dyes or conjugated polymers have been shown to have potential in light-emitting applications, as the dendrimer acts as a shield preventing undesirable interactions of the chromophores. Here, rigid shape-persistent polyphenylene dendrimers have proved to be more efficient shields than more flexible dendrimers.

Dendrimers with chromophores attached to the surface have been used as models for probing interchromophore interactions as the unique, rigid, shape-persistence of these materials means that the functional groups attached to them are essentially fixed with respect to each other in very well-defined spatial relationships. As a result the processes of energy or electron transfer between chromophores in such dendrimers can be studied by single molecule spectroscopic techniques, and insight gained into similar processes in other systems such as the photosynthetic centres in plants. Multifunctional systems with different dyes at the core, scaffold and surface are efficient light-harvesters.

Dendrimers with polar groups and their composites with gold nanoparticles show potential as chemical vapour sensors, while materials with biologically active units display potential in bioassay and RNA delivery applications. Attachment of PEO chains gives rise to core–shell materials, which can be used as supports for catalysts in olefin polymerisation.

In all these applications, the rigidity and shape-persistence of the polyphenylene structure has proved to be of crucial importance. The proven ability to control the shape, size, density and substitution pattern of these materials means that new and even more exciting prospects await them. In particular the incorporation of biologically active units, e.g. sugars, into the core and/or scaffold as well as the surface offers the prospect for new biological applications, as such materials would contain well-defined cavities which could selectively bind biologically active molecules through molecular recognition, with obvious applications in bioassay or drug delivery. Another possibility that offers promise for new applications is the self-assembly of polyphenylene dendrimers.

Acknowledgements We would like to thank all our many colleagues and collaborators who have contributed to this work. Funding for research into functional dendrimers from the European Union, the Bundesministerium für Bildung und Forschung, the Deutsche Forschungsgemeinschaft, the Fonds der Chemischen Industrie, and the Volkswagenstiftung is also gratefully acknowledged.

References

1. Buhleier E, Wehner W, Vögtle F (1978) Synthesis 2:155
2. Tomalia DA, Baker H, Dewald J, Hall M, Kallos G, Martin S, Roeck J, Ryder J, Smith P (1986) Macromolecules 19:2466
3. Fischer M, Vögtle F (1999) Angew Chem Int Ed 38:884
4. Scherrenberg R, Coussens B, van Vliet P, Edouard G, Brackman J, de Brabander E, Mortensen K (1998) Macromolecules 31:456
5. Sheiko SS, Moller M (2001) Top Curr Chem 212:137
6. Hart H (1993) Pure Appl Chem 65:27
7. Xu ZF, Moore JS (1993) Angew Chem Int Ed Engl 32:1354
8. Xu ZF, Kahr M, Walker KL, Wilkins CL, Moore JS (1994) J Am Chem Soc 116:4537
9. Wiesler UM, Weil T, Müllen K (2001) Top Curr Chem 212:1
10. Zhang H, Grim PCM, Foubert P, Vosch T, Vanoppen P, Wiesler UM, Berresheim AJ, Müllen K, De Schryver FC (2000) Langmuir 16:9009
11. Zhang H, Grim PCM, Vosch T, Wiesler UM, Berresheim AJ, Müllen K, De Schryver FC (2000) Langmuir 16:9294
12. Wind M, Saalwachter K, Wiesler UM, Müllen K, Spiess HW (2002) Macromolecules 35:10071
13. Wind M, Wiesler UM, Saalwachter K, Müllen K, Spiess HW (2001) Adv Mater 13:752
14. Lescanec RL, Muthukumar M (1990) Macromolecules 23:2280
15. Rosenfeldt S, Dingenouts N, Pötschke D, Ballauff M, Berresheim AJ, Müllen K, Lindner P (2004) Angew Chem Int Ed 43:109
16. Degennes PG, Hervet H (1983) J Phys Lett 44:L351
17. Liu DJ, Zhang H, Grim PCM, De Feyter S, Wiesler UM, Berresheim AJ, Müllen K, De Schryver FC (2002) Langmuir 18:2385
18. Pope M, Swenberg CA (1998) Electronic processes in organic crystals. Oxford University Press, Oxford
19. Herrmann A, Weil T, Sinigersky V, Wiesler UM, Vosch T, Hofkens J, De Schryver FC, Müllen K (2001) Chem Eur J 7:4844
20. Qu J, Pschirer NG, Liu D, Stefan A, De Schryver FC, Müllen K (2004) Chem Eur J 10:528
21. Qu J, Zhang J, Grimsdale AC, Müllen K, Jaiser F, Yang X, Neher D (submitted) Macromolecules
22. Liu D, De Feyter S, Cotlet M, Wiesler U-M, Weil T, Herrrmann A, Müllen K, De Schryver FC (2003) Macromolecules 36:8489
23. Meisel A, Herrmann A, Miteva T, Nothofer H-G, Scherf U, Müllen K, Neher D (manuscript in preparation) Macromolecules
24. Grebel-Koehler D, Liu DJ, De Feyter S, Enkelmann V, Weil T, Engels C, Samyn C, Müllen K, De Schryver FC (2003) Macromolecules 36:578
25. Hampel C (2001) PhD thesis, Johannes-Gutenberg-University, Mainz
26. Loi S, Butt HJ, Hampel C, Bauer R, Wiesler UM, Müllen K (2002) Langmuir 18:2398
27. Schlüter AD, Rabe JP (2000) Angew Chem Int Ed 39:864
28. Sato T, Jiang D-L, Aida T (1999) J Am Chem Soc 121:10658

29. Marsitzky D, Vestberg R, Blainey P, Tang BT, Hawker CJ, Carter KR (2001) J Am Chem Soc 123:6965
30. Setayesh S, Grimsdale AC, Weil T, Enkelmann V, Müllen K, Meghdadi F, List EJW, Leising G (2001) J Am Chem Soc 123:946
31. Pogantsch A, Wenzl FP, List EJW, Leising G, Grimsdale AC, Müllen K (2002) Adv Mater 14:1061
32. List EWJ, Pogantsch A, Wenzl FP, Kim C-H, Shinar J, Loi MA, Bongiovanni G, Mura A, Setayesh S, Grimsdale AC, Nothofer HG, Müllen K, Scherf U, Leising G (2001) Mater Res Soc Symp Proc 665: C5.47.1
33. Lupton JM, Schouwink P, Keivanidis PE, Grimsdale AC, Müllen K (2003) Adv Funct Mater 13:154
34. Pogantsch A, Gadermeier C, Cerullo G, Lanzani G, Scherf U, Grimsdale AC, Müllen K (2003) J Chem Phys 119:6904
35. Pogantsch A, Gadermeier C, Cerullo G, Lanzani G, Scherf U, Grimsdale AC, Müllen K, List EJW (2003) Synth Met 139:847
36. Jacob J, Oldridge L, Zhang J, Gaal M, List EJW, Grimsdale AC, Müllen K (in press) Curr Appl Phys 4
37. Grimsdale AC, Müllen K (2001) Chem Rec 1:243
38. Wu J, Fechtenkötter A, Gauss J, Watson MD, Kastler M, Fechtenkötter C, Wagner M, Müllen K (2004) J Am Chem Soc 126:11311
39. Grimsdale AC, Bauer R, Weil T, Tchebotareva N, Wu JS, Watson M, Müllen K (2002) Synthesis 9:1229
40. Schlupp M, Weil T, Berresheim AJ, Wiesler UM, Bargon J, Müllen K (2001) Angew Chem Int Ed 40:4011
41. Weil T, Wiesler UM, Herrmann A, Bauer R, Hofkens J, De Schryver FC, Müllen K (2001) J Am Chem Soc 123:8101
42. Bauer R, Müllen K (manuscript in preparation)
43. Oesterling I, Müllen K (manuscript in preparation)
44. Bernhardt S, Müllen K. Personal communication
45. Weil T, Reuther E, Müllen K (2002) Angew Chem Int Ed 41:1900
46. Mihov G, Bauer RE, Tchebotareva N, Knoll W, Schmidt M, Thüneman AF, Müllen K (2004) Polym Mater Sci Eng 90:41
47. Gensch T, Hofkens J, Heirmann A, Tsuda K, Verheijen W, Vosch T, Christ T, Basche T, Müllen K, De Schryver FC (1999) Angew Chem Int Ed 38:3752
48. Karni Y, Jordens S, De Belder G, Schweitzer G, Hofkens J, Gensch T, Maus M, De Schryver FC, Hermann A, Müllen K (1999) Chem Phys Lett 310:73
49. Karni Y, Jordens S, De Belder G, Hofkens J, Schweitzer G, De Schryver FC, Herrmann A, Müllen K (1999) J Phys Chem B 103:9378
50. Hofkens J, Maus M, Gensch T, Vosch T, Cotlet M, Kohn F, Herrmann A, Müllen K, De Schryver F (2000) J Am Chem Soc 122:9278
51. Hofkens J, Latterini L, De Belder G, Gensch T, Maus M, Vosch T, Karni Y, Schweitzer G, De Schryver FC, Hermann A, Müllen K (1999) Chem Phys Lett 304:1
52. Hofkens J, Schroeyers W, Loos D, Cotlet M, Kohn F, Vosch T, Maus M, Herrmann A, Müllen K, Gensch T, De Schryver FC (2001) Spectrochim Acta Part A-Mol Biomol Spectrosc 57:2093
53. Vosch T, Hofkens J, Cotlet M, Köhn F, Fujiwara H, Gronheid R, Van Der Biest K, Weil T, Herrmann A, Müllen K, Mukamel S, Van der Auweraer M, De Schryver FC (2001) Angew Chem Int Ed 40:4643
54. Gronheid R, Hofkens J, Kohn F, Weil T, Reuther E, Müllen K, De Schryver FC (2002) J Am Chem Soc 124:2418
55. Zhang J, Grimsdale AC, Müllen K. Personal communication

56. Lor M, Thielemans J, Viaene L, Cotlet M, Hofkens J, Weil T, Hampel C, Müllen K, Verhoeven JW, Van der Auweraer M, De Schryver FC (2002) J Am Chem Soc 124:9918
57. Wu J, Qu J, Tchebotareva N, Müllen K (submitted) Tetrahedron Lett
58. Freeman AW, Koene SC, Malefant PRL, Thompson ME, Fréchet JMJ (2000) J Am Chem Soc 122:12385
59. Bauer RE, John H, Sonar P, Heinze J, Müllen K (manuscript in preparation)
60. Taubert A, Wiesler UM, Müllen K (2003) J Mater Chem 13:1090
61. Vossmeyer T, Guse B, Besnard I, Bauer RE, Müllen K, Yasuda A (2002) Adv Mater 14:238
62. Krasteva N, Besnard I, Guse B, Bauer RE, Müllen K, Yasuda A, Vossmeyer T (2002) Nano Lett 2:551
63. Hernandez-Lopez JL, Bauer RE, W-S C, Glasser G, Grebel-Koehler D, Klapper M, Kreiter M, Leclaire J, Majoral J-P, Mittler S, Müllen K, Vasilev K, Weil T, Wu J, Zhu T, Knoll W (2003) Mater Sci Eng C 23:697
64. Eichman JD, Bieleinska AU, Kurowska-Latallo JF, Donovan BW, Baker Jr JR (2001) In: Fréchet JMJ, Tomalia DA (eds) Dendrimers and other dendritic polymers. Wiley, West Sussex, p 441
65. Herrmann A, Mihov G, Vandermeulen GWM, Klok H-A, Müllen K (2003) Tetrahedron 59:3925
66. Mihov G, Herrmann A, Prawitt D, Brixel L, Hengsler JG, Müllen K (submitted) Chem Commun
67. Mihov G, Chelmecka M, Gröhn F, Müllen K (manuscript in preparation)
68. Minard-Basquin C, Weil T, Hohner A, Rädler JO, Müllen K (2003) J Am Chem Soc 125:5832
69. Langhauser F, Kerth J, Kersting M, Kölle P, Lilge D, Müller P (1994) Angew Makromol Chem 223:155
70. Koch M, Falcou A, Nenov N, Klapper M, Müllen K (2001) Macromol Rapid Commun 22:1455
71. Atanasov V, Sinigersky V, Klapper M, Müllen K (manuscript in preparation)
72. Müller M, Kübel C, Mullen K (1998) Chem Eur J 4:2099
73. Simpson CD (2003) PhD Thesis, Johannes-Gutenberg-University, Mainz
74. Simpson CD, Martin K, Gherghel L, Räder H-J, Müllen K (in press) Angew Chem Int Ed
75. Simpson CD, Mattersteig G, Martin K, Gherghel L, Bauer RE, H-J R, Müllen K (2004) J Am Chem Soc 126:3139

Nanoscale Objects: Perspectives Regarding Methodologies for Their Assembly, Covalent Stabilization, and Utilization

Karen L. Wooley[1] (✉) · Craig J. Hawker[2] (✉)

[1] Department of Chemistry and Center for Materials Innovation, Washington University in St. Louis, One Brookings Drive, Saint Louis, Missouri, 63130 mo, USA
klwooley@artsci.wustl.edu
[2] Materials Research Laboratory and Department of Chemistry & Biochemistry, University of California, Santa Barbara, CA 93106, USA
hawker@mrl.ucsb.edu

1	Introduction	287
2	Nanoscale Assemblies by Supramolecular Interactions	289
3	Covalent Stabilization of Supramolecular Assemblies Leading to Nanoscale Objects	294
4	Utility of Supramolecular Assemblies and Nanoscale Objects	300
5	Conclusions	303
	References	304

Abstract In this review, we highlight some of the most recent advances in the design and utilization of organic nanoscale objects. Initially discussed is the preparation of well-defined nanoscale assemblies, typically the precursors of nanoscale objects, by programmed supramolecular interactions, and the subtle interplay between molecular structure and self-assembly is highlighted. The covalent stabilization of these supramolecular structures to produce robust nanoscale objects is then addressed from both intramolecular and intermolecular perspectives. Finally, the evolving field of the utilization of these nanoscale objects is described.

Keywords Nanostructured organic materials · Covalently-crosslinked supramolecular assemblies

1
Introduction

While the fashioning of organic molecules has only a short history, continuous developments in synthetic methodology have had a tremendous impact on the ability to prepare well-defined and complex molecular structures. The sophistication now available to the modern synthetic chemist permits amazingly complex architectures to be prepared, which can contain several levels of structural information. In this review, we will limit the discussions to organic systems, for which organic transformations [1, 2] with high degrees of regiochemical and stereochemical control have evolved to allow the syntheses of elaborate natural and synthetic products. This ability to mimic nature in the design of molecules is one of the driving forces behind the rapidly evolving field of nanotechnology, whereby synthetic organic chemistry is being extended to targets of increasing dimensions. As the size of the targets increases, new approaches to their design and new synthetic methodologies are being developed. Successful examples of these have most often involved a bottom-up supramolecular assembly of molecules or macromolecules that are compositionally and structurally programmed for assembly according to the rules of nature, involving a balance of intermolecular attractive and repulsive forces. Moreover, supramolecular assembly is applied in an iterative fashion, along with templating strategies and covalent stabilization to produce materials of high orders of complexity.

Equally important to both nanotechnology and the continued evolution of synthetic chemistry has been the development and further refinement of a range of analytical tools which, for the first time, has permitted a thorough investigation of the properties and characteristics of structures less than 50 nm in size. Using combinations of techniques, such as atomic force microscopy (AFM), transmission electron microscopy (TEM), and neutron-scattering, together with traditional spectroscopic tools, a direct correlation can now be obtained between minor structural changes at the atomic level and molecular organization on the 10–100 nm scale. This unprecedented ability has led to the concept of designer materials that have enormous potential in commercially important fields ranging from molecular/microelectronics to biocompatible surfaces. Arguably, one of the most exciting areas of designer materials to emerge as a result of these advances in molecular imaging is the concept of nanoscale objects – organic structures which are shape-persistent and have overall dimensions of between 5 and 100 nm.

In this review, we highlight some of the most recent advances in the design and utilization of nanoscale objects. Initially, the preparation of well-defined nanoscale assemblies, typically the precursors of nanoscale objects, by programmed supramolecular interactions is discussed and the subtle interplay between molecular structure and self-assembly highlighted. The covalent stabilization of these supramolecular structures to produce true nanoscale objects is then addressed from both intramolecular and intermolecular perspectives. Finally, the evolving field of the utilization of these nanoscale objects is described.

2
Nanoscale Assemblies by Supramolecular Interactions

Hierarchical assembly of supramolecular systems is ubiquitous in biological systems [3] and this level of sophistication is being approached by a number of recent studies where there is a direct correlation between molecular structure and self-assembly. Ghadiri et al. [4] have shown the power of hydrogen-bond-directed self-assembly of cyclic structures to give open-ended hollow tubular nanoscale assemblies in which the specific design of the macrocyclic ring leads to efficient stacking. The stacked assemblies are subsequently stabilized primarily by hydrogen-bonding. While the majority of the early work [5] in this area involves cyclic peptides, the range of possible structures has been extended recently to include heterocyclic triazole rings, **1**, prepared by Click Chemistry [6], and oligo(phenylethynyl) derivatives [7–9], which are primarily stabilized by π-π and hydrophobic interactions (Fig. 1). In each of these cases, assembly-active ends remain, which theoretically allow for the lengthwise growth to continue until a capping unit is encountered.

Nanoscale objects can also be constructed as limited growth, discrete entities, by the molecular recognition and self-assembly of small molecules that are designed for directional assembly into a closed structure. An elegant recent example of this is the construction of a molecular cage composed of three calix[4]arene end units, **2**, and six connecting barbiturate groups, **3**, reported by Crego-Calama, Reinhoudt et al. [10], where the structures and conformations of the small molecules direct and limit the assembly processes to the formation of enclosed cages. This provides a central cavity that is ca. 3 nm in diameter and 0.7 nm in height, which has demonstrated dynamic complexation and release of noncovalent guests (Fig. 2). Atwood and Szumna have utilized cationπ and hydrogen-bonding interactions geometrically placed within a molecular capsule to directionalize electrostatic interactions for the production of

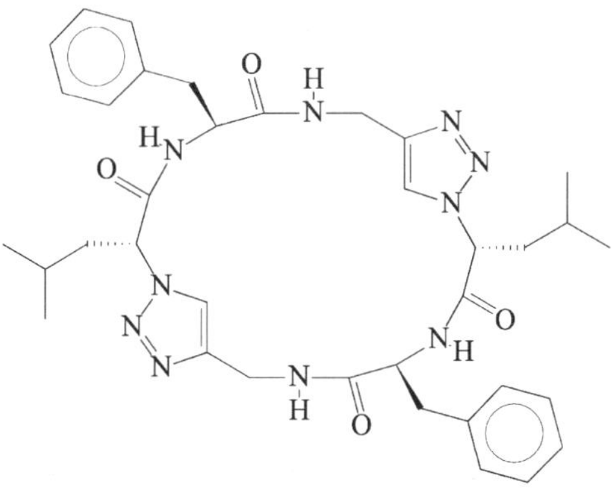

Fig. 1 Structure of self-assembling heterocyclic cyclic peptide, **1**

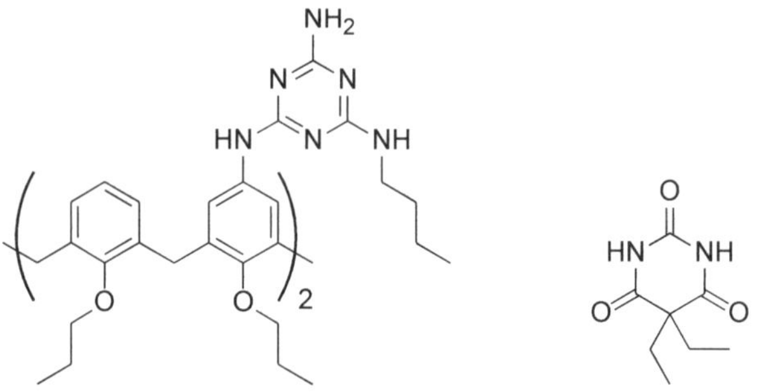

Fig. 2 Structure of calix[4]arene (**2**) and barbiturate (**3**) used in the 3:6 self-assembly of a molecular cage for dynamic guest complexation

anion-sealed single-molecule capsules [11]. This placement of the entire ion-pair within a molecular framework is a new approach to the preparation of anion receptors, based upon non-covalent packaging of the cationic tetramethylammonium salt within the cavity of the organic capsule molecule, **4**. In a recent report, a second-generation anion receptor (Fig. 3) was designed to optimize the electrostatic and hydrogen-bonding interactions, to contain the tetramethylammonium cation within the cavity, to attract a halide anion to seal the vessel, and also to provide for bulky groups along the upper rim to stabilize the directionally templated ion-pair. These systems represent extremely high levels of control over the self-assembly processes and illustrate the predictability that can be now achieved with the manners by which small molecules undergo well defined assembly to afford discrete structures. Progressing to large, polymeric self-assembling systems introduces a range of new difficulties as well as added opportunities.

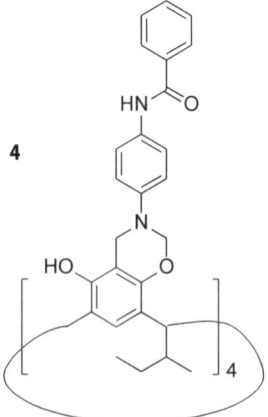

Fig. 3 Atwood's second generation molecular capsule (**4**), which acts as a receptor for the directional non-covalent capture of ion-pairs

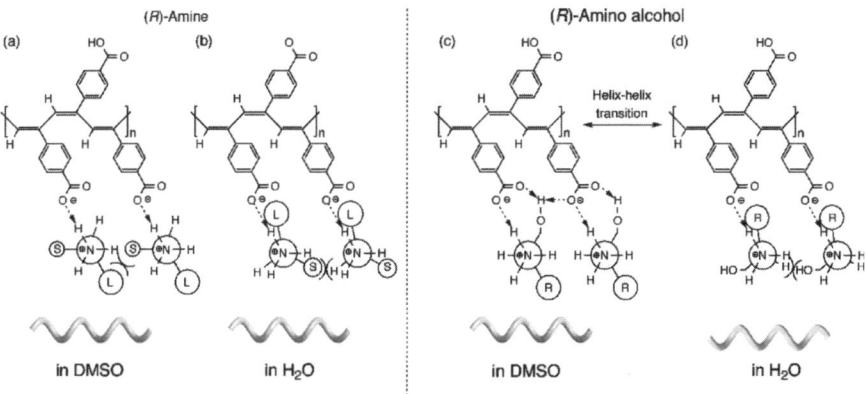

Fig. 4a–d Solvent induced helical transition in linear poly(*p*-carboxyphenyl)acetylenes

By extending the self-assembly process to polymeric systems, organizational control can also be induced by the addition of external agents and with reliance upon the increased conformational degrees of freedom of macromolecules, in effect mimicking biological systems such as chaperonins. One of the most dramatic illustrations of this ability to manipulate structural conformation is the solvent-induced switching of macromolecular helicity by the addition of optically active amino alcohols to linear poly(*p*-carboxyphenyl)acetylene chains [12]. The helical sense of the polymers exhibited opposite Cotton effect signs, as observed by circular dichroism, by simply changing the solvent from water to dimethylsulphoxide. As illustrated in Fig. 4, the model that has been proposed to explain this phenomenon involves a switching in the acid-base interactions under different solvent conditions. This example demonstrates the ability to alter molecular conformation and chirality by appropriate choice of the achiral or chiral external conditions (Fig. 4). The induction of chirality through a self-assembling event has also been observed during catenation and may lead to a new switching system where chirality appears on catenation but disappears on decatenation [13].

The intermolecular self-organization of polymeric or oligomeric units can also lead to a range of different nanoscale assemblies. A recent review by Discher and Eisenberg [14] clearly demonstrates the rich array of structures that can be obtained from the solution-mediated assembly of polymers, and further underscores the fact that the structural diversity possible with synthetic systems leads to a significantly greater array of chemical and physical possibilities when compared to biological systems, such as lipids. This diversity naturally leads to improved properties and functionality, critical criteria in many applications. Moreover, the ability to control the shapes of the assembling polymers, by control over their structure and topology [15], leads to further diversity in the supramolecular materials.

By a combination of coordination chemistry and electrostatic interactions, the fullerene-triggered unidirectional self-assembly of an acyclic zinc por-

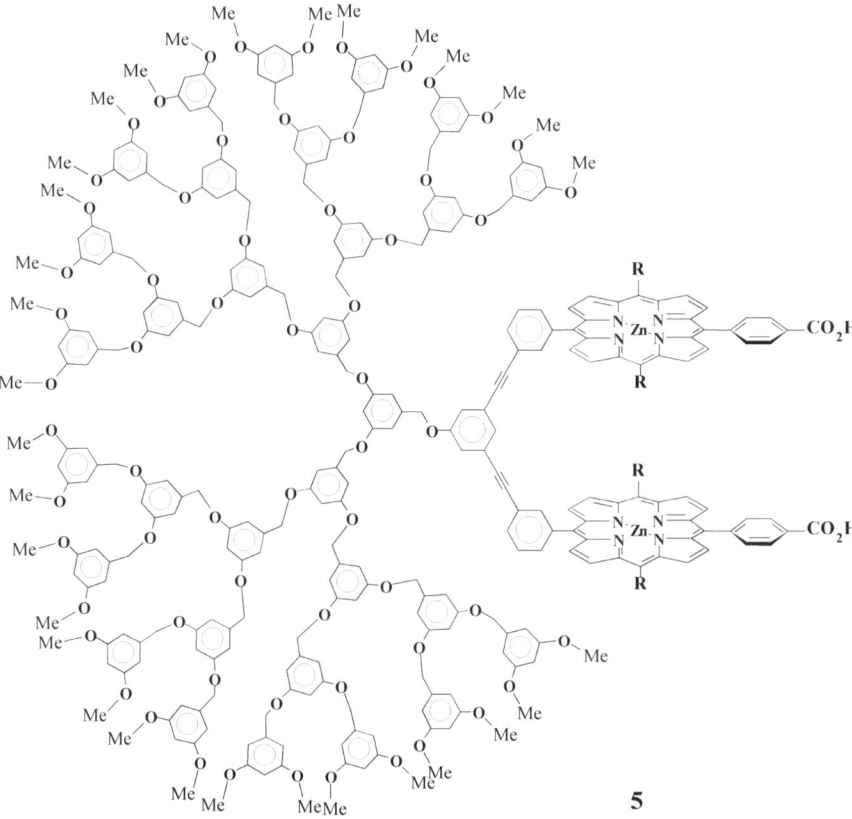

Fig. 5 Structure of dendritic bis(porphyrin), **5**, which undergoes specific self-assembly in the presence of C_{60} to give well-defined nanorods

phyrin dimer bearing a large fourth-generation poly(benzyl ether) dendritic wedge, **5**, has recently been reported by Aida et al. (Fig. 5) [16]. No discrete organization of the dendrimer was observed in either the absence of C_{60} or when the six acid groups (R=-Ph-CO_2H) were esterified. However, in the presence of the fullerene, the carboxylic acid derivative initially formed an inclusion-like complex with the fullerene coordinated between the two porphyrin rings. This complex then induced one-dimensional aggregation through the dimerization of the carboxylic acid side group to give untangled, discrete nanoscale objects, having very high aspect ratios and diameters of 12 nm, in agreement with that estimated from molecular modeling of two dendritic porphyrins.

These selective interactions, built into the structures on the Ångström level and drawing inspiration from biomolecular interactions, can also be employed to create other nanoscale assemblies. Illustrative examples have involved the use of quaternary hydrogen-bonding units [17], nucleobases [18], or biotin–streptavidin interactions [19]. An intriguing variation on this theme, only demonstrated on a microscopic scale as yet, involves the recent use of magnetic interactions in thin nanorods (diameter~0.4 µm and length 2.2 µm) composed of alternating sections of Au and Ni which gives rise to three-dimensional self-assembly into highly stable nanoobjects [20]. These nanoobjects are essentially "bundles" of individual rods in which the alternating layers of Au and Ni align and defects in the packing or alignment of the rods were rare (<1%). While not organic in nature, the process used for manufacture of the rods could easily be applied to polymeric materials and suggests that a combination of magnetic forces with specific surface chemistries for interaction may be powerful tools in future self-assembling strategies.

In concert with the potential use of external fields to induce supramolecular assembly, other relatively new approaches to controlling molecular organization have been developed. One of the most exciting involves architectural differences, for example the self-assembly of diblock or triblock copolymers in which one of the segments is a rigid rod with a great tendency to order and the others are flexible coils that have low tendencies toward ordering. In addition to this ordering difference, variations in molecular architecture have also been shown to have a powerful effect on the ability of these rod-coil block copoly-

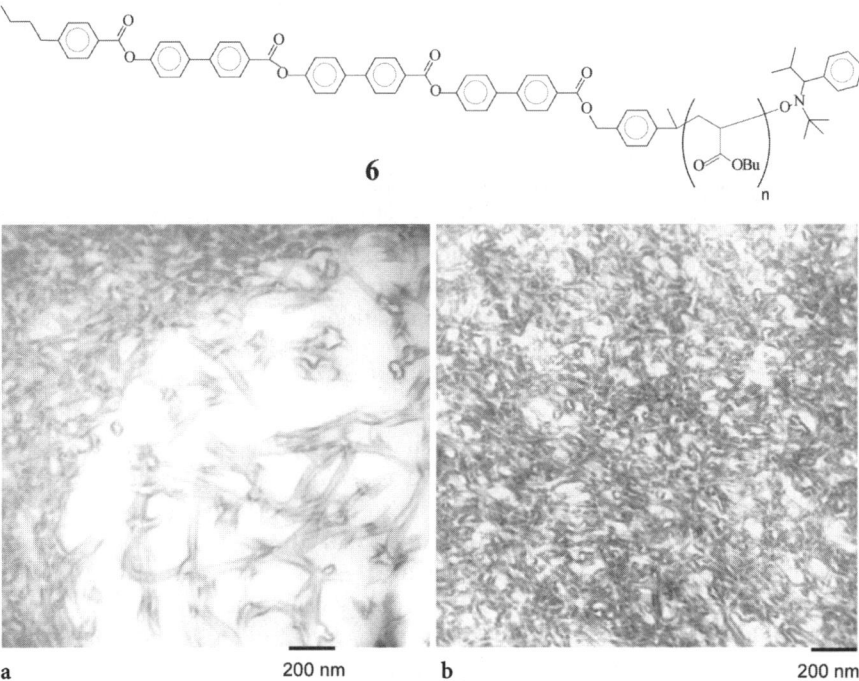

Fig. 6a, b Structure of rod-coil block copolymer, **6**, and transmission electron microscopy images of a RuO_4-stained film cast from THF and annealed at 125 °C for 48 h. **a** The edge of the film showing cylindrical aggregates lying parallel and perpendicular to the substrate. **b** The center of the film

mers to assemble into well-defined nanostructures. The pioneers in this new direction have been Stupp et al. [21–23], who have developed a range of structures, each programmed to assemble into one-dimensional nanoribbons, two-dimensional sheets or three-dimensional mushroom-like structures by simply varying the chemical structure and architecture of the block copolymers. A recent example [24] building upon this work involved the synthesis of rod–coil polymers consisting of an aromatic ester rod and an amorphous vinyl polymer coil, **6**. In both the solid state and in solution, **6** revealed a range of unusual nanostructures, the formation of which was driven by the stacking of the aromatic polyester unit (Fig. 6). It is also interesting to speculate that the high degree of definition observed for the nanostructures is a result of the monodisperse nature of the rod units, in both this work and that of Stupp, which further promotes and enhances crystallization/stacking.

3
Covalent Stabilization of Supramolecular Assemblies Leading to Nanoscale Objects

One of the key attributes of supramolecular assemblies is the dynamic nature of the molecular organization, which leads to exchange of individual molecules between molecular clusters. This has been exploited by Meijer et al. [25] in the design of a novel family of telechelic polymers bearing hydrogen-bonding termini, which exhibit unique physical and mechanical properties when compared to their counterparts that lack the terminal molecular recognition elements. These materials illustrate the influence that can be exerted by single, intermolecularly attractive units. Owing to the ability of the small concentration of sophisticated chain end groups to alter the properties of commodity polymer chains, this approach and these materials are expected to dramatically impact a variety of commercially important applications, where external stimuli such as heat or dilution trigger the polymers to behave like a fluid that is easy to process by the manufacturer or consumer. However, in many other applications it is highly desirable to have a static structure that is not affected by reversibility. To this end, significant effort has been devoted in recent years to the covalent stabilization of supramolecular assemblies, which leads to discrete nanoscale objects that have fundamentally different properties and structural characteristics when compared to the individual constituents of the assembly.

Arguably, one of the first examples of covalent stabilization involves the core-crosslinking of polymer micelles to give a crosslinked nanoparticle surrounded by a corona of linear polymer chains [26–29]. This concept has been revolutionized in recent years by the concept of shell-crosslinked nanoparticles in which it is the coronal chains of the polymer micelle that are crosslinked, leaving the central core comprised of linear chains that are covalently linked to the inner surface of the crosslinked membrane capsule. The shell-crosslinked nanoparticle possesses a degree of structural organization, having a well-defined external shell the nature and crosslink density of which can be controlled, an interfacial layer and a central core composed of linear chains. Armes et al. [30] has effectively combined both concepts, shell-crosslinking and core-crosslinking, to afford an intermediate structure in which a central layer is crosslinked. Using a triblock copolymer, for which the central block contains reactive groups that can be used for crosslinking, a hybrid nanoobject is created in which a central core of linear chains is surrounded by a crosslinked shell/layer that is, itself, surrounded by a corona of linear polymer chains. The broad nature of these concepts allow a wide variety of polymeric blocks and copolymer architectures to be used and results in tremendous structural diversity for these nanoobjects [31], leading in turn to tremendous potential for applications.

Such a covalently stabilized nanostructure can be manipulated physically and/or chemically under varying conditions while still retaining its individuality and altering the discrete size, shape and structure, each of which is dictated originally by the supramolecular assembly of linear polymer chains. This is an exciting and emerging area of current interest – the regiocontrolled physical and chemical manipulation of well-defined nanostructures preformed by supramolecular assembly to produce, ultimately, nanomaterials through iterative processes that could not be obtained via supramolecular assembly alone. A recent example of the structural diversity that can be exercised for the subsequent thermal reshaping of the nanostructures involved the preparation of nanodroplets of polyisoprene fluid contained within a poly(acrylic acid-*co*-acrylamide) shell from polyacrylic acid-*b*-isoprene block copolymers (Fig. 7) [32, 33].

Comparative TEM and AFM analyses of the shell-crosslinked nanostructures revealed discrete particles, the size and shape of which were dictated by the glass transition temperatures (T_gs) of the inner core domains. The core T_g values for shell-crosslinked nanoparticles containing polyisoprene within their internal region were controlled by both the microstructure [32] and chemical nature [33] of the core polymer chains. Polyisoprene prepared by anionic polymerization in hexane or by living radical polymerization exhibited a low T_g

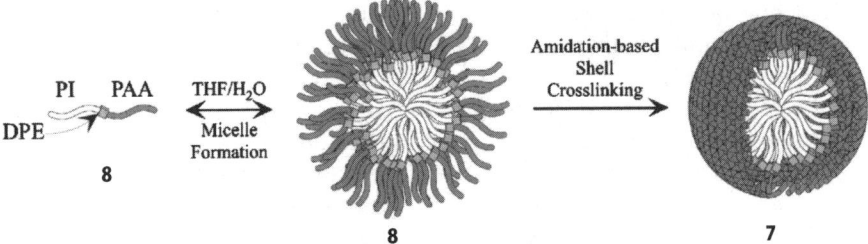

Fig. 7 Synthetic approach for the formation of shell-crosslinked nanoparticles, 7, from poly(acrylic acid-*b*-isoprene) block copolymers, 8. *DPE* Diphenylethylene, *PI* polyisoprene, *PAA* poly(acrylic acid)

value (ca. −70 °C), which allowed for significant flattening of the nanoparticles upon deposition from water onto the substrate at room temperature. In contrast, the preparation of polyisoprene by anionic polymerization conducted in tetrahydrofuran (THF) or conversion to a hydrochlorinated polyisoprene afforded core materials that exhibited T_gs in excess of room temperature, and these materials remained spheres when deposited onto substrates and imaged at room temperature. Thermal treatment of the nanostructures containing the higher T_g core materials resulted in shaping of the structures, whereby their shapes conformed to that of the flat substrate. Moreover, complete chemical degradation and extraction of the core polymer chains, for example by ozonolysis of polyisoprene [34] or hydrolysis of poly(ε-caprolactone) [35], led to an even more flattened structure, essentially by removing the core and creating a "hollow" nanocage. It should be noted that these manipulations were not possible for the precursor polymer micelles due to complete loss of structure, which further demonstrates the validity and utility of the covalent stabilization approach.

Similar "hollowed-out" structures have recently been reported by Zimmerman et al. from either dendritic or star polymer precursors [36]. Instead of using a supramolecular assembly, covalent macromolecules are first prepared containing a cleavable core and crosslinkable groups at the chain ends; in these cases homoallyl groups were crosslinked by ring-closing metathesis reactions using Grubbs' Type 1 catalyst. Extremely efficient crosslinking of the chain ends was observed, possibly owing to proximity effects. Subsequently, the functionalized core could be removed to give a hollow, crosslinked macromolecule. The critical feature of structural retention has been demonstrated in these systems and, as in the shell-crosslinked nanoparticle case, degradation of the core for the non-crosslinked case results in disintegration of the macromolecule into smaller fragments. Furthermore, as a result of the chain termini connectivity for the crosslinked dendrimers, molecular imprinting with high to moderate degrees of selectivity can be achieved using guests that resemble the core fragment [37, 38].

The versatility inherent in this regioselective crosslinking approach has recently been utilized by a variety of groups to prepare well-defined and functional nanoobjects. Using small-molecule amphiphiles that contain a crosslinkable double bond, **9**, McQuade et al. [39] have been able to form reverse micelles at low water content and then covalently stabilize these dynamic structures to give crosslinked spheres, **10**, whose radii (ca. 3–5 nm) are similar to those of the pre-crosslinked reverse micelles (Fig. 8). Significantly, these nanoobjects were shown to be permeable and the carboxylate counterions on the interior could be exchanged with both fluorescent dyes and transition metal ions, **11**. In a similar vein to the dendritic catalysts developed by Fréchet et al. [40], these nanoobjects can act as catalysts in which reactants can enter the nanoparticle, undergo a chemical transformation in the interior and then exit.

A similar strategy has been used by Emrick et al. [41] and Nolte et al. [42] to stabilize interfacial assemblies of block copolymers leading to hollow spher-

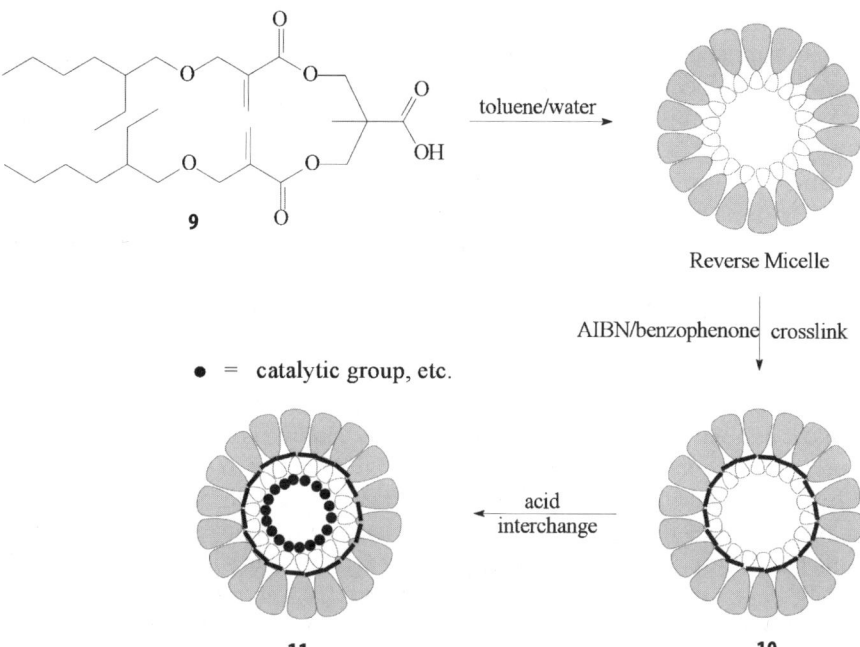

Fig. 8 Crosslinking reactions performed between polymerizable double bonds placed at the interface of amphiphilic small molecule assemblies provided for the preparation of stabilized reverse micelle nanostructures. *AIBN* Azobisisobutyronitrile

ical nanoobjects. By assembling graft copolymers of poly(octadiene)-*g*-poly-(ethylene glycol) at the interface of toluene/water, discrete spherical assemblies can be obtained that have the poly(ethylene glycol) (PEG) chains directed into the bulk water phase and the poly(olefin) backbone located at the toluene/water interface [41]. These dynamic structures are then stabilized by crosslinking with a bis(cyclooctene)-functionalized PEG derivative to give toluene-filled nanocapsules that can be isolated and are now stable to a wide variety of solvent treatments. Other examples of semipermeable capsules prepared by self-assembly can also be found in a recent highlight article by Bergbreiter [43].

In contrast, Nolte, Rowan et al. [42] have employed a rod–coil block copolymer derived from a polystyrene block and a thiophene-functionalized polyisocyanate block and examined its aggregation behavior in water/THF (5:1). Following equilibration, vesicles were obtained with diameters of 100–200 nm and a shell thickness of 10–20 nm. To stabilize the structure, the thiophene groups in the "skin" of the aggregate could be electrochemically polymerized and the resulting hollow nanocapsules were capable of including enzymes into their interior, thus affording an alternative route to stable, nanoscale reactors. Additionally, the use of a crosslinked thiophene stabilization strategy permits the construction of conducting vesicles with potentially tunable properties.

A common feature of the above examples is the use of a fluid/solvent-filled core to template the nanoscopic object. However it is also possible to use a solid, nanoscopic object as a support, thereby leading to a more uniform hollow capsule of predictable and tunable size after removal of the core. The major challenge in this strategy is designing the chemistry of the organic shell to be compatible with the conditions necessary for removal of the template. As a result, silica nanoparticles have been widely used since treatment with HF results in facile removal under conditions that are typically unreactive towards the organic shell. An initial report [44] on the use of surface-initiated, living, free-radical polymerization from silica nanospheres, coupled with benzocyclobutene crosslinking chemistry, has shown the importance of covalent stabilization in this strategy. As shown in Fig. 9, the growth of polystyrene from the surface of silica nanoparticles, followed by the removal of the silica template produces hollow nanostructures (Fig. 9a). However treatment of these particles with a good solvent such as THF leads to complete dissolution and a continuous thin film is obtained with no individual structures still recognizable (Fig. 9b). In direct contrast, if the styrene is copolymerized with 4-vinylbenzocyclobutene and then the polymer-coated silica nanoparticles crosslinked by heating at 250 °C before dissolution of the silica core with HF, a significantly more robust structure is obtained. Treatment with THF does not lead to dissolution of the individual polymer capsules; instead, individual particles are still obtained that have simply undergone collapse due to plasticization of the crosslinked polystyrene shell (Fig. 9c). This collapse process is reversible and the capsules can be "reinflated" under the appropriate conditions.

The versatility of this covalent stabilization of nanostructures is not restricted to spherical objects: a variety of other three-dimensional structures can also be produced using this methodology. By changing the hydrophilic/hydrophobic balance of di- or triblock copolymers, rod-like structures can initially be formed during the solution-mediated assembly process and, as described above, the structures may be stabilized by crosslinking of functional groups placed in either the shell [45, 46] or the core [47]. The resulting struc-

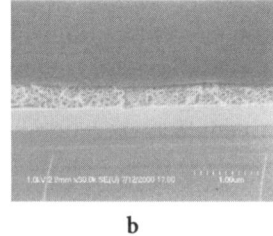

a b c

Fig. 9a–c Scanning electron microscopy images of **a** hollowed polystyrene capsules templated upon silica, **b** destruction of the polystyrene capsules by treatment with organic solvent, and **c** polystyrene capsules crosslinked via benzocyclobutene units and collapsed following treatment with organic solvent

tures have a diameter similar to that of the corresponding micelles (ca. 20–50 nm); however, they can be up to hundreds of micrometers in length, with persistent lengths from 500 nm to tens of micrometers.

The idea of stabilizing monolayers and bilayer vesicles of small molecule amphiphiles was introduced more than 20 years ago and has generated a significant amount of literature for both lamellar and nonlamellar phases [48, 49]. The success of these initial small-molecule studies has recently been translated to synthetic polymers or peptides, and the increased stability of the polymeric structures, combined with the greater statistical probability of efficient crosslinking owing to the large number of reactive groups, opens up a broader range of possible applications [50]. Recent examples include poly(ethylene glycol)-*b*-poly(butadiene) crosslinked assemblies and blending of these structures with non-crosslinkable poly(ethylene glycol)-*b*-poly(lactide) can be used to control their stability, porosity and mechanical properties [51]. Proceeding from simple diblock copolymers to more sophisticated structures such as triblock rod–coil block copolymers leads to a concomitant increase in the complexity of the macromolecular object obtained after crosslinking. Stupp et al. [21] has employed thermal crosslinking of butadiene units in low molecular weight triblock systems to create well-defined nanoscopic objects with a narrow size distribution by first controlling the self-assembly process. Of particular interest is the highly anisotropic nature of the objects obtained, 2×8 nm, which have a characteristic shape that resembles a mushroom and are able to self-assemble themselves to form materials that exhibit a liquid crystalline (LC) state. Proceeding even further from classical block copolymer substrates is the work of Emrick, Russell et al. [52] who prepared CdSe nanoparticles stabilized by ligands containing reactive vinylbenzene moieties. The crosslinkable nanoparticles can then be dispersed in toluene and an aqueous solution of a free radical initiator added. The nanoparticles are then able to assemble at the toluene/water interface, and heating of the mixture to 60 °C resulted in covalent linkage of the CdSe nanoparticles giving a robust two-dimensional sheet of up to several centimeters square, which can be manipulated and transferred to different solvents without physical destruction of the film.

In a slightly different application, the concept of covalent stabilization of a supramolecular assembly can also be used to trap linear polymer chains in a collapsed state [53]. In effect, this involves the controlled intramolecular collapse and crosslinking of a variety of different linear polymer structures (i.e., random, **12**, block, graft, star copolymers) to give the corresponding nanoparticles, **13** [54]. By the use of free radical or benzocyclobutene crosslinking chemistry, no change in molecular weight occurs, however gel permeation chromatography (GPC) and light-scattering analysis of the collapsed structure shows a substantial decrease in molecular volume which is fully consistent with the transition of a random coil chain to a collapsed three-dimensional nanoparticle (Fig. 10).

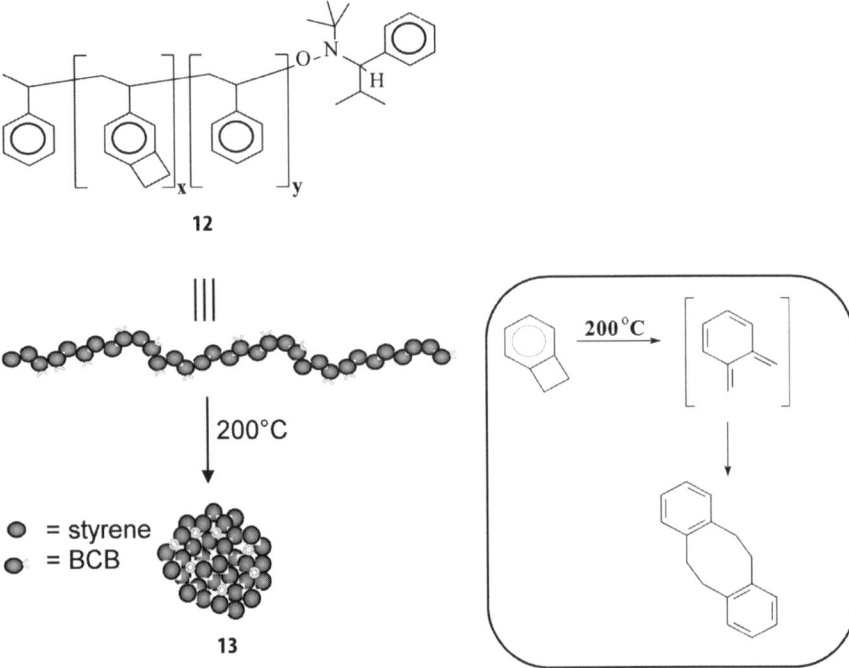

Fig. 10 Collapse of random copolymer, 12, using benzocyclobutene chemistry (see *insert*) to give the nanoparticle, 13. *BCB* benzocyclobutene

4
Utility of Supramolecular Assemblies and Nanoscale Objects

With the continued improvement in analytical techniques and instrumentation for examining physical properties on ever-diminishing size scales, it is now possible to investigate materials on the nanometer size scale. These dimensions are only slightly larger than the molecules themselves, and so bulk materials properties may no longer be observed. This leads to many physical and mechanical properties that are not only different to those observed for traditional systems, but in select cases are not even predicted by long-standing theories. This fascinating and exciting possibility is one of the fundamental reasons for developing strategies leading to well-defined nanoscale objects. To further promote the study of nanoscopic materials and to justify the increased funding levels and scientific interest, it is critical that the unique properties associated with these systems, which cannot be accessed with bulk materials or conventional structures, be exploited. Some of the most promising initial targets for nanoscale organic materials include studies of quantum-confined behavior, and use as templates for inorganic nanostructures, as novel delivery diagnostic systems for biomolecules, and as additives for enhancing the properties of traditional bulk materials.

An intriguing recent example of a diagnostic system based on supramolecular assembly is the temperature-sensitive zinc porphyrin complex of Tsuda, Aida et al. [55], which displays a stepwise color change from green to yellow to red on heating from 0 to 50 to 100 °C. Over this temperature range, the pyridyl-substituted zinc porphyrin complex adopts a cyclotetrameric structure in solution. This temperature-driven assembly process is affected by the substitution pattern of the porphyrin and shows the dramatic influence of molecular structure on physical properties induced by nanometer scale processes.

The influence of nanostructure on physical properties has also been recently demonstrated in the breakdown at nanometer size scales of a classical theory of Einstein from 1906 [56], in which he predicts that the addition of particles to a fluid should increase the fluid's viscosity. Experimentally, this has been demonstrated for an extremely diverse range of systems from normal liquids such as water to viscous polymer melts. However, available synthetic techniques only allowed relatively large particles (diameter >100 nm) to be prepared and studied as viscosity modifiers [57]. Recent work by Mackay, Hawker et al. [58] using nanoscopic objects prepared by the covalent collapse and stabilization of linear polystyrene chains has shown that the addition of nanoparticles with a size similar to that of the fluid molecules (linear polystyrene molecules) causes an unexpected decrease in viscosity, directly opposite to that predicted by the theory of Einstein. The exact mechanism for this anomalous behavior is not yet known, although it is clear that the "quantum-like" size of the nanoparticles causes severe distortion of the linear polystyrene chains of the surrounding fluid in contrast to larger particles, which do not cause any distortion. This change in conformation and free volume of the polymer results in the viscosity decrease. This leads to the important conclusion that Einstein's original theory is not wrong, but that it does not address the use of nanoparticles and, rather, at these small dimensions different physics come into play.

Similar nanoscopic confinement leading to enhanced properties has been observed by Kato [59] in the construction of highly efficient nanostructured ion-conducting films. By specifically designing a monomer containing a mesogen, a polymerizable methacrylate unit and an imidazolium ionic liquid moiety, LC smectic structures can be assembled in which the orientation induced by the mesogen causes a molecular, layered structure to be formed. The inherent thermal instability of such an LC system limits its usefulness; however, the presence of the polymerizable double bonds allows the nanostructure to be covalently stabilized. As shown in Fig. 11, this results in a two-dimensional sheet structure in which alignment and covalent stabilization of the imidazole groups in discrete planes gives rise to significantly enhanced ion conduction. In addition, the presence of ion-insulating layers leads to an increased conductance (ca. two to three orders of magnitude) in the direction parallel to the smectic layer compared to the perpendicular direction.

The challenge of forming aligned nanoscale objects has also been addressed by University of Toronto researchers [46] who were able to form shell-crosslinked cylinders of poly(isoprene)-b-poly(ferrocenyldimethylsilane) by

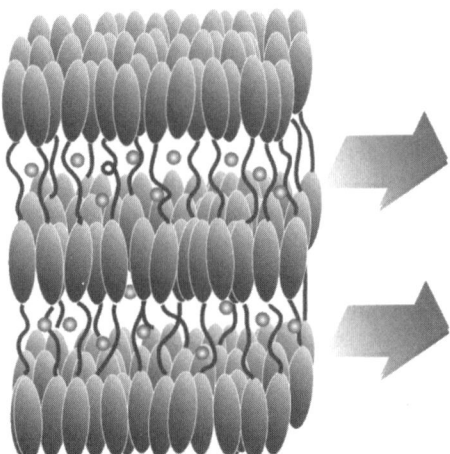

Fig. 11 Schematic representation of the 2-dimensional confinement of ion flow in a crosslinked smectic liquid crystal

Pt-catalyzed hydrosilylation of the pendent vinyl group of the coronal poly(isoprene) chains. Variation in the relative block lengths of the isoprene and ferrocenyldimethylsilane units allowed cylindrical self-assembled micelles to be obtained in contrast to spherical structures [45], with TEM analysis of the crosslinked structures giving a core diameter of 20 nm and lengths of 50–400 nm. While these resemble the values obtained for the dynamic self-assembled structure, the covalent stabilization is apparent in the stability of the nanostructures. The shell-crosslinked cylinders are stable in solvents common to both blocks, and from a technology viewpoint the ability to manipulate, align and pattern the covalently stabilized nanostructures is an enormous benefit. For example, complex micrometer-sized patterns of aligned nanocylinders can be prepared by capillary forces inside microchannels, and in a separate experiment, it was possible to convert the organic shell-crosslinked structures into ceramic materials by pyrolysis, possibly leading to magnetic properties.

The use of organic nanostructures as templates for the formation of inorganic materials with features' sizes in the nanometer size range is also of current technological interest in fields ranging from ultra-low dielectric constant thin films for advanced microelectronics [60] to artificial bone mimics. Stupp et al. [61] and, more recently, Sommerdijk et al. [62] have shown that the self-assembly of tailor-made amphiphiles allows the construction of a nanostructured fibrous scaffold which mimics the extracellular matrix. The high degree of sophistication in the design of the amphiphile, **14**, is evidenced by the five key structural components; a hydrophobic hexadecyl alkyl chain (region 1), a crosslinkable tetra(cysteine) fragment (region 2), a flexible tri(glycine) unit (region 3), a single phosphorylated serine residue that is designed to bind calcium ions and act as a nucleating site for the deposition of hydroxyapatite (region 4), and a RGD tripeptide cell adhesion ligand (region 5). Each subunit plays a spe-

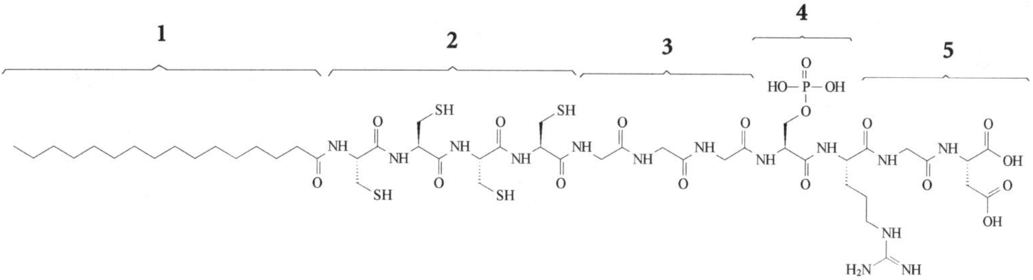

Fig. 12 Structure of the amphiphile, **14**, used in the production of material mimicking bone showing the five different structural regions, which are critical for performance

cific role in the biomineralization process and, as has been shown in the numerous examples above, the crosslinkable groups are essential to provide for covalent stabilization of the self-assembled complex (Fig. 12). Interestingly, the crosslinked fibers of **14** were able to direct the mineralization of hydroxyapatite, which gave rise to a similar alignment between the inorganic crystals/long axis of the fibers to that found for naturally occurring collagen fibers/hydroxyapatite.

5
Conclusions

The preparation of macromolecules having well defined topologies and presenting molecular recognition units in predetermined directions is maturing to the point that rational design strategies can lead to predictable supramolecular complexes that exhibit interesting structures and properties. This concept is not unique; rather, supramolecular chemistry is a rich and diverse field of study, yet there are many new advances that can be made with the preparation and study of molecules of increasing degrees of sophistication. Unique aspects arise when covalent stabilization is imparted within selective regions of the assemblies, especially following subsequent physical or chemical manipulation of the materials. Research directions often strive to mimic natural materials; however, highly novel and complex materials can be produced via the combination of biological design, synthetic materials, and creative design parameters, several examples of which have been highlighted here. This is a blossoming field of study, which will witness over the coming years many significant advances in fundamental knowledge, available materials and technological applications. These advances will be facilitated by the development of new synthetic methodologies and enhancements in analytical tools. This is certainly an exciting time for polymer chemists, with the opportunity to participate in the evolution of synthetic polymer chemistry for the preparation and characterization of elaborate and uniform nanostructured materials, in much the same way as organic chemistry has evolved over the past two centuries.

References

1. Smith MB, March J (2001) March's advanced organic chemistry: reactions, mechanisms and structure. Wiley Interscience, New York
2. Larock RC (1999) Comprehensive Organic Transformations. Wiley Interscience, New York
3. Elemans JAAW, Rowan AE, Nolte RJM (2003) J Mater Chem13:2661
4. Horne WS, Stout CD, Ghadiri MR (2003) J Am Chem Soc 125:9372
5. Bong DT, Clark TD, Granja JR, Ghadiri MR (2001) Angew Chem Int Ed 40:988
6. Rostovtsev VV, Green LG, Fokin VV, Sharpless KB (2002) Angew Chem Int Ed 41:2596
7. Zhao DH, Moore JS (2002) J Am Chem Soc 124:9996
8. Zhao DH, Moore JS (2003) Macromolecules 36:2712
9. Blatchly RA, Tew GN (2003) J Org Chem 68:8780
10. Kerckhoffs JMCA, van Leeuwen FWB, Spek AL, Kooijman H, Crego-Calama M, Reinhoudt DN (2003) Angew Chem Int Ed 42:5717
11. Atwood JL, Szumna A (2003) Chem Commun: 940
12. Maeda K, Morino K, Yashima E (2003) J Polym Sci Part A: Polym Chem 41:3625
13. Hori A, Kataoka H, Akasaka A, Okano T, Fujita M (2003) J Polym Sci Part A: Polym Chem 41:3478
14. Discher DE, Eisenberg A (2002) Science 297:967
15. Bosman AW, Janssen HM, Meijer EW (1999) Chem Rev 99:1665
16. Yamaguchi T, Ishii N, Tashiro K, Aida T (2003) J Am Chem Soc 125:13934
17. Hirschberg JHKK, Koevoets RA, Sijbesma RP, Meijer EW (2003) Chem Eur J 9:4222
18. Rowan SJ, Suwanmala P, Sivakova S (2003) J Polym Sci Part A: Polym Chem 41:3589
19. Caswell KK, Wilson JN, Bunz UHF, Murphy CJ (2003) J Am Chem Soc 125:13914
20. Love JC, Urbach AR, Prentiss MG, Whitesides GM (2003) J Am Chem Soc 125:12696
21. Zubarev ER, Pralle MU, Li L, Stupp SI (1999) Science 283:523
22. Zubarev ER, Stupp SI (2002) J Am Chem Soc 124:5762
23. Lecommandoux S, Klok H-A, Sayar M, Stupp SI (2003) J Polym Sci Part A: Polym Chem 41:3501
24. Gopalan P, Li X, Li M, Ober CK, Gonzales CP, Hawker CJ (2003) J Polym Sci Part A: Polym Chem 41:3640
25. Folmer BJB, Sijbesma RP, Versteegen RM, van der Rijt JAJ, Meijer EW (2000) Adv Mater 12:874
26. Prochazka K, Baloch MK, Tuzar Z (1979) Makromol Chem 180:2521
27. Wilson DJ, Riess G (1988) Eur Polym J 24:617
28. Saito R, Ishizu K, Fukutomi T (1992) Polymer 33:1712
29. Henselwood F, Liu GJ (1997) Macromolecules 30:488
30. Bueten V, Wang X-S, de Paz Banez MV, Robinson KL, Billingham NC, Armes SP, Tuzar Z (2000) Macromolecules 33:1
31. Clark CG, Wooley KL (2001) Covalently-stabilized discrete polymer nanoscale assemblies from solution. In: Encyclopedia of materials: science and technology. Elsevier, Oxford, p 1727
32. Huang H, Kowalewski T, Wooley KL (2003) J Polym Sci Part A: Polym Chem 41:1659
33. Murthy KS, Ma Q, Remsen EE, Kowalewski T, Wooley KL (2003) J Mater Chem 13:2785
34. Huang H, Remsen EE, Kowalewski T, Wooley KL (1999) J Am Chem Soc 121:3805
35. Zhang Q, Remsen EE, Wooley KL (2000) J Am Chem Soc 122:3642
36. Schultz LG, Zhao Y, Zimmerman SC (2001) Angew Chem Int Ed 40:1962
37. Mertz E, Zimmerman SC (2003) J Am Chem Soc 125:3424
38. Zimmerman SC, Zharov I, Wendland MS, Rakow NA, Suslick KS (2003) J Am Chem Soc 125:13504

39. Jung HM, Price KE, McQuade DT (2003) J Am Chem Soc 125:5351
40. Piotti ME, Rivera F, Jr., Bond R, Hawker CJ, Frechet JMJ (1999) J Am Chem Soc 121:9471
41. Breitenkamp K, Emrick T (2003) J Am Chem Soc 125:12070
42. Vriezema DM, Hoogboom J, Velonia K, Takazawa K, Christianen PCM, Maan JC, Rowan AE, Nolte RJM (2003) Angew Chem Int Ed 42:772
43. Bergbreiter DE (1999) Angew Chem Int Ed 38:2870
44. Blomberg S, Ostberg S, Harth E, Bosman AW, Van Horn B, Hawker CJ (2002) J Polym Sci Part A: Polym Chem 40:1309
45. Ma Q, Remsen EE, Clark CG Jr, Kowalewski T, Wooley KL (2002) Proc Natl Acad USA 99:5058
46. Wang X-S, Arsenault A, Ozin GA, Winnik MA, Manners I (2003) J Am Chem Soc 125:12686
47. Dalhaimer P, Bates FS, Discher DE (2003) Macromolecules 36:6873
48. Regen SL, Czech B, Singh A (1980) J Am Chem Soc 102:6638
49. Dorn K, Klingbiel RT, Specht DP, Tyminski PN, Ringsdorf H, O'Brien DF (1984) J Am Chem Soc 106:1627
50. Xu H, Goedel WA (2002) Langmuir 18:2363
51. Ahmed F, Hategan A, Discher DE, Discher BM (2003) Langmuir 19:6505
52. Lin Y, Skaff H, Boeker A, Dinsmore AD, Emrick T, Russell TP (2003) J Am Chem Soc 125:12690
53. Mecerreyes D, Lee V, Hawker CJ, Hedrick JL, Wursch A, Volksen W, Magbitang T, Huang E, Miller RD (2001) Adv Mater 13:204
54. Harth E, Van Horn B, Lee VY, Germack DS, Gonzales CP, Miller RD, Hawker CJ (2002) J Am Chem Soc 124:8653
55. Tsuda A, Sakamoto S, Yamaguchi K, Aida T (2003) J Am Chem Soc 125:15722
56. Einstein A (1906) Ann Phys (Leipz) 19:371
57. Glotzer SC (2003) Nat Mater 2:713
58. Mackay ME, Dao TT, Tuteja A, Ho DL, Van Horn B, Kim H-C, Hawker CJ (2003) Nat Mater 2:762
59. Hoshino K, Yoshio M, Mukai T, Kishimoto K, Ohno H, Kato T (2003) J Polym Sci Part A: Polym Chem 41:3486
60. Hawker CJ, Hedrick JL, Miller RD, Volksen W (2000) MRS Bull 25:54
61. Hartgerink JD, Beniash E, Stupp SI (2001) Science 294:1684
62. Donners JJJM, Nolte RJM, Sommerdijk NAJM (2003) Adv Mater 15:313